W9-BFU-786

Basic ^{1}H- and ^{13}C-NMR Spectroscopy

Basic ^{1}H- and ^{13}C-NMR Spectroscopy

Metin Balcı

Department of Chemistry
Middle East Technical University
Ankara, Turkey

2005

ELSEVIER

Amsterdam – Boston – Heidelberg – London – New York – Oxford – Paris
San Diego – San Francisco – Singapore – Sydney – Tokyo

ELSEVIER B.V.	ELSEVIER Inc.	ELSEVIER Ltd	ELSEVIER Ltd
Sara Burgerhartstraat 25	525 B Street, Suite 1900	The Boulevard, Langford Lane	84 Theobalds Road
P.O. 211, 1000 AE Amsterdam	San Diego, CA 92101-4495	Kidlington, Oxford OX5 1GB	London WC1X 8RR
The Netherlands	USA	UK	UK

First edition 2005

Library of Congress Cataloging in Publication Data
A catalog record is available from the Library of Congress.

British Library Cataloguing in Publication Data
A catalogue record is available from the British Library.

ISBN: 0-444-51811-8 (hardbound)

⊗ The paper used in this publication meets the requirements of ANSI/NISO Z39.48-1992 (Permanence of Paper). Printed in The Netherlands

To
My wife Jale and
children Gülşah and Berkay

Preface

Although there are already a number of excellent books on NMR spectroscopy, the progressive growth of the field has compelled me to write this book. As a synthetic chemist as well as a teacher at the undergraduate and graduate level, I know which problems students encounter when using NMR spectroscopy. Therefore, I have concentrated especially on problems such as coupling modes, simple spin systems and their treatments etc., whilst shielding the students from the complicated mathematical treatments that are applied within NMR spectroscopy.

The basic theory with an oversimplified model of the NMR phenomenon is provided in Chapter 2 along with a description of the Continuous Wave and Fourier Transform NMR spectrometer. A more realistic description of the NMR phenomenon is described in the ^{13}C-NMR spectroscopy section. I did not want to hinder the students at the onset, and so I chose to start with an uncomplicated introduction, which makes the book much more readable. Two nuclei, the proton (^{1}H) and carbon (^{13}C), are covered in this book, and these two important elements are intentionally discussed separately. Over many years I have noticed that it is much more useful to first cover proton NMR rather well, and subsequently proceed to carbon NMR. After learning proton NMR in detail, the students can learn the carbon NMR in a rather short timeframe. The homonuclear and heteronuclear couplings are discussed in detail and supported with some applications. Very simple second-order spectra that are encountered very frequently in the spectra are also discussed in detail and their importance is emphasised with original examples. The fundamental principles of the pulse NMR experiment and 2D NMR spectra are discussed in the second part of this book.

The aim of this book is to provide an introduction to the principles and applications of NMR spectroscopy at a level that is suitable for undergraduate as well as graduate students. Most of the spectra shown in the text were obtained in the author's research laboratory.

A number of companies have been kind enough to permit me to use some spectra. The author gratefully acknowledges the many helpful comments and critical readings that his students provided at the Atatürk and Middle East Technical universities. My sincere gratitude especially goes to my PhD student Cavit Kazaz at Atatürk University in Erzurum for spending a tremendous amount of time obtaining most of the NMR spectra that are presented in this book. I also would like to thank Fatoş Doğanel at the Middle East Technical University for recording spectra. Last but not least, my wife Jale and my children Gülşah and Berkay offered unwavering patience and support. There is no adequate way to express my appreciation.

Metin Balcı
Ankara, June 2004

Contents

Part I

¹H-NMR Spectroscopy

– 1 –

Introduction

1.1 STRUCTURE ELUCIDATION AND NMR

Today several million organic compounds are known, and thousands of new ones are prepared each year. Industry and academia account for more than 90% of the organic compounds synthesized in the laboratory, while the rest are derived from natural resources. Spectral analysis is used to identify structural features of these compounds. The advent of spectral analysis has allowed chemists to confirm whether a reaction has resulted in the desired product or not. Presently, several analytical techniques are available. Depending on the area of study (organic, inorganic, physical chemistry, or biochemistry), the analytical methods may vary. The most commonly applied methods in organic chemistry are as follows:

(1) infrared (IR) spectroscopy;
(2) ultraviolet (UV) spectroscopy;
(3) mass spectrometry;
(4) X-ray spectroscopy;
(5) nuclear magnetic resonance (NMR) spectroscopy;
(6) elemental analysis.

Now let us focus on the kind of molecular structure information that can be obtained using these methods. The compound ethyl benzoate (**1**) will be used as an example to illustrate these methods.

1

1.1.1 IR spectroscopy

IR spectroscopy provides information about the functional groups present in a molecule such as COR, COOR, CN, NO_2, etc. The IR spectrum of ethyl benzoate will reveal the existence of an ester and double bond. This method of spectroscopy does not provide information regarding the location or connectivity of these functional groups.

1.1.2   UV spectroscopy

The principal use of UV spectroscopy is to identify structures containing conjugation. The existence of a benzene ring in conjugation with another system is evident from the UV spectrum of ethyl benzoate. In some instances, it is difficult to interpret this spectrum completely by UV and IR spectroscopy.

1.1.3   Mass spectrometry

Mass spectrometry provides information regarding molecular weight, or the formula weight of a compound. In the case of ethyl benzoate, analysis reveals the formula weight of 150 amu. Also, high-resolution mass spectroscopy (HRMS) may be used to determine the empirical formula of the molecule. In the case of ethyl benzoate, the empirical formula is $C_9H_{10}O_2$. However, there may be hundreds of constitutional isomers of the same empirical formulae, making it difficult to distinguish these isomers. Even though the ethyl benzoate can be detected amongst these isomers, it is almost impossible to do so for molecules having a more complex structure.

1.1.4   X-ray spectroscopy

X-ray spectroscopy is an excellent method to determine the structure of a compound. However, this technique requires the availability of a compound as a single crystal. Most chemists find this process very tedious, time consuming and it requires a skillful hand. In the event when other spectral methods fail to reveal a compound's identity, X-ray spectroscopy is the method of choice for structural determination where the other parameters such as bond lengths and bond angles are also determined. However, advances in technology have made it possible to have an NMR spectrum complete in as little as one minute.

1.1.5   NMR spectroscopy

NMR spectroscopy is a useful technique for identifying and analyzing organic compounds. This extremely important experimental technique is based on magnetic nuclear spin of ^1H, ^{13}C, ^{15}N, ^{19}F, ^{31}P, and so forth. Only ^1H and ^{13}C will be considered in this example. Proton and carbon NMR spectra of ethyl benzoate (Figure 1) contain more detailed and definitive information:

(1) ^1H-NMR provides information about the number of protons in the molecule.
(2) ^1H-NMR reveals the existence of an aromatic moiety that is monosubstituted. This is indicative of 5 protons attached to an aromatic ring.
(3) The existence of a methyl group and its attachment to a CH_2 group is indicated.
(4) Analysis of the signals provides information about a CH_3 group being adjacent to a methylene group that is attached to electronegative atom (oxygen).
(5) ^{13}C NMR provides information about the number of carbons in the molecule.

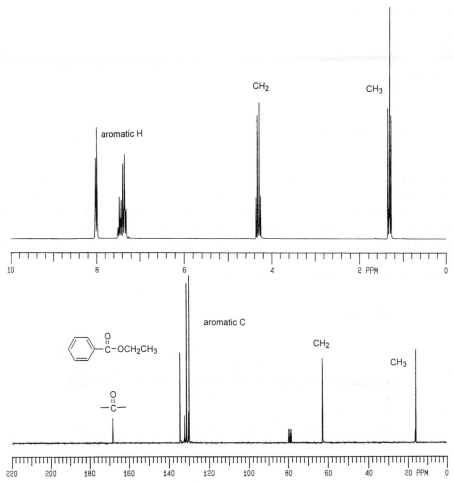

Figure 1   200 MHz ^{1}H NMR and 50 MHz ^{13}C NMR spectra of ethyl benzoate in CDCl$_3$.

(6) The existence of a carbonyl carbon and four different carbon atoms on an aromatic ring (monosubstituted benzene ring) is revealed from the ^{13}C NMR.

The combination of all analytical information shows that the molecule is indeed ethyl benzoate. Hence, when information obtained from NMR and other analytical methods is compared, it is apparent that the most detailed and necessary information to recognize a structure is obtained from NMR spectra. However, this does not mean that NMR is always sufficient for structural analysis. For absolute analysis of structures, it is necessary to apply all spectroscopic methods. If a new compound is synthesized, the chemist who does scientific research should have all the spectroscopic data related to this chemical compound. The information provided by an NMR spectrum is useful only if it can be interpreted well.

In addition to providing information related to constitutional analysis, NMR spectra provide detailed information pertinent to configurational and conformational analysis. Illustrations of the use of NMR in distinguishing these aspects will be explored. In the classical reaction of nitration of acetophenone (**2**), the nitro group attaches to the *meta* position.

Conventionally, it is accepted that nitration occurs at the *meta* position, but how can this be proved? There must be a method to determine the position of the nitro group (the constitutional analysis of the molecule). Indeed, this question is easily answered by the use of NMR.

Most chemical reactions produce configurational isomers. For instance, there may be two products (**5** and **6**) which represent the *endo* and *exo* isomers in the epoxidation of benzonorbornadiene with *m*-chloroperbenzoic acid.

Which product is formed from the reaction or do both occur? Which isomer is the major product and what is the ratio of the formation? These questions can be answered absolutely by means of NMR spectra, thus making NMR spectroscopy a definitive tool for the determination of the correct configuration of the molecule.

The cyclohexane ring contains two types of protons, equatorial and axial. When a substituent is attached to the cyclohexane ring, it is known that generally the substituent prefers to be in an equatorial position. Bromocyclohexane will be used to demonstrate how NMR spectra can be used to identify which conformation the bromine atom prefers. Additional information can be obtained in reference to the ratio of the population and all thermodynamic parameters (activation energy, activation entropy, etc.) of the dynamic process.

NMR spectroscopy is not only a method for the determination of the structure of the molecule, but also a very functional spectroscopic method that explains the dynamic process (by recording the NMR spectra at different temperatures) in the molecule, approximate bond length, the bond angles, etc.

The development of modern spectroscopic methods has made the structural analysis of chemical compounds faster and easier. Today, the structural elucidation of very complicated chemical compounds can be done in a very short period of time. Before these methods were known, structural analysis was done by chemical methods. Let us consider the reaction in which cyclobutene is formed, for instance.

process (by recording the NMR spectra at different temperatures) in the molecule, approximate bond length, the bond angles, etc.

If the compound is really a cyclobutene, it should have a double bond. Therefore the alkene (**9**) is reacted with ozone in order to break the double bond to form dialdehyde (**10**) and then the dialdehyde is oxidized to carboxylic acid by $KMnO_4$. The acid obtained from oxidation is compared with the structurally known succinic acid (**11**). If both the compounds have the same properties, it is understood that succinic acid occurs as a result of the reaction. This finding reveals that the structurally unknown compound is cyclobutene (**9**).

This example shows that the identification of the structure is not easy using chemical methods. Besides the fact that the determination of structure by chemical methods is both difficult and time consuming, another drawback is that the compound needs to be broken into known compounds in order to analyze the structure. Furthermore, it is difficult to use chemical methods for the determination of structure of compounds which have a complex structure. Most of the time it is almost impossible to obtain a result.

1.2   DEVELOPMENT OF NMR SPECTROSCOPY

The first NMR signal was observed independently in 1945 by two physicists, Bloch and Purcell. Due to their accomplishment, both received the 1952 Nobel prize in physics. Physicists were mainly interested in resonance and usually did not devote much attention to structural analysis. Soon after the finding of an NMR signal, chemists realized the importance of this technique and started using NMR spectroscopy in the structural analysis of molecules. In early 1953, the first low resolution NMR instrument was released on the market. NMR was introduced in Europe in 1960 and the developments in this area have been continuing at an incredible rate. The first NMR instruments consisted of the magnet for which the magnetic fields (T) were 1.41 (60 MHz), 1.87 (80 MHz), 2.20 (90 MHz) and 2.35 (100 MHz). After the 1970s, superconducting magnets with high resolution which work at the temperature of liquid helium (4 K) were produced. Ernst received the Nobel prize in chemistry in 1993 for his important work on these instruments. Today, many chemistry departments have a couple of 200, 300, 400, 600 NMR instruments. Currently, a 900 MHz NMR instrument is under construction.

Besides chemists, NMR instruments are also utilized by physicists, biochemists, pharmaceutical chemists and medical doctors. These professionals make use of this technique to elucidate structural information and the physical properties of a molecule. The use of NMR analysis in the field of medicine is becoming commonplace for the identification and medical treatment of certain diseases. As a result, the use of X-rays will soon become obsolete in the near future. Reasons for the popularity of NMR stem from the fact that:

1. It works with radiofrequency waves; therefore, it is not hazardous to peoples' health.
2. It can distinguish between soft and hard tissues.
3. Besides anatomical information, physiological information can be obtained.

In recent years, the routine measurements of NMR spectra of elements such as phosphorous, sodium, potassium and nitrogen have been studied. It is likely that these techniques will find application in the medical area rather soon, for instance, sketching the phosphorous distribution of a tissue slice.

The importance of NMR should now be obvious. It is the goal of this book to highlight the indispensability of NMR and its use in structural analysis. The first part of the book concentrates on explaining the concepts of resonance and NMR interpretation at a level that a novice can understand. The second part of the book will focus on resonance in a more detailed fashion.

– 2 –

Resonance Phenomena

2.1 MAGNETIC PROPERTIES OF ATOMIC NUCLEI

Nuclear magnetic resonance spectroscopy is based on the magnetic properties of atomic nuclei. Let us first discuss the magnetism of atomic nuclei. Atoms consist of a dense, positively charged nucleus surrounded at a relatively large distance by negatively charged electrons. At this stage we will ignore the electrons and concentrate on the nucleus. We can consider it to be a spherical spinning object with all of its charge spread uniformly over its surface. This model is, of course, considerably oversimplified. Since atomic nuclei are positively charged, they can also be thought of as being much like tiny magnets. In some nuclei, this charge spins on the nuclear axis. The circulation of nuclear charge generates a magnetic dipole along the axis of rotation (Figure 2).

The intrinsic magnitude of the dipole generated is expressed in terms of *magnetic moment* μ. The arrow on μ (Figure 2) means that the magnetic moment is a vector and has values. We know from classical physics that any kind of spinning body has an angular momentum. Consequently, a spinning atomic nucleus also has an angular momentum. Since it is well established that the magnetic moment μ and the angular momentum P behave as parallel vectors, the relation between these quantities is expressed by the following equation:

$$\mu = \gamma P \tag{1}$$

where μ is the magnetic moment, P is the angular momentum, and γ, gyromagnetic ratio, is a constant characteristic of the particular nucleus. The gyromagnetic ratios γ of some important elements in NMR spectroscopy and their natural abundance are given in Tables 2.1 and 2.2.

2.2 SPIN QUANTUM NUMBERS OF ELEMENTS

Before we discuss the behavior of a magnetic dipole (of any kind of element) in a magnetic field, we should consider the concept of electron spin. The spin quantum numbers of an electron are $+1/2$ and $-1/2$. These numbers indicate that an electron can have only one of the two possible spin orientations. We usually show these orientations

9

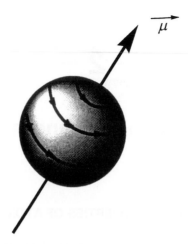

Figure 2 The spinning proton resembles a tiny bar magnet.

by arrows pointing either up ↑ or down ↓. As in the case of electrons, spinning nuclei will then also have nuclear spin quantum numbers since they are a charged species. As we will show later, whether a given element is active in NMR spectroscopy or not solely depends on the magnitude of its nuclear spin quantum number I.

The actual value of the nuclear spin quantum number of any given nucleus depends on the mass number and the atomic number. Consequently, the isotopes of an element have different spin quantum numbers. However, the spin quantum numbers cannot be determined using the numbers of protons and neutrons in the nuclei. There are some very useful generalizations between the spin quantum numbers of elements and the numbers of protons and neutrons. These can be summarized as follows:

(1) *Even–even elements*: The elements whose mass numbers and atomic numbers are even belong to this group, for example, $^{12}_{6}C$, $^{16}_{8}O$, and $_{16}^{34}S$. $^{12}_{6}C$ has 6 protons and 6

Table 2.1

Gyromagnetic ratios of some nuclei

Nucleus	Gyromagnetic ratio, γ (10^8 rad s^{-1} T^{-1})
^1H	2.674
^2H (D)	0.410
^{13}C	0.672
^{14}N	0.913
^{15}N	−0.271
^{17}O	−3.620
^{19}F	2.516
^{31}P	1.083

Table 2.2

Spin quantum numbers and natural abundance (%) of some useful
nuclei

Nucleus	Spin quantum number, I	Natural abundance
^{1}H	1/2	99.98
^{2}H (D)	1	0.016
^{10}B	3	18.83
^{11}B	3/2	81.17
^{13}C	1/2	1.108
^{14}N	1	99.635
^{15}N	1/2	0.365
^{17}O	5/2	0.037
^{19}F	1/2	100.00
^{29}Si	3/2	4.70
^{31}P	1/2	100.00
^{33}S	3/2	0.74

neutrons; $^{16}_{8}$O has 8 electrons and 8 neutrons. The nuclear spin quantum number of all elements (isotopes) that belong to this group is zero ($I = 0$). *These elements ($I = 0$) are inactive in NMR spectroscopy.* We will explain the reason for this later.

(2) (a) *Odd–odd elements*: The elements whose mass numbers and atomic numbers are odd belong to this group, for example, $^1_1$H and $^{19}_9$F. $^1_1$H has only one proton and does not have any neutrons. Fluorine has 9 protons and 10 neutrons.

(b) *Odd–even elements*: The elements whose mass numbers are odd and atomic numbers are even belong to this group, for example, $^{13}_6$C and $^{17}_8$O. $^{13}_6$C has 6 protons and 7 neutrons. The nuclear spin quantum numbers I of all elements that belong to this group (odd–odd and odd–even) are integral multiples of 1/2:

$$I = \frac{1}{2}, \frac{3}{2}, \frac{5}{2}, \frac{7}{2}, \frac{9}{2}$$

(3) *Even–odd elements*: The elements whose mass numbers are even and atomic numbers are odd belong to this group, for example, $^2_1$H (D) and $^{14}_7$N. $^2_1$D has 1 proton and 1 neutron. The nuclear spin quantum number I of all elements that belong to this group are integral multiples of 1:

$$I = 1, 2, 3, 4, 5$$

In summary, nuclei such as ^{12}C, ^{32}O and ^{34}S have no spins ($I = 0$) and these nuclei do not yield NMR spectra. Other nuclei with spin quantum numbers $I > 0$ are active and do in fact yield NMR spectra. In this book, we shall deal primarily with the spectra that arise from proton (^{1}H) and from carbon (^{13}C), both of which have $I = 1/2$.

2.3 BEHAVIOR OF AN ATOMIC NUCLEUS IN A MAGNETIC FIELD

2.3.1 The relation between the magnetic moment and angular momentum

We have already discussed the fact that nuclei of certain isotopes possess an intrinsic spin, which is associated with the angular momentum. In classical physics, the angular momentum can have any value. However, in quantum physics, angular momentum is quantized, which means that angular momentum can have certain values. This fact cannot be explained by an argument based on classical physics. This is the most important difference between classical and quantum physics. In quantum physics, angular momentum is defined by the following relation:

$$P = \frac{h}{2\pi} m \tag{2}$$

where h is Planck's constant and m is the magnetic quantum number. According to this equation, the angular momentum P of a given element depends directly on the magnetic quantum number m. Therefore, to calculate the angular momentum of any element, we have to know the value of the magnetic quantum number m.

   These magnetic quantum numbers are related to the nuclear spin quantum numbers, I, of the respective nucleus:

$$m = (2I + 1) \tag{3}$$

If we know the spin quantum number I, we can calculate the magnetic quantum number m. For example, the spin quantum number of a proton is $I = 1/2$. When we replace I in eq. 3 by $I = 1/2$:

$$m = \left(2\frac{1}{2} + 1 \right) = 2$$

we obtain $m = 2$. Now we have to explain the meaning of this number '2'. It is not that the magnetic quantum number of the proton is equal to 2, but rather that the proton has *two different magnetic quantum numbers*. As we will show later, the proton has two different alignments in the presence of a static magnetic field.

   We have determined that the proton has two different magnetic numbers. Now we have to define these numbers. The magnetic quantum number can take any of the integral values from $-I$ to $+I$, including zero ($\Delta m = 1$):

$$m = -l, (-l + 1), (-l + 2), ..., 0, ..., (l - 1), +l \tag{4}$$

We determine the magnetic quantum numbers for the proton as follows:

$$m = -l, ..., +l$$

$$m = -\frac{1}{2}, ..., +\frac{1}{2}$$

$$m_1 = +\frac{1}{2} \quad \text{and} \quad m_2 = -\frac{1}{2}$$

The proton has two different magnetic quantum numbers, which are given above. Let us determine the magnetic quantum numbers of an element whose spin quantum number is

$I = 2$. If I is equal to 2, then the magnetic quantum number m will be 5. From eq. 4 we obtain the following magnetic quantum numbers for the representative element. The magnetic quantum number m can have any value between -2 and $+2$ where $\Delta m = 1$:

$$m = -2, ..., +2$$

$$m = -2, -1, 0, +1, +2$$

$$m_1 = +2, \quad m_2 = +1, \quad m_3 = 0, \quad m_4 = -1, \quad m_5 = -2$$

After showing the relation between the magnetic quantum number m and the angular momentum P, we can define the relation between the magnetic moment μ and the magnetic quantum number m:

$$P = \frac{h}{2\pi} m \tag{2}$$

$$\mu = \gamma P \tag{1}$$

When we add these two equations, we obtain a new equation (eq. 5), which shows us that the magnetic moment μ of a given element depends only on the magnetic quantum numbers m:

$$\mu = \gamma \frac{h}{2\pi} m \tag{5}$$

We can conclude that the nuclear magnetic moment μ is also quantized and can only have certain values. These values are determined by the magnetic quantum numbers m.

2.3.2 The energy of a magnetic dipole in a magnetic field

When we place a magnetic dipole (of any nucleus) in a static homogeneous magnetic field (H_0), we can define the potential energy of this dipole. The potential energy E of a dipole is given by

$$E = \mu H_0 \tag{6}$$

where μ is the magnetic moment of the nucleus. We have already calculated the magnetic moment μ (see eq. 5). Adding eqs. 5 and 6 will give us the new energy formula:

$$E = \frac{\gamma h H_0}{2\pi} m \tag{7}$$

Let us attempt to explain this energy formula. In a given magnetic field, the strength of the magnetic field H_0 is not varied. The other parameters, h and γ, are constant values, which means that the energy of a magnetic dipole in a static magnetic field depends only on the magnetic quantum numbers m. At this point, we can draw the following important conclusion: the energy of a dipole in a static magnetic field is quantized and it can only have certain values. If we know the magnetic quantum numbers of the element, we can calculate the different energies of a dipole in a magnetic field. Since we will deal with the proton (^1H) and carbon (^{13}C) NMR spectroscopy, let us apply this equation to the proton

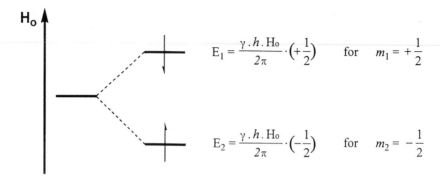

Figure 3   Energy level schemes of proton by parallel and antiparallel alignments.

and carbon nuclei. The spin quantum number of the proton and carbon nucleus is 1/2. The corresponding magnetic quantum numbers are

$$m_1 = +\frac{1}{2} \quad \text{and} \quad m_2 = -\frac{1}{2}$$

Since the proton (or carbon) has two different magnetic quantum numbers, it has two different energy states in the presence of a static magnetic field, i.e. two different alignments in the magnetic field. These two different energy states can be calculated by replacing m in eq. 7. Then we can obtain the energies of the two different states:

$$E_1 = \frac{\gamma h H_0}{2\pi}\left(+\frac{1}{2}\right) \qquad \text{for } m_1 = +\frac{1}{2} \tag{8}$$

$$E_2 = \frac{\gamma h H_0}{2\pi}\left(-\frac{1}{2}\right) \qquad \text{for } m_2 = -\frac{1}{2} \tag{9}$$

Now let us try to explain qualitatively the energy separation of the protons in the presence of a magnetic field (Figure 3).

In the absence of a strong external magnetic field, the nuclear spins of protons are randomly oriented (Figure 4a). When a sample containing protons is placed in a static, homogeneous magnetic field, the magnetic moment μ of the protons will interact with

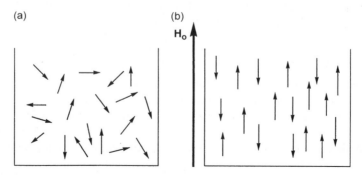

Figure 4   Orientation of magnetic moments of protons (a) in the absence of a magnetic field and (b) in the presence of a magnetic field.

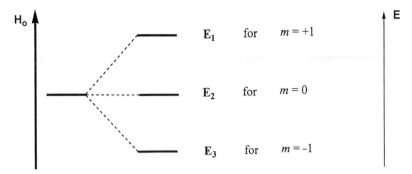

Figure 5   Energy level schemes for deuterium (generally for nuclei with spin quantum number $I = 1$).

the external magnetic field, and the nuclei will adopt a specific orientation (Figure 4b), much as a compass needle orients itself in the earth's magnetic field. In summary, some of the nuclear spin will align parallel with the external magnetic field and some antiparallel. The energy of the nuclear spins that are parallel oriented is lower than that of those spins that are antiparallel oriented.

Generally, the spin quantum numbers I of nuclei determine the number of different energy states of a nucleus in a magnetic field. We have already explained the energy separation of the nuclei whose spin quantum number $I = 1/2$. Let us predict the energy levels of an element with the spin quantum number $I = 1$. For example, the spin quantum number of deuterium, D, is 1. The magnetic quantum number of D is then 3 (eq. 3). This means that when deuterium is placed in a magnetic field, the energy will be split into three different states, which are determined by the magnetic quantum numbers (Figure 5). In other words, deuterium nuclei have three different orientations in a magnetic field.

2.3.3   Resonance phenomena and resonance condition

We have explained that all nuclei with the spin quantum number $I > 0$ are split into different energy levels when they are placed in a static and homogenous magnetic field. The energy difference between two energy states (in the case of proton or carbon we have two different energy states) is very important in NMR spectroscopy. If the oriented nuclei are now irradiated with electromagnetic radiation of the proper frequency, energy absorption occurs, and the lower energy state flips to the higher energy state. For that reason, in NMR spectroscopy, we have to know the exact energy difference between these states:

$$E_{\text{parallel}} = \frac{\gamma h H_0}{2\pi}\left(-\frac{1}{2}\right), \qquad E_{\text{antiparallel}} = \frac{\gamma h H_0}{2\pi}\left(+\frac{1}{2}\right)$$

The energy difference can be easily determined by subtracting the above equations:

$$\Delta E = E_{\text{antiparallel}} - E_{\text{parallel}} = \frac{\gamma h H_0}{2\pi}\left(+\frac{1}{2}\right) - \frac{\gamma h H_0}{2\pi}\left(-\frac{1}{2}\right) = \frac{\gamma h H_0}{2\pi} \qquad (10)$$

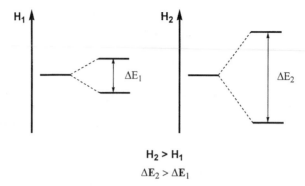

$$H_2 > H_1$$
$$\Delta E_2 > \Delta E_1$$

Figure 6   The energy difference ΔE between two adjacent energy levels as a function of applied magnetic field.

This equation provides us with the exact energy difference between the two energy states. As one can see from eq. 10, the energy separation of a given nucleus produced in the presence of an external magnetic field is proportional only to the strength of the applied magnetic field H_0, since h, γ, and π are constants. There is only one variable parameter, the magnetic field H_0, in eq. 10. Figure 6 illustrates the fact that the energy difference between the states increases with the increased strength of the magnetic field. Increasing of the energy, ΔE, will be reflected in the sensitivity of the NMR instruments, which will be discussed later.

If a nucleus is not placed in a magnetic field ($H_0 = 0$), there will be no different energy states. According to eq. 10, ΔE will be equal to zero. The energy states will be degenerated. The exact amount of energy is necessary for the resonance. Once the two energy levels for the proton have been established, it is possible to introduce energy in the form of radiofrequency (rf) radiation to effect the transition between these energy levels in a static homogeneous magnetic field of a given strength H_0. The introduced radiofrequency ν is given in megahertz (MHz).

$$E = h\nu \tag{11}$$

If we combine eqs. 10 and 11, the frequency ν at which the absorptions occur can readily be calculated.

$$E = h\nu = \frac{\gamma h H_0}{2\pi} \tag{12}$$

Modification of this equation, to include the gyromagnetic ratio γ, gives the fundamental NMR equation:

$$\nu = \frac{\gamma H_0}{2\pi} \tag{13}$$

This equation shows us the necessary condition for the resonance in the NMR experiments. Now let us summarize the NMR experiments on the basis of this equation.

When a sample containing protons is placed in an external homogeneous magnetic field, the protons will have two orientations either parallel to or against the external field. The two different orientations do not have the same energy. The parallel orientation is slightly lower in energy. Energy is required to 'flip' the proton from the lower energy level to its higher energy level. In the NMR spectrometer this energy is supplied by electromagnetic radiation in the rf region. When the system is irradiated with the proper frequency (see eq. 13), energy absorption occurs. Absorption of energy results in the excitation of the nucleus to the higher spin state. The nuclei are said to be in *resonance* with the applied radiation. Therefore, this process is called *nuclear magnetic resonance*.

The amount of applied energy has to be exactly the same as the energy difference between the two states. However, if a stronger magnetic field is applied, according to eq. 13, higher energy is required in order to bring the nuclei into resonance (Figure 6).

To carry out an NMR experiment, the sample (containing a proton) is placed between the poles of a strong magnet. The sample will then be irradiated. A radiofrequency oscillator or transmitter is used as the source of exciting radiation. Experimentally, there are two different ways to achieve resonance. The NMR equation

$$\nu = \frac{\gamma H_0}{2\pi} \tag{13}$$

shows that there are two variable parameters: (i) frequency and (ii) magnetic field. The most obvious way to achieve resonance is to vary the frequency at a fixed magnetic field strength. This technique is called the *frequency sweep method* (Figure 7a). Alternatively, the oscillator frequency may be held constant as the magnetic field strength is varied. This technique is called the *field-sweep method* (Figure 7b). Both methods are used, but the *field-sweep technique* is preferred. Field-sweep spectrometers possess at the pole pieces of the magnet special coils, which allow continuous variation of the magnetic field strength. In both cases, the different protons (nuclei) are brought into resonance one by one.

Since in both techniques one parameter is kept constant while the other parameter is varied, this mode is called *continuous-wave* (CW) spectrometry. The CW method was the basis of all NMR instruments constructed up to about the end of the 1960s. CW is still used in some lower resolution instruments. However, CW instruments have been entirely superseded by pulsed Fourier transform (FT) instruments. For the interpretation of NMR spectra it is not important whether the NMR spectrum is recorded by the CW or FT technique. Therefore, we will discuss the FT NMR technique in the second part of this

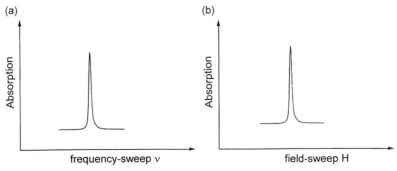

Figure 7   Energy absorption and formation of an NMR peak (a) by frequency-sweep technique and (b) field-sweep technique.

book (see Chapter 11). We do not want to bother beginners with the details of the FT technique at this stage.

We have discussed resonance phenomena with an oversimplified model. Students who are interested in the details of resonance can now go to Chapter 10 and learn about actual resonance. In the second part of this book, the carbon nucleus is discussed. The basic theories are applied to all kinds of elements, to where it is not important as to which nucleus is discussed.

As we have shown above, the resonance frequency of the protons depends on the strength of the magnetic field applied (see eq. 13). The first commercial NMR spectrometer used permanent magnets with fields of 1.41, 1.87, 2.20 or 2.35 T corresponding to 60, 80, 90 or 100 MHz. When describing NMR instruments we always state the frequency at which they operate. For example, if we talk about a 60 MHz NMR instrument, it means that the strength of the magnetic field is 1.41 T (14,100 G). The demand for NMR instruments has resulted in the wide use of 200–600 MHz instruments. A 900 MHz instrument (21.1 T) is currently in production.

2.3.4 Populations of the energy levels

As we mentioned in the introduction, when a sample containing protons is placed in a static magnetic field, the nuclei will adopt different orientations. The magnetic moments of some nuclei will align parallel with the external magnetic field and some against. Now we are interested in the question of which factors can influence this distribution of the nuclei. Since the nuclei flip from the lower energy state to the higher energy state during the resonance, the number of nuclei populating the lower energy level (parallel orientation) has to be greater than the number populating the higher energy level (antiparallel orientation) in order to obtain an NMR signal. According to the laws of thermodynamics, if there are two different energy levels, the population of these levels has to be different. Therefore, when the sample is placed in the magnetic field, the nuclei will distribute themselves between the different energy states in thermal equilibrium. This distribution of nuclei between two states is given by the Boltzmann relation:

$$\frac{N_\alpha}{N_\beta} = e^{-\Delta E/kT} \tag{14}$$

where N_α is the number of nuclei at excited state, N_β is the number of nuclei at ground state, ΔE is the energy difference between them, k is the Boltzmann constant, and T is the temperature.

When we replace the energy ΔE by eq. 12, we obtain the following equation, which can be further modified:

$$\frac{N_\alpha}{N_\beta} = e^{-\gamma hH_0/2\pi kT} \approx 1 - \frac{\gamma hH_0}{2\pi kT} \tag{15}$$

The energy difference between the two energy states built up in a magnetic field is very small. For example, we can calculate the energy difference ΔE between the different levels of protons in a magnetic field of 14,100 G (1.4 T) using eq. 10. We obtain a value of $\Delta E = 0.024$ cal/mol, which shows a small separation of the energy states. In the case of a magnetic field of 23,500 G, the energy gap is then increased to $\Delta E = 0.04\ kcal/mol$. The population excess can be determined using eq. 14. For room temperature ($T = 25\ °C$), we obtain the following ratio:

$$\frac{N_\alpha}{N_\beta} = 0.999984$$

Let us explain this number with a numerical example. When we assume that we have 400,003 molecules, 200,000 of them populate the excited state and 200,003 the ground state. The population excess in the ground state is only 3. This excess determines the probability of a transition and thereby the sensitivity of the experiment. If the populations of the two levels were exactly the same (this does occur under certain conditions and is referred to as saturation), the numbers would be equal and there would be no net absorption of energy. In this case we do not observe an NMR signal.

Irrespective of the measurement techniques (CW or FT) used, sensitivity (peak intensity) is a very important aspect of NMR spectroscopy. Sensitivity specifications for NMR spectrometers are a prime area of competition between the instrument manufacturers. Nowadays, it is possible to obtain a spectrum on 1 μg of a compound under favorable conditions. The signal intensity is directly proportional to the ratio of the protons populating the two energy levels. The greater the population excess at lower energy level is, the greater the probability for the resonance. If the number of the resonating protons is in some way increased, the signal intensity is also increased. In order to interpret the NMR spectra better, the signal intensity has to be increased. As we have discussed above, we have to increase the number of protons populating the lower energy levels in order to obtain an intense signal. The sample concentration at a given volume can increase the peak intensity. However, it is limited by limited solubility. In other words, we have to attempt to change the population difference. Analysis of eq. 15, Boltzmann's Law, indicates that there are only two variable parameters that can directly affect the population difference: temperature T and the strength of the magnetic field H_0. Sensitivity can be slightly increased by lowering the sample temperature, since T appears in the denominator of the exponent of the Boltzmann function. By decreasing the temperature, the relative population of the lower energy level will increase, which will then be reflected in the increase in signal intensity. Care must be taken with temperature-dependent spectra. There are two main reasons why we do not use this low-temperature measurement to increase the sensitivity. We always encounter two main problems. The first is the solubility of the sample, which will decrease at lower temperatures. The sample can precipitate out of the solution. Furthermore, an increase in the viscosity of the solution will cause line broadening in the signals. The second is that the freezing of some dynamic processes within the compound (bond rotation, ring inversion, valence isomerization, etc.) will cause a change in the NMR spectra (see Chapter 8). Therefore, decreasing the temperature can never be used in order to increase the peak intensity (sensitivity of the instruments) in NMR spectroscopy. On the other hand, the NMR spectrum of a given sample can also be recorded at higher temperatures. According to eq. 15, the sensitivity of signals will then be decreased. Temperature measurements in NMR spectroscopy are used only to study the above-mentioned dynamic processes, and not to affect the sensitivity.

The remaining parameter affecting the population ratio is the strength of the magnetic field H_0. When the magnetic field strength is increased, the energy gap between the two energy states will also increase (eq. 10). The increased energy gap will directly influence the population difference, i.e. the number of protons populating the lower energy level will increase. Consequently, resonance probability will increase. Finally, we will obtain a more intense signal. We can conclude that the signal intensity can be achieved by increasing the strength of the magnetic field. Therefore, researchers are always interested in buying NMR instruments with higher magnetic fields. NMR instruments with higher magnetic fields can also simplify the NMR spectra, as well as increase the sensitivity. In this case, NMR spectra can be interpreted much more easily. This subject is dealt with in Chapter 6.

2.4 RELAXATION

As we have discussed in detail, the absorption of energy results in the excitation of the nucleus from the lower to higher energy level. This process is called resonance. After resonance, of course, the population of the different energy levels will change. The number of nuclei at the lower energy level is then reduced. In order to maintain the slight excess of nuclei at the lower energy level, which is the condition for resonance, the nuclei at the higher energy level must somehow be able to lose their energy and return to the lower energy level. The various processes that tend to restore the original equilibrium condition among the energy levels are referred to as *relaxation phenomena* (Figure 8).

As long as this equilibrium is not reestablished, a second NMR spectrum from the same sample cannot be recorded. This process of *relaxation* is important because the time spent in the excited state determines the width of the absorption peaks, pattern of the peaks, and ease with which the nucleus is observed.

The relaxation time is the time needed to relax the nuclei back to their equilibrium distribution. Relaxation processes can be divided into two categories:

(1) spin–lattice relaxation (longitudinal relaxation) T_1;
(2) spin–spin relaxation (transverse relaxation) T_2.

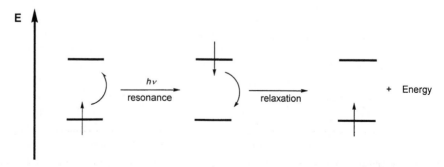

Figure 8 Simple presentation of resonance phenomenon and relaxation process.

The probability of spontaneous emission of the excess energy in the form of electro-magnetic radiation is negligible, although the excitation occurs by the adsorption of electromagnetic radiation. Before we start to discuss spin–lattice relaxation, the term *lattice* should be defined. Lattices are defined as all kinds of aggregates of atoms or molecules in the solution. Solvent molecules and dissolved gases are also considered to be lattices. The spin–lattice relaxation designated by the time T_1 involves the transfer of energy from the excited protons to the surrounding lattice. For the moment we will confine our attention to liquids and gases which undergo random translational and rotational motion. Since some or all of these atoms and molecules contain magnetic nuclei, such motions will be associated with fluctuating magnetic fields. The fluctuating lattice fields can be regarded as being composed of a number of oscillating components. If a component matches the precessional frequency of the magnetic nuclei, the transfer of energy from the excited proton to the surrounding atoms or molecule can take place that are tumbling at the appropriate frequencies. The energy will be transferred into kinetic and thermal energy. The magnitude of T_1 varies to a con-siderable extent depending on the nucleus and its environment. Liquids usually have a T_1 of 10^{-2}–10^2 s, while in solids this value is much larger. For many organic compounds, T_1 is less than 1 s. See Section 10.2 for more detailed descriptions of these processes.

The spin–spin relaxation, characterized by the time T_2, involves the transfer of energy among the processing protons. The most important feature of spin–spin relaxation is that it determines the natural width of the lines in the spectra. For more information see Section 10.2.

From the foregoing discussion, certain precautions must be taken in the preparation of the sample. The solvent must be free of magnetic impurities and paramagnetic species.

We have already mentioned the importance of relaxation times in controlling the width of peaks. If the relaxation time is very short, we observe a broadening of the peaks. In the case of longer relaxation times, there are more intense peaks (Figure 9).

This relationship is described by the *Heisenberg uncertainty principle*, which states that the product of the uncertainty of the energy ΔE and the uncertainty of the life time

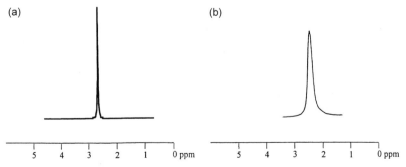

Figure 9   Influence of relaxation time on peak shape: (a) longer relaxation time; (b) decreased relaxation time.

Δt interval is a constant:

$$\Delta E \Delta t = \frac{h}{2\pi} \tag{16}$$

Substituting ΔE by ($\Delta E = h\nu$) gives

$$h \Delta \nu \Delta t = \frac{h}{2\pi}$$

$$\Delta \nu = \frac{1}{\Delta t 2\pi} \tag{17}$$

This equation shows the relation between the peak width and relaxation time. Thus, if the life is very short the uncertainty in energy will be large. This uncertainty in energy is, in turn, manifested in an uncertainty in the frequency. In short, if Δt is small, then $\Delta \nu$ is large, and the peak is therefore broad. Any kind of paramagnetic impurities existing in the NMR tube will decrease the relaxation time. Consequently, line broadening will occur so that we cannot see fine splitting, which provides very important information about the structures. Therefore, the removal of fine suspended matter by filtering through a pad of celite is recommended. The increased viscosity will also affect the relaxation times. One also has to pay attention to the concentration of the solution. The solution prepared should not be highly concentrated.

2.5 NMR INSTRUMENTATION

Two types of NMR spectrometer, based on different designs, are now used: CW and FT spectrometers. The basic requirements for all high-resolution NMR spectrometers are as follows:

(1) a magnet capable of producing a very strong static and homogeneous field;
(2) a stable radiofrequency generator;
(3) a radiofrequency receiver;
(4) a detector.

A schematic diagram of a classical NMR spectrometer is given in Figure 10.

The magnet is necessary to produce the different energy levels. The remaining components then have analogs in other methods of absorption spectroscopy (IR, UV, etc.). Mainly permanent or electromagnets are employed in low-field NMR spectrometers (1–2 T). In sweep NMR spectrometers, the compound is irradiated with electromagnetic energy of a constant frequency while magnetic field strength is varied or swept. Electronically, it is easier to maintain a constant radiofrequency and vary the magnetic field strength. By this technique the protons are brought into resonance one by one. Recording of a single spectrum takes approximately 5 min. CW spectrometers were employed in the early instruments. CW is still used in some of the lower resolution instruments (60 MHz), but CW instruments have been almost completely superseded by FT spectrometers.

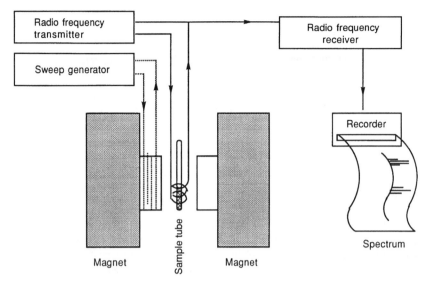

Figure 10 Schematic diagram of a classical NMR spectrometer.

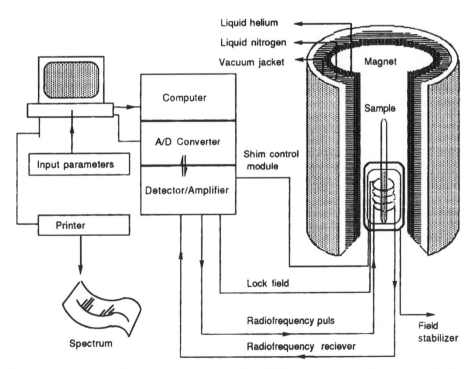

Figure 11 Schematic diagram of a Fourier transform NMR spectrometer with a superconducting magnet.

An important innovation in the field of NMR spectroscopy was the introduction of superconducting magnets that have a much higher magnetic field strength than their predecessors (up to 21 T). A superconducting magnet consists of a coil and solenoids made from a special alloy. The coil is immersed in liquid helium. The magnet has to be kept permanently at the temperature of liquid helium (4 K). To minimize the helium consumption and increase the periods between refills, the helium tank is cooled by liquid nitrogen, which has additional thermal insulation. Contrary to the CW instruments, the samples in the FT spectrometers are irradiated with a short pulse of rf radiation. This pulse excites all the nuclei at once. The most important difference between the CW and FT instruments is that the sweep method takes 5 min to record a complete spectrum, whereas the FT technique can produce a spectrum in 2–3 s. The FT technique then allows the accumulation of hundreds or thousands of spectra in a very short time by computer in order to increase the sensitivity. At this stage, we additionally do not want to perplex beginners with the details of FT NMR spectrometers. It should be noted that the technique by which NMR spectra are recorded has no bearing on their interpretation. Therefore, detailed information about the FT technique is reserved for the second part of this book (see Chapter 11). A schematic diagram of an FT NMR spectrometer is given in Figure 11.

To record an NMR spectrum of a given compound, the sample is placed in a homogeneous magnetic field. In all experiments, the sample tube is rapidly spun around its long axis by an air turbine in order to improve the field homogeneity. Furthermore, the magnetic field is always shimmed before every measurement. The shims are small magnetic fields used to cancel out the errors in the static field. During measurements the ratio of magnetic field strength to transmitter frequency must be held constant. This is usually accomplished by electronically 'locking on' to a strong, narrow signal and compensating a tendency for the signal drift. A deuterium signal is used for long-term stability of the magnetic field. Therefore, for FT instruments, deuterated solvent always has to be used.

– 3 –

Chemical Shift

3.1 LOCAL MAGNETIC FIELDS AROUND A NUCLEUS

In the previous chapter we explained that the following criterion has to be met in order to bring a nucleus into resonance:

$$\nu = \gamma \frac{H_0}{2\pi} \tag{13}$$

As we have seen before, the gyromagnetic ratio, γ, is a constant characteristic of a particular nucleus. The proton has a definitive magnitude for the gyromagnetic ratio (see Table 2.1). The different chemical environment does not have any effect on the gyromagnetic ratio. Consequently, only a single proton peak is to be expected from protons of all kinds in a given magnetic field, in accordance with the basic NMR equation (eq. 13). In this case, the nuclear magnetic resonance spectroscopy would allow us to provide the information about the type of nucleus, the elemental composition (C, N, O, etc.), which are characterized by their individual resonance frequencies. Then, proton NMR spectroscopy would provide information whether a proton is present in a compound or not. However, there are much simpler techniques for determining the presence of protons, such as elemental analysis. The fact that all protons do not have a single resonance frequency according to eq. 13 was the main factor leading to the development of NMR spectroscopy. Since different protons have different resonance frequencies, we have to try to understand this phenomenon in relation to eq. 13.

Protons resonate at different frequencies according to their chemical environments and locations. Even in a homogeneous magnetic field, the strengths of the local magnetic fields that influence the protons differ slightly from the external magnetic field. Generally speaking, all protons of a given compound located in 'different' chemical environments are under the influence of different magnetic fields. Conversely, protons located in the same chemical environment are under the influence of the same magnetic field. According to the basic NMR equation (eq. 13), if the protons are under the influence of different magnetic fields, they will naturally also have different resonance frequencies. This fact will help us to recognize different protons in NMR spectroscopy.

Now we can raise the question of why the magnetic fields around the protons are different from the applied external magnetic field. When we were discussing resonance, we were concerned only with the external magnetic field, always ignoring the effect of the electrons surrounding the protons. Electrons under the influence of an external magnetic field generate their own magnetic fields, either increasing or decreasing the influence of the external magnetic field. For example, methyl acetate

$$H_3C-\overset{\displaystyle \overset{O}{\|}}{C}-O-CH_3$$

has two different methyl groups. One of these methyl groups is bonded to the carbon atom whereas the other is bonded to the more electronegative oxygen atom. In the ^{1}H-NMR spectrum (Figure 12) we can see different signals for each methyl group.

Electrons are charged particles. Therefore, the applied external magnetic field induces circulations in the electron cloud surrounding the nucleus causing the electrons to generate their own magnetic fields. In accordance with Lenz's law, the induced magnetic field (*the secondary magnetic field*), opposed to the external magnetic field, is produced. The circulation of electron clouds and the direction of the generated magnetic field are shown in Figure 13.

Since the induced magnetic field opposes the external magnetic field, the strength of the external magnetic field around the nucleus is then reduced. The reduction of the external magnetic field around a nucleus is called *shielding*. In summary, when a compound is placed in a homogeneous magnetic field, the protons will be under the influence of different magnetic fields because of tiny magnetic fields induced when opposing the external magnetic field. Let us explain this phenomenon using some numbers. For example, assume that a proton resonates in the presence of a magnetic field of 14,000 G. We bring this proton in an external magnetic field of 14,000 G. Now, the electrons around this proton will generate a secondary magnetic field. Let us assume that the strength of this induced magnetic field is 50 G. The proton is no longer under the influence of a field of 14,000 G. Since the induced magnetic field opposes the external magnetic field, the proton is then under the influence of a total magnetic field of 13,950 G. This value is not sufficient to bring the proton into resonance, which would require the strength of the external magnetic field to be increased to 14,050 Hz. The electrons are said to shield the proton.

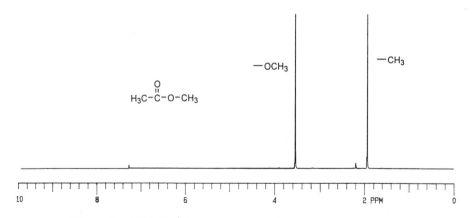

Figure 12   200 MHz ^{1}H-NMR spectrum of methyl acetate in CDCl$_3$.

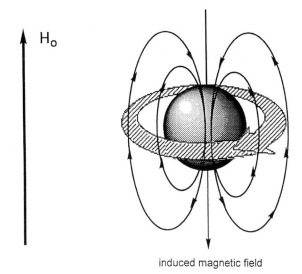

induced magnetic field

Figure 13   The circulation of the electrons under the influence of an external magnetic field. The electron circulation generates an induced magnetic field, which opposes the external magnetic field.

In certain cases, the proton may be under the influence of an increased magnetic field. Consequently, to fulfill the resonance condition, the strength of the external magnetic field has to be reduced. This is called *deshielding*. We will deal with this in the chapter on aromatic compounds.

Since the secondary magnetic field is generated by the external magnetic field, the magnitude of the secondary magnetic field depends on the strength of the external magnetic field. This is expressed by

$$H_{\text{sec}} = \sigma H_0 \tag{18}$$

where H_{sec} is the secondary magnetic field, H_0 is the external magnetic field, and σ is the shielding constant.

The term σ is referred to as the diamagnetic shielding constant, and its value depends on the electron density around the protons. We now come to the conclusion that the protons are under the influence of the external magnetic field as well as of the magnetic field induced by the electrons. This means that the effective magnetic field H_{eff} sensed by the proton is slightly less than the external field by an amount σH_0:

$$H_{\text{eff}} = H_0 - H_{\text{sec}} \tag{19}$$

$$H_{\text{eff}} = H_0 - \sigma H_0 = H_0(1 - \sigma) \tag{20}$$

Consequently, the resonance frequencies of the protons in a given molecule will be different as long as their chemical environments are different. The shielding constant will determine the resonance frequency. It has already been mentioned that the magnetic shielding constant of a nucleus is determined mainly by the electrons around the nucleus. However, there are other factors that influence the magnetic shielding constant σ.

The shielding of nuclei can be expressed as the sum of four terms:

$$\sigma = \sigma_{\text{dia}} + \sigma_{\text{para}} + \sigma_{\text{neig}} + \sigma_{\text{sol}} \tag{21}$$

where σ_{dia} is the diamagnetic shielding, σ_{para} is the paramagnetic shielding, σ_{neig} is the neighboring group shielding, and σ_{sol} is the solvent shielding.

We have explained that the diamagnetic shielding arises from the circulation of the electrons at the nucleus and that it reduces the strength of the external magnetic field. Paramagnetic shielding effects arise only for heavier nuclei and are observed in nonspherical molecules. The electrons in p-orbitals do not have a spherical symmetry, and produce large magnetic fields at the nucleus. Since the proton does not have p-orbitals, no contribution of the paramagnetic shielding in the proton is observed (see Section 12.1). Shielding arising from the neighboring groups also plays an important role in determining the chemical shifts. We will discuss these effects later. Concentration and solvent molecules also have very important contributions to the shielding constant, the most important of which are diamagnetic shielding and neighboring group shielding.

All protons will have different shielding, since the electron density around the protons and the chemical environments are different. Consequently, the effective magnetic field around the protons will vary from one proton to another. In accordance with the NMR equation (eq. 13), the protons will have different resonance frequencies. Shielding and deshielding effects cause the absorption of the protons to shift from the position at which a bare proton would absorb. These shifts, the resonance of different protons at different frequencies, are called *chemical shifts*. If two protons are located in the same chemical environment, they will be under the influence of the same magnetic field and then their resonance signals will overlap.

Inspection of the ^{1}H-NMR spectrum of methyl acetate (Figure 12) reveals two methyl resonances, which show that these methyl groups are in different chemical environments. One of these groups is bonded to a carbon atom and the other to an oxygen atom. A single

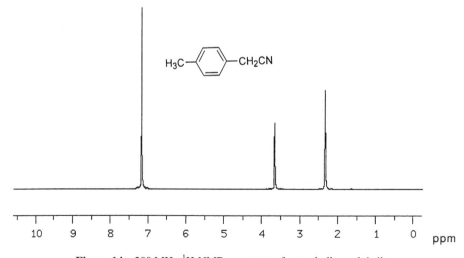

Figure 14   200 MHz ^{1}H-NMR spectrum of *p*-methylbenzylnitrile.

resonance for a methyl group furthermore supports the equivalency of the three methyl protons.

Figure 14 shows the ^{1}H-NMR spectrum of *p*-methylbenzylnitrile showing three distinct resonances. This means that the compound contains three different protons: aromatic, methylenic and methyl protons. One would expect three types of aromatic protons. The appearance of one signal indicates that the aromatic protons do not differ greatly from each other.

3.2   THE UNIT OF THE CHEMICAL SHIFT

After the discussion of the differential shielding of individual protons in a given magnetic field, we have to define the position of a resonance signal in an NMR spectrum. Chemical shifts are measured with reference to the absorptions of protons of reference compounds. The most generally used reference compound is tetramethylsilane (TMS). In practice, a small amount of tetramethylsilane is usually added to the sample so that a standard reference absorption line is produced. The distances between the signals of the sample and the reference signal are given in hertz.

$$
\begin{array}{c}
CH_3 \\
| \\
H_3C-Si-CH_3 \\
| \\
CH_3
\end{array}
$$

There are many reasons why tetramethylsilane is used as a reference.

1. The TMS signal, which is at the right-hand side of the spectrum, is clearly distinguished from most other resonances. Since the carbon atom is more electronegative than silicon, the methyl groups bonded directly to the silicium atom are shielded more, and therefore resonate at the high field (on the right-hand side of the spectrum) (Figure 15). There are also some compounds whose protons resonate on the right side of the TMS signal. These signals mostly arise from the protons located in the strong shielding area of the aromatic compounds. We will discuss this kind of compound later (see Section 3.4.1).
2. TMS is a cheap and readily available compound.
3. TMS is chemically inert. There is no reaction between TMS and the sample.
4. TMS has a low boiling point (27 °C). It can be easily removed from the sample by evaporation after the spectrum is recorded.

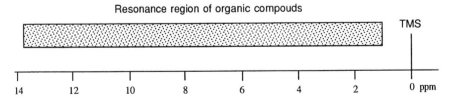

Figure 15   Approximate proton chemical shifts region of organic compounds.

5. The 12 protons of TMS produce a sharp signal. Even at lower TMS concentrations, the reference signal can be easily recognized.

Before recording a spectrum, TMS is added to the sample whose spectrum is being measured. The distance between the TMS signal and the sample signal is called the *chemical shift* of the corresponding proton.

As you will recall from the resonance condition equation

$$\nu = \gamma \frac{H_0}{2\pi} \tag{13}$$

the resonance frequency of a proton depends on the strength of the magnetic field. Since many different kinds of NMR spectrometers operating at many different magnetic field strengths (1.4–21 T) are available, chemical shifts given in Hz will vary from one instrument to another. Finally, the reported chemical shifts of a given compound will vary between different research groups working with different sized NMR spectrometers. This system would lead to chaos. Therefore, it is necessary to develop a scale to express chemical shifts in a form that is independent of the strength of the external magnetic field.

We have already defined the chemical shift as the distance between the reference signal and the sample signal:

$$\Delta \nu = \nu_{sample} - \nu_{standard(TMS)}$$

This difference varies with the size of the operating NMR spectrometer. The frequency difference (chemical shift) $\Delta \nu$ will increase with the increased magnetic field strength. To remove the effect of the magnetic field on chemical shift, we divide the chemical shift (in Hz) by the frequency of the spectrometer (in Hz) and multiply it by 10^6:

$$\delta = \frac{\nu_{sample} - \nu_{standard}}{\nu_{spectrometer}} \times 10^6 \tag{22}$$

where δ is the chemical shift, ν_{sample} is the resonance frequency of the sample, $\nu_{standard}$ is the resonance frequency of the standard, and $\nu_{spectrometer}$ is the spectrometer frequency in Hz.

The resonance frequency for the reference compound TMS is set to zero, by definition. Thus, the chemical shift equation (eq. 22) will have the following form:

$$\delta = \frac{\Delta \nu}{\nu_{spectrometer}} \times 10^6 \tag{23}$$

Since the chemical shifts are always very small (typically < 6000 Hz) compared with the total field strength (commonly the equivalent of 100, 400 and 600 million Hz), the factor 10^6 is introduced in order to simplify the numerical values, and thus the δ-values are always given in parts per million (ppm). Now the chemical shifts of NMR absorptions expressed in ppm or δ units are constant, regardless of the operating frequency of the spectrometer. Let us give an example in order to illustrate this point.

Figure 16 shows the ^{1}H-NMR spectra of *p*-methylbenzylnitrile recorded on 60 and 100 MHz NMR spectrometers. There are three sets of signals belonging to the methyl, methylenic and aromatic protons. The signal appearing at the right-hand side is the TMS absorption signal. The distances between the TMS signal and the sample signals are given in Hz. As shown in the spectra, the distances vary with the magnetic field strength.

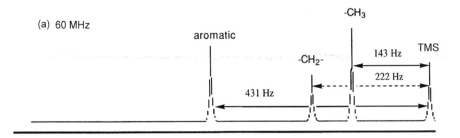

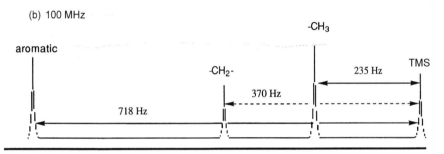

Figure 16 Schematic presentation of ^{1}H-NMR spectrum of p-methylbenzylnitrile recorded (a) on a 60 MHz NMR instrument and (b) on a 100 MHz NMR instrument. The distances in Hz between the standard signal TMS and resonance signals vary with the strength of the applied magnetic field.

To determine the chemical shifts, we apply eq. 23. For example, the methyl resonance absorbs 143 Hz downfield from the TMS signal on a 60 MHz instrument. We find a chemical shift of 2.35 ppm. The methyl proton resonance on a 100 MHz instrument appears 235 Hz downfield from the TMS signal. The calculation of the chemical shift gives the same δ-value of 2.35 ppm.

60 MHz instrument	100 MHz instrument
$\delta_{CH_3} = \dfrac{143 \text{ Hz}}{60,000,000 \text{ Hz}} \times 10^6 = 2.35 \text{ ppm}$	$\delta_{CH_3} = \dfrac{235 \text{ Hz}}{100,000,000 \text{ Hz}} \times 10^6 = 2.35 \text{ ppm}$
$\delta_{CH_2} = \dfrac{222 \text{ Hz}}{60,000,000 \text{ Hz}} \times 10^6 = 3.70 \text{ ppm}$	$\delta_{CH_2} = \dfrac{370 \text{ Hz}}{100,000,000 \text{ Hz}} \times 10^6 = 3.70 \text{ ppm}$
$\delta_{arom} = \dfrac{431 \text{ Hz}}{60,000,000 \text{ Hz}} \times 10^6 = 7.18 \text{ ppm}$	$\delta_{arom} = \dfrac{718 \text{ Hz}}{100,000,000 \text{ Hz}} \times 10^6 = 7.18 \text{ ppm}$

Now the chemical shifts can be expressed in dimensionless units, independent of the applied frequency. The resonance signal of TMS is set to zero. The range in which most NMR absorption occurs is quite narrow. Almost all ^{1}H-NMR absorptions occur $0-10\delta$ downfield from the absorption of TMS.

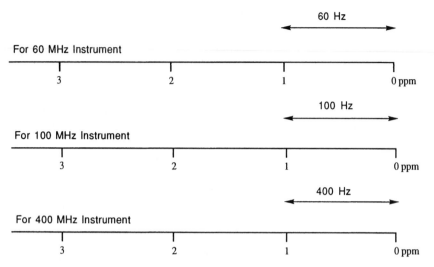

Figure 17 Presentation of ppm values in Hz on different NMR instruments.

It has been demonstrated that the above-mentioned eq. 23 provides chemical shift values regardless of the operating frequency of the spectrometer. A proton absorbs at 3.0δ on a 60 MHz instrument and its resonance signal is 180 Hz away (downfield) from the TMS signal. This means that 1.0 ppm is equal to 60 Hz on a 60 MHz instrument. The same proton absorbs also at 3.0δ on a 100 MHz instrument. In this case, the resonance signal is 300 Hz downfield from the TMS signal, which means that 1.0 ppm is equal to 100 Hz on a 100 MHz instrument. Let us assume that a sample contains two different protons. These proton resonances in ppm will not vary on different NMR instruments. However, *the difference between the chemical shift values in hertz varies from one instrument to another*. For example, we assume that our NMR spectrum consists of two singlets appearing at 1.0 and 4.5 ppm. These values cannot vary with magnetic field strength. If the spectrum is recorded on a 60 MHz instrument, the chemical shift difference between these signals is 210 Hz for the 60 MHz instrument and 350 Hz for the 100 MHz instrument (Figure 17):

$$\Delta\delta = 4.5 - 1.0 = 3.5 \text{ ppm} \qquad 3.5 \text{ ppm} = 3.5 \times 60 = 210 \text{ Hz for 60 MHz}$$
$$\Delta\delta = 4.5 - 1.0 = 3.5 \text{ ppm} \qquad 3.5 \text{ ppm} = 3.5 \times 100 = 350 \text{ Hz for 100 MHz}$$

The advantage of using an instrument with high rather than low field strength is that the different NMR absorptions are more widely separated (in hertz). This will affect the appearance of the NMR spectra and simplify the interpretation of the spectra (see Section 6.1, second-order spectra).

Electronic integration of the ^{1}H-NMR peak areas reveals the relative number of protons responsible for these protons.

3.3   SAMPLE PREPARATION

The preparation of the sample for the measurements of NMR spectra is as important as the recording of the spectra. Therefore, it is desirable to pay particular attention to the preparation of the samples. The sample has to be dissolved in the chosen solvent. A routine sample for proton NMR on a CW-NMR instrument should be around 30–50 mg, depending on the size of the molecule, in about 0.5 ml of solvent. However, for high-field NMR spectra, 1 mg of sample in 0.5 ml would be a strong solution. Under favorable conditions, it is possible to obtain a spectrum on 1 μg of a compound of modest molecular weight. An NMR tube is a cylindrical tube made of special glass 18 cm in length and 5 mm in external diameter (Figure 18).

High-quality NMR spectra can only be obtained on solutions that are completely free from suspended dust or fibers, which may affect the homogeneity of the magnetic field, which causes the broadening of NMR lines. To achieve this it is always recommended to filter the solution for NMR directly into the sample tube through sintered glass. The solution should fill the tube to a height of 4–5 cm. The detection coil in the NMR probe only receives signals from a finite volume, and any sample outside this volume may as well not be there. There is, naturally, a practical limit to the extent to which the sample volume can be decreased since the reduction causes a concurrent loss of signal intensity. Likewise, when there is an excess of solvents, the concentration of the sample will be reduced, which will directly influence the intensity of lines, and expensive NMR solvent

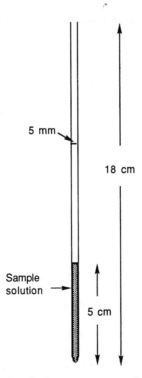

Figure 18   NMR sample tube for measurement with an internal standard.

will be wasted. Furthermore, increased concentration and solvent volume will affect the spin rate of the NMR tubes. In this case, desirable high-resolution NMR spectra cannot be obtained. In order to partially compensate for the inhomogeneity of the magnetic field, the NMR tube is spun during measurements with the help of air pressure. The ideal spin rate has to be adjusted. Lower and higher spin rates will directly affect the appearance of the signals. The choice of a suitable solvent for the determination of NMR spectra largely depends on the solubility in the selected solvent of the compound to be studied. The amount of solvent used is approximately 500 mg. The sample to be recorded can be 1.0–10.0 mg. In any case, the concentration of the solvent molecules is 50–500 times higher than the sample concentration. If the solvent molecules contain protons, their resonance signals will be 50–500 times more intense than those of the sample molecules, and the sample signals will be lost. Therefore, the ideal solvent should contain no protons (such as is in carbon tetrachloride, tetrachlorethylene, or carbondisulphide) or the solvent protons have to be replaced by deuterium. Since pulsed instruments depend on deuterium in the field-frequency lock, deuterated solvents are required. Carbon tetrachloride, which is the cheapest solvent, can only be used on CW instruments. Furthermore, its use as a solvent is limited because of its low polarity. In order to solve a polar compound, polar solvents have to be used. The recommended solvent is deuterochloroform, $CDCl_3$ is used whenever circumstances permit, which is in fact most of the time. The replacement of protons of solvent by deuterium atoms does not mean that deuterated solvents are inactive in NMR spectroscopy. Since the spin quantum number of deuterium is equal to $I_D = 1$, deuterium is active in the NMR spectroscopy. According to the NMR equation (eq. 13) deuterium resonances appear outside the range where proton resonances are observed (the gyromagnetic ratio of deuterium is different from that of this proton). Consequently, deuterium resonances are not observed in the proton NMR spectra. The solvents used have to be inert, low boiling, and inexpensive. For poorly soluble samples a series of deuterated solvents are available on the market: acetone-d_6, methanol-d_4, dichloro-methane-d_2, pyridine-d_5, benzene-d_6, dimethyl sulphoxide-d_6, acetonitrile-d_3, and cyclohexane-d_{12} (Table 3.1). These solvents increase in price according to the difficulty of preparation in deuterated form. It should be noted that the measurements in pyridine and benzene will slightly influence the chemical shifts. These changes (0.1–0.4 ppm) will never be so great as to affect the interpretation of spectra. By reporting the chemical shift values, the solvent that is used always has to be described. The viscosity of the chosen solvent can have a considerable effect on the observed line-width. For example, measurements in dimethyl formamide or dimethyl sulphoxide can cause line broadening, especially at high concentrations of the sample. The highest resolution can only be obtained in nonviscous solvents. Some solvents, such as water, methanol, and acetic acid, have exchangeable protons, which will prevent the observation of other exchangeable protons in the sample. Typically, the commercially available solvents do not contain 100% deuterium atoms. The isotopic purity of the deuterated solvents ranges from 99.5 to 99.99%. Routinely used chloroform has a purity of 99.5%. Of course, the high purity grades are considerably more expensive.

To study the dynamic processes by NMR spectroscopy, the spectra have to be measured in a range of -185 to $+200\,°C$ depending on the system. The choice of solvent is therefore very important. The melting and boiling points of solvents have to be taken into consideration. For high temperatures, 1,1,2,2-tetrachloroethane or dimethyl

Table 3.1

Physical properties of solvents important for NMR spectroscopy

Solvent	Dielectric constant	mp (°C)	bp (°C)	δ_{1H}	δ_{13C}
CCl$_4$	2.24	−2.3	77	−	96.7
CS$_2$	2.64	−112	40	−	193.1
D$_2$O	78.5	3.8	101.4	4.63	−
CDCl$_3$	4.80	−64	61	7.27	77
CD$_2$Cl$_2$	8.90	−97	40	5.32	53.8
Dioxane-d$_8$	2.20	12	100	3.53	66.5
THF-d$_8$	7.60	−106	65	3.58−1.73	67.4−25.2
Benzene-d$_6$	2.28	6.8	79.1	7.16	128
Pyridine-d$_5$	12.40	−42	114.4	8.71−7.55, 7.19	149.9, 135.5, 123.5
Acetone-d$_6$	20.70	−93.8	55.5	2.05	206.7−30.7
CD$_3$COOD		15.8	115.5	11.5−2.03	178.4−20.0
CF$_3$COOD		−15	75	11.3	163.8−115.7
CD$_3$OD	32.7	−98	65.4	3.31	49.0
CD$_3$CN	37.50	−48	80.7	1.93	118.2, 39.5
DMSO-d$_6$	46.7	18	190	2.50	167.7−35.2
DMF-d$_7$	36.7	−61	153	8.01, 2.91, 2.74	30.3

sulphoxide is often used. For low temperatures, dichloromethane, acetone, methanol or solvent mixtures can be used.

3.4 FACTORS INFLUENCING THE CHEMICAL SHIFT

Before we begin a discussion of the factors that influence the chemical shifts, let us give a short summary of the absorption regions of some characteristic functional groups. The chemical shifts of some functional groups depending on the hybridization of carbon atoms are presented in Figure 19 and Table 3.2.

As these data show, proton chemical shifts fall within the rather narrow range of 0−14δ. Saturated hydrocarbons absorb generally between 1.0 and 4.0 ppm, whereas the resonances of olefinic protons appear in the region of 5.0−6.5 ppm and the resonances of aromatic protons in the region of 7.0−8.5 ppm. These values are never exact

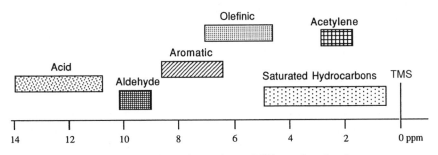

Figure 19 General absorption regions of different functional groups.

Table 3.2

^{1}H-NMR chemical shifts of some characteristic functional groups

Functional group		Chemical shift, δ
Cyclopropane		0.2
Primary hydrocarbons	$R-CH_3$	0.9
Secondary hydrocarbons	R_2CH_2	1.3
Tertiary hydrocarbons	R_3CH	1.5–1.6
Allylic	$-C=C-CH_3$	1.7
Amine	$R-NH_2$	1.0–5.0
Alcohol	$R-OH$	1.0–5.5
Carbonyl	$-CO-CH-$	2.0–2.7
Acetylene	$-C\equiv CH-$	2.0–3.0
Benzylic	$Ar-CH-$	2.2–3.0
I	$-CH_2-I$	2.0–4.0
Br	$-CH_2-Br$	2.5–4.0
Cl	$-CH_2-Cl$	3.0–4.0
F	$-CH_2-F$	4.0–4.5
Alcohol	$-CH-OH$	3.4–4.0
Ether	$RO-CH-$	3.3–4.0
Ester	$RCOOCH-$	3.7–4.1
Olefine	$-C=CH-$	4.5–6.5
Aromatic	$Ar-H$	6.0–8.5
Aldehyde	$R-CHO$	9.0–10.0
Acid	$R-COOH$	10.0–14.0

ranges. Resonances outside these given ranges can be encountered depending on the electronegativity and the number of the attached substituents. The values presented in Figure 19 are values for monosubstituted functional groups. As an aromatic proton can resonate in the range of olefinic protons, so can an olefinic proton resonate in the range of aromatic protons. Increased crowding of alkyl groups or electron-withdrawing substituents at a carbon atom causes a successive downfield shift of the proton next to the saturated center. We will analyze these shifts in detail.

There are two main factors that determine the position of proton absorptions:

(1) The local magnetic field generated by the circulation of the electrons around the nucleus.
(2) The magnetic anisotropy of neighboring groups.

Let us first discuss the influence of electron density (charge density) on chemical shift.

3.4.1   Influence of electron density on chemical shift

As mentioned before, the circulation of the electrons induces a secondary magnetic field, which opposes the external magnetic field according to Lenz's law. Consequently, the strength of the external magnetic field is reduced, and thus we need a higher magnetic field for the resonances of the individual protons. In the NMR spectra, the strength of

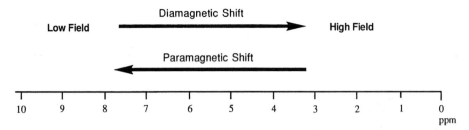

Figure 20   Presentation of diamagnetic and paramagnetic shifts on NMR scale.

the magnetic field increases from left to right. The shift of the resonances to a higher field (to the right-hand side of the chart) is called *diamagnetic shift*. A shift in the opposite position, to the lower field, is called *paramagnetic shift* (Figure 20).

The degree of shielding (inducing of secondary magnetic field) depends on the density of the circulating electrons. The degree of shielding of a proton on a carbon atom will depend on the inductive effect of other groups attached to the carbon atom. As a consequence, the increased electron density will generate a stronger secondary magnetic field, which will further decrease the strength of the external magnetic field. The chemical shifts will then move to a higher field. Conversely, less electron density will generate less shielding and the resonance will appear at low field.

Increased electron density	Decreased electron density
↓	↓
Strong secondary magnetic field	Weak secondary magnetic field
↓	↓
Strong shielding	Less shielding
↓	↓
Resonance at high field	Resonance at low field

The chemical shift of a methine group (CH) attached to three functional groups depends entirely on the electronic nature of substituents such as X, Y, and Z.

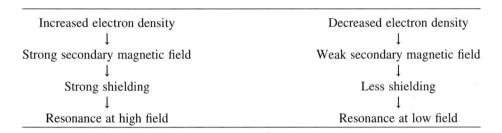

The substituents with electron-withdrawing ability will decrease the electron density around the proton and cause less shielding. Then, the chemical shift will appear at a lower field. However, if the substituents are electron-releasing groups, the shielding will be increased. The proton resonances will appear at high field. If an electron-withdrawing substituent and an electron-releasing substituent are attached to the same carbon atom, their effects will be compensatory and the proton resonance will appear at either high or low field depending on which substituent has a stronger effect.

The chemical shifts of mono-, di- and trisubstituted alkyl halides clearly demonstrate the effect of electron density on the chemical shift (Table 3.3).

Table 3.3

^{1}H-NMR chemical shifts of alkyl halides (δ) [1]

Alkyl halide	Substituent X			
	F	Cl	Br	I
CH$_3$X	4.3	3.0	2.7	2.2
CH$_2$X$_2$	5.4	5.3	5.0	3.9
CHX$_3$	6.2	7.3	6.8	4.9

Since fluorine is the most electronegative element, the electron density around the protons in CH$_3$F decreases more than in the other methyl halides, and therefore the protons of CH$_3$F resonate at a lower field ($\delta = 4.3$ ppm), i.e. increasing δ_H values. An increased number of electronegative substituents shift the chemical shifts further to low field. Table 3.3 shows that the shifts of the substituents are additive.

The proton NMR chemical shifts of the following charged nonbenzenoid compounds clearly reflect a linear correlation between the electron density and the chemical shifts. Benzene has 6 π-electrons and 6 protons. The number of electrons per H atom is $\frac{6}{6} = 1$. The increase or decrease of electron density is directly reflected in the chemical shifts.

$$\text{Electron density} = \frac{\text{Number of electrons}}{\text{Number of protons}}$$

Aromatic compounds

	13	14	15	16	17	18
π-Electron density	0.676	0.85	1.0	1.11	1.2	1.25
^{1}H chemical shift, δ	10.9	9.17	7.27	6.75	5.37	5.65

Benzene resonates at 7.27 ppm, whereas tropylium cation **14**, which is an aromatic compound, resonates at 9.17 ppm due to the decreased electron density (electron density: 0.85). Cyclopropenyl cation **13** resonance appears at a much lower field, 10.9 ppm because of the more decreased electron density (0.676) around protons. On the other hand, negatively charged aromatic compounds **16**, **17** and **18** resonate at high field, compared to benzene, because of the increased shielding caused by the electrons.

The effect of oxygen substituents on proton chemical shifts has also been investigated in detail (Table 3.4). Oxygen atoms behave like halides, decreasing the electron density around protons and causing low field shift. Alkoxy protons in ether, alcohol, and esters resonate in the range of approximately 3.0–4.5 ppm. Proton NMR resonance data of some saturated hetero compounds are also presented in Table 3.4.

Table 3.4

^{1}H-NMR chemical shifts in selected compounds containing heteroatoms (δ)

Alcohols

H$_3$C—OH	H$_3$C—H$_2$C—OH	H$_3$C—HC—OH (CH$_3$)	H$_3$C—C—OH (CH$_3$, CH$_3$)
3.39	1.18 3.59	1.16 3.94	1.22

HO—H$_2$C—CH$_2$—OH
3.58

HO—H$_2$C—CH$_2$—O—CH$_2$—CH$_2$—OH
3.7   3.7

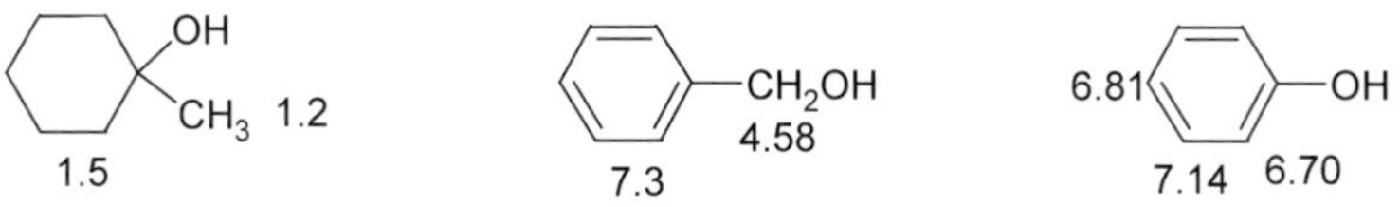

OH
CH$_3$ 1.2
1.5

-CH$_2$OH
4.58
7.3

6.81 ⟨⟩ -OH
7.14 6.70

Ethers

H$_3$C—O—CH$_3$
3.3

H$_3$C—CH$_2$—O—CH$_2$—CH$_3$
1.3   3.7

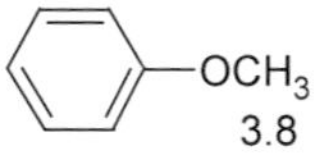

-OCH$_3$
3.8

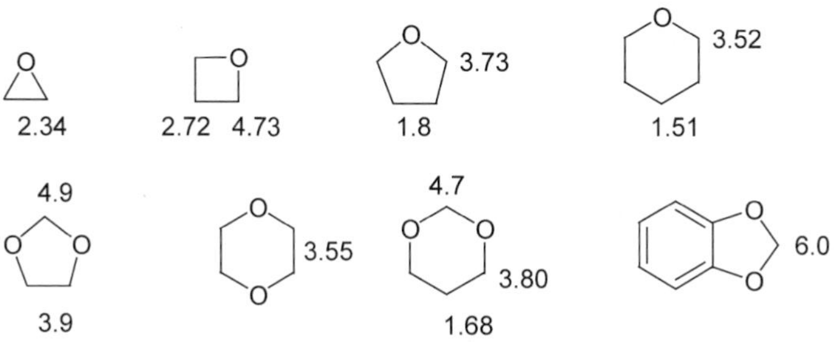

2.34

2.72 4.73

3.73
1.8

3.52
1.51

4.9

3.9

3.55

4.7
3.80
1.68

6.0

Esters

O‖—C—O—CH$_3$
3.7

O‖—C—O—CH$_2$—CH$_3$
4.2   1.3

O‖—C—O—CH$_3$
3.7

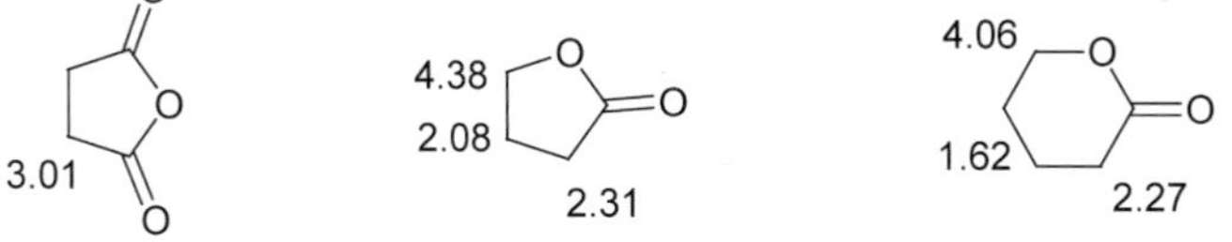

3.01

4.38
2.08
2.31

4.06
1.62
2.27

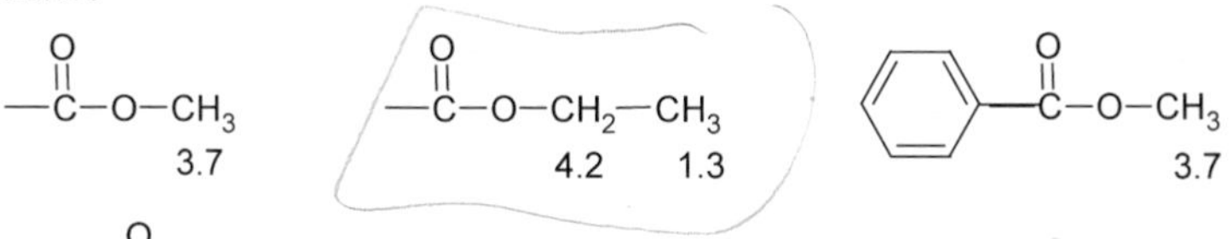

Saturated heterocycles

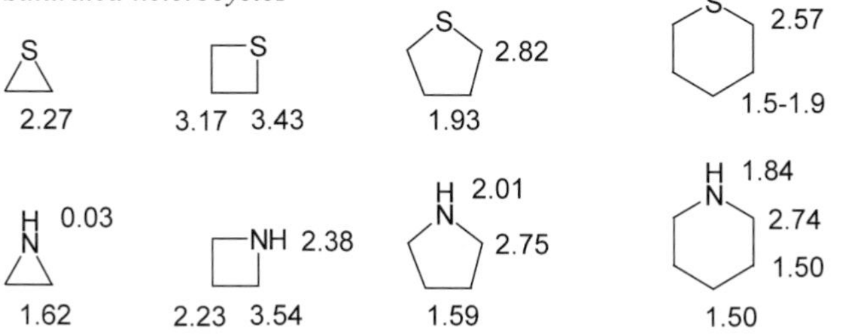

S
2.27

S
3.17 3.43

S
2.82
1.93

S
2.57
1.5-1.9

H—N 0.03
1.62

—NH 2.38
2.23 3.54

H—N 2.01
2.75
1.59

H—N 1.84
2.74
1.50
1.50

The chemical shifts of the proton in saturated compounds can be calculated by means of substituent constants (σ-values). The σ-values of different substituents are illustrated in Table 3.5.

With the help of the substituent constants we can estimate the chemical shifts of given compounds. The proton whose chemical shift has to be calculated can be a primary, secondary or a tertiary proton. The chemical shift of a methylene group attached to two substituents can be calculated by means of the substituents constants using the following equation (Table 3.5):

$$\delta_{CH_2} = 1.25 + \sigma_1 + \sigma_2$$

For example, the resonance frequency of the methylene protons of the following compound is to be determined:

$$H_3CCH_2 - CH_2 - O - \bigcirc$$

$$\delta_{CH_2} = 1.25 + 2.3 + 0.4 = 3.95 \text{ ppm}$$

The result is in good agreement with the experimental data of 3.86 ppm.

Table 3.5

Substituent constants (σ) for estimating ^{1}H chemical shifts of methylene and methine protons (δ) [2]

$R_1-\mathbf{CH_2}-R_2$ $\delta_{CH_2} = 1.25 + \sigma_1 + \sigma_2$	$R_1-\overset{\overset{\displaystyle R_3}{\vert}}{\mathbf{CH}}-R_2$ $\delta_{CH} = 1.50 + \sigma_1 + \sigma_2 + \sigma_3$
Substituent	Substituent constants (σ)
−Cl	2.0
−Br	1.9
−I	1.4
−Alkyl	0.4
−C=C−	0.8
−C≡C−	0.9
−C≡N−	1.2
−Ph	1.3
−CHO, −COR	1.2
−COOH	0.8
−NH$_2$	1.0
−NO$_2$	3.0
−SH	1.3
−OH	1.7
−OR	1.5
−O−CO−Ph	2.9
−O−CO−R	2.7

The chemical shift for the methine proton of the following compound is calculated from the σ-values in Table 3.5:

$$\text{NO}_2$$
$$\text{Cl}-\overset{|}{\text{CH}}-\text{CH}_2\text{CH}_3$$

$$\delta_{CH} = 1.5 + 1.2 + 1.3 + 2.7 = 6.70 \text{ ppm}$$

The value found is 6.66 ppm.

Ring current effect (diamagnetic ring current)

Aromatic protons resonate at a field 1–2 ppm lower than olefinic protons (Table 3.2 and Figure 19). The low-field shift of the protons directly bonded to an aromatic ring is caused by the circulation of the π-electrons (Figure 21). The presence of a ring current is characteristic of all Hückel aromatic molecules and serves as an excellent test of aromaticity.

Let us explain the ring current with a benzene ring. When a benzene ring (or an aromatic compound) is placed in an external magnetic field, the π-electrons of the ring circulate around the ring and generate a ring current. This ring current in turn induces a local secondary magnetic field (Figures 21 and 22), as we have seen in the circulation of the electrons around a nucleus. This leads to two current loops: one above and one below the plane of the σ-bonds. The induced field opposes the applied field in the middle of the ring (above and below), but reinforces the applied field outside the ring. This ring current generated by the π-electrons is called the *diamagnetic ring current*.

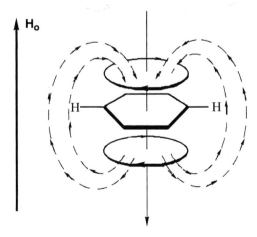

Figure 21 The schematic presentation of aromatic ring current. Aryl protons are deshielded by the induced magnetic field caused by π-electrons circulating in the molecular orbitals of aromatic ring.

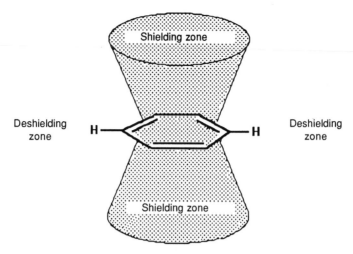

Figure 22   Schematic presentation of shielding and deshielding zones around benzene ring.

   This ring current decreases the strength of the external magnetic field in the middle of the ring (above and below) and increases the strength of the external magnetic field outside the ring. As a result, protons in the molecular plane and outside the ring are under the influence of an increased magnetic field and are deshielded. The resonances of these protons shift to low field. Conversely, protons in the region above and below the plane are under the influence of a decreased magnetic field and are shielded. Therefore, resonances of these protons appear at high field. The benzene ring does not contain any protons in the middle of the ring or above and below the ring. All benzene protons are located in the molecular plane and outside the ring. Chemical shifts of the aromatic protons therefore appear at low field. We will discuss later in this section some aromatic compounds that possess protons in the middle of the ring.

   We have already mentioned that the low field shift of the aromatic proton resonances is the best criterion for aromaticity. As a result, aromatic protons resonate at a field 1–2 ppm lower than olefinic protons. Figure 23 shows the ^{1}H-NMR spectra of some aromatic compounds. Since the six benzene protons are equal, they resonate at 7.27 ppm as a singlet. Naphthalene has two different protons. However, the ^{1}H-NMR spectrum of naphthalene looks complex and shows further peak splitting. The splitting in the peaks will be explained later. At this stage, we are interested only in the fact that naphthalene protons resonate as two sets of multiplets. One of these belongs to the α-protons and the other to the β-protons. Naphthalene protons resonate at a field *ca.* 0.5 ppm lower than benzene protons. Thus, the α-protons in naphthalene absorb at a lower field than β-protons because the contributions of the two benzene rings (diamagnetic ring currents) are more important at the α-position since the latter is closer to both rings.

   The third spectrum in Figure 23 shows the ^{1}H-NMR spectrum of 1,6-methano[10] annulene. The 1,6-methano[10]annulene (**20**) has 10 π-electrons, obeys the Hückel rule and is an aromatic compound.

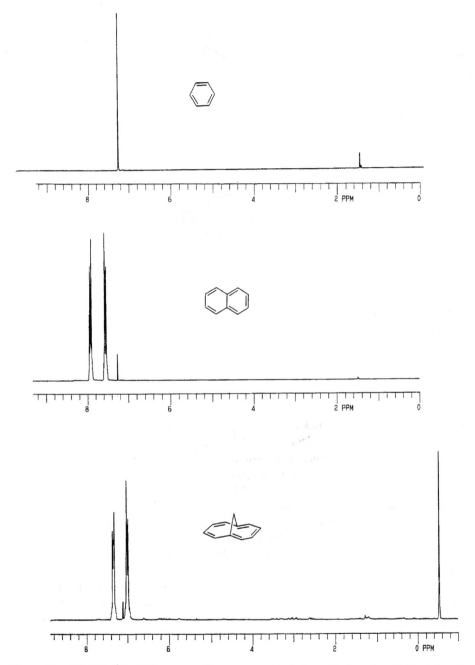

Figure 23 200 MHz ^{1}H-NMR spectra of benzene, naphthalene, and 1,6-methano[10]annulene in CDCl$_3$.

15   **19**   **20** H$_a$

δ_H = 7.27 ppm δ_{H_a} = 7.81 ppm δ_{H_a} = 7.27 ppm

                               δ_{H_b} = 7.46 ppm δ_{H_b} = 6.95 ppm

                                                               δ_{H_c} = −0.5 ppm

The chemical shifts observed in the ^{1}H-NMR spectra clearly indicate whether a given compound possesses an aromatic structure (inducing a strong diamagnetic ring current) or not. The 1,6-methano[10]annulene (**20**) has two types of protons: (i) protons attached to the ring and (ii) methylenic bridge protons located above the aromatic ring. The aromatic protons are deshielded and absorb at a field lower than normal (olefinic protons) and the methylenic protons are shielded and absorb at a field higher (− 0.5 ppm) than normal [3]. This resonance appears on the right side of the TMS signal. The methylenic protons are bonded to two vinyl groups as shown below:

$$-CH=C-CH_2-C=CH-$$

According to the substituent constants, such a −CH$_2$− group should resonate at approximately 3 ppm. However, the resonance signal of the methylenic protons is shifted around 3.5 ppm upfield. This observation indicates the presence of a strong diamagnetic ring current in **20** and the location of the −CH$_2$− protons in the shielding cone. The 1,6-methano[10]annulene is therefore a strong aromatic compound. Such molecules are called *diatropic*.

The cyclooctatetraene dianion (**25**) has a planar structure and is an aromatic compound. The protons of the cyclooctatetraene dianion have a resonance frequency of 6.4 ppm (Table 3.6). This value is higher than the resonance frequency of the other aromatic protons. The existence of two (−) charges in the molecule contributes to the further shielding of the cyclooctatetraene dianion protons, which causes high field resonance. Even this high field shift shows the strong diamagnetic ring current present in **25**. In [14]-annulene (**26**) it is noteworthy that the signal of the methyl protons (located in the middle of the ring current) is found at $\delta = -4.25$ ppm. In the absence of the specific shielding effects of the ring current, one would expect a signal for these protons at about $\delta = 1.0$ ppm. This dramatic shift of the methyl protons of around 5.5 ppm supports the strong aromaticity and the location of the methyl group in the most effective region of the shielding cone.

[18]-Annulene (**27**) has 12 outside protons and 6 inside protons. In spite of the fact that all of these protons are attached to sp^2-hybridized carbon atoms, the inner protons are strongly shielded and resonate at − 2.99 ppm, whereas the outer protons resonate at

Table 3.6

^{1}H-NMR chemical shifts in selected aromatic compounds (δ)

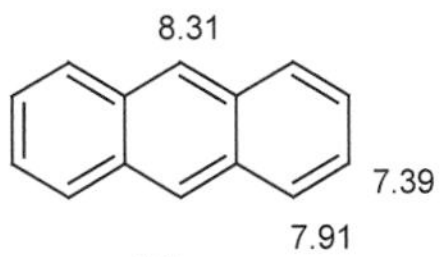

21

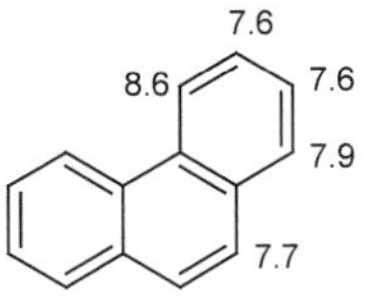

22

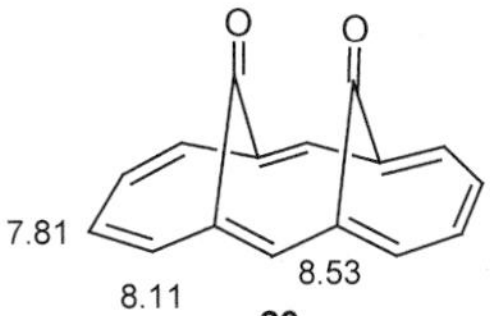

23

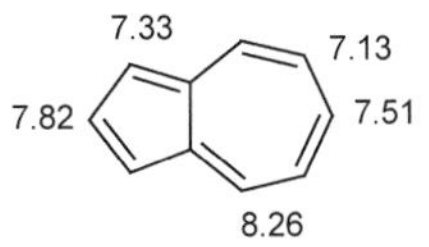

24

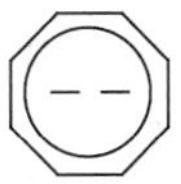

25 [10]–Annulene[4]

δ_H = 6.4 ppm

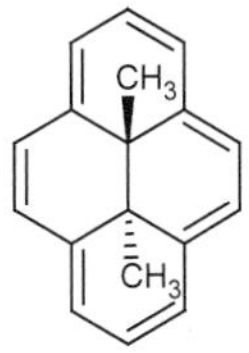

26   [14]–Annulene[5]

δ_{ring} = 7.95–8.67 ppm

δ_{CH_3} = −4.25 ppm

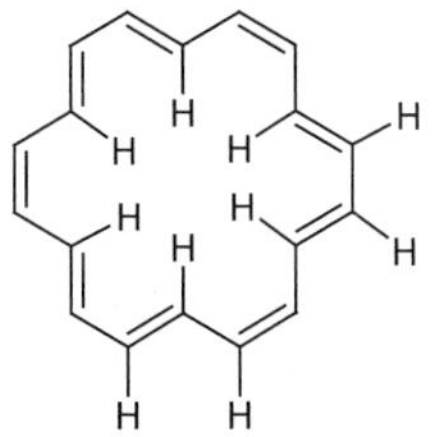

27 [18]–Annulene[6]

δ_{outer} = 9.3 ppm

δ_{inner} = −2.99 ppm

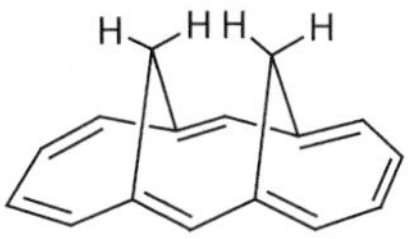

28 [18]–Annulene[7]

$\delta_{aromatic}$ = 7.0–7.9 ppm

δ_{bridge} = 0.9 and −1.2 ppm

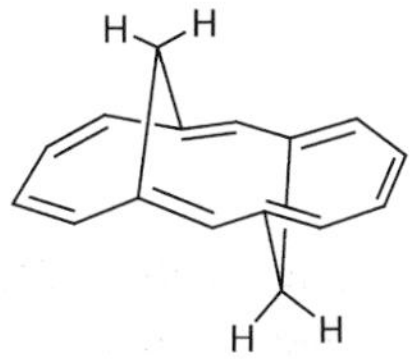

29 [18]–Annulene[8]

$\delta_{olefinic}$ = 5.7–6.6 ppm

δ_{bridge} = 1.9 and 2.6 ppm

9.3 ppm. The unusual chemical shift difference of 12.3 ppm between the inner and outer protons clearly shows the effect of the strong diamagnetic ring current in **27**. In particular, the high-field resonance of the inner proton supports the presence of a strong shielding effect.

The presence of a strong ring current and deshielding requires the aromatic compound to have a planar structure. For the effective overlapping of the carbon $2p_z$ orbitals, the orbitals must have parallel orientation, which is only the case in complete planar structures. Thus, the [14]-annulenes **28** and **29** have different structures with regard to the orientation of the bridge methylene groups. The spectra of the two compounds show distinct differences in the resonance frequencies of the bridge as well as of the perimeter protons. *Anti*-[14]-annulene **29** behaves like an olefine, because of the extensive twisting of the carbon–carbon bonds, which hinders the formation of the ideal conformation for the effective overlap of the carbon $2p_z$ orbitals. The resonance frequency of the perimeter protons of **29** ($\delta = 5.7$–6.4 ppm) shows that the expected ring current does not exist. On the other hand, the observed resonance frequency for perimeter protons of *syn*-isomer **28** ($\delta = 7.0$–7.9 ppm) and high-field absorption of the bridge protons ($\delta = 0.9$ and -1.2 ppm) indicate the existence of a strong diamagnetic ring current.

[10]-Paracyclophane (**30**) is one of the best examples of organic compounds, proficiently reflecting the presence of a ring current and distribution of magnetic anisotropy inside and outside the benzene ring.

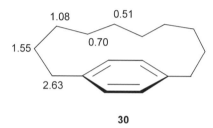

30

[10]-Paracyclophane (**30**) contains a carbon chain of 10 –CH$_2$– units, which is attached to the benzene ring at 1- and 4-positions. Let us analyze the chemical shifts of these methylene groups. Those directly adjacent to the benzene ring resonate at 2.63 ppm, and are deshielded [9]. The methylene proton resonances next to these appear at 1.55 ppm. Comparison of this value with that of alkane chain methylene protons (1.3–1.4 ppm) shows that these protons are slightly deshielded (0.2–0.3 ppm). This value indicates that the deshielding effect of the ring current at the position of these methylene protons is decreased. The third methylene protons resonate at 1.08 ppm, which means that these protons are located in the shielding zone. The resonance frequencies of the remaining methylene protons appear at a higher field (0.5 and 0.7 ppm) because of stronger shielding. The different δ_H values of various methylene groups reflect their positions with respect to the aromatic ring and distribution of the shielding and deshielding zones around the benzene ring.

Changes in the chemical shift due to the aromatic ring current can be calculated at any point around the benzene ring. The zone is limited where the ring current is effective. Shielding and deshielding decrease with the distance from the middle of the benzene ring. The simplest method is to calculate the chemical shift differences ($\Delta\delta$) for any point P with the following equation:

$$\Delta\delta = \mu \frac{(1 - 3\cos^2\Theta)}{r^3} \tag{24}$$

where r and Θ are shown in Figure 24 [10]. As one can easily see from eq. 24 and Figure 24, proton chemical shifts are strongly affected by the distance r to the point dipole's center and the angle Θ between the line joining this center to the observed nucleus and the axis A.

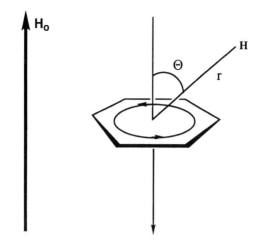

Figure 24   Graphical presentation of a proton and the aromatic ring current of benzene.

Paramagnetic ring current: ^{1}H-NMR spectra of antiaromatic compounds

We have explained the magnetic anisotropy generated by the aromatic compounds in a given magnetic field. Furthermore, we have seen that the resonances of protons outside the aromatic ring shift to lower field, whereas the resonances of protons located above (below) and in the ring shift to higher field.

The ^{1}H-NMR spectra of the antiaromatic compounds are completely different from those of the aromatic compounds. Quantum mechanical calculations predict that the antiaromatic compounds generate a paramagnetic ring current effect with just the opposite consequences for the proton resonance frequencies. Therefore, protons outside and inside the plane of the ring are shielded and those within the perimeter are now deshielded. This variant behavior of antiaromatic compounds can be rationalized with the help of a simple quantum mechanical model. At this stage, one should not suppose that the antiaromatic compounds generate a similar ring current effect with the opposite direction.

In order to understand this variant behavior of aromatic and antiaromatic compounds, we have to analyze the electronic structures of benzene and cyclobutadiene, which are typical members of $4n + 2$ and $4n$ π-electron systems. The energy level diagrams of these compounds and populations of the orbitals are shown in Figure 25. In the case of benzene, all bonding orbitals are filled and the energy gap between the bonding and antibonding orbitals is high. Consequently, in an aromatic system a closed shell results (Figure 25) and the occupied orbitals produce a diamagnetic contribution to the magnetic susceptibility. However, cyclobutadiene has two degenerated highest nonbonding orbitals that contain only one electron each, the spins of which are unpaired (Hund's population rule). According to the Jahn–Teller theorem, the degeneracy of the highest occupied orbitals can be destroyed by a slight perturbation of the molecular symmetry, and this allows both electrons to occupy a single lower lying energy level. The resulting energy level diagram (Figure 25a) shows that there is only a small energy gap between the highest occupied molecular orbital (HOMO) and the lowest unoccupied molecular orbital (LUMO).

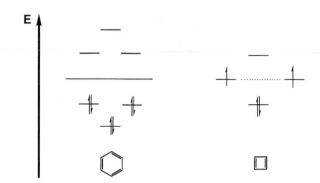

Figure 25   Energy level diagram for benzene and cyclobutadiene.

This difference is much smaller than the corresponding energy difference in the case of benzene (aromatic compounds).

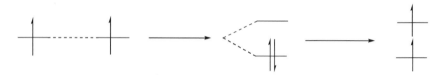

An interaction with the magnetic field H_0 will cause an electron jump from the HOMO to the LUMO, since this energy gap is very small. These unpaired electrons produce a paramagnetic contribution to the shielding constant σ. This is much larger in magnitude than the diamagnetic contribution to the shielding constant σ.

Therefore, the ^{1}H-NMR spectra of the antiaromatic compounds are different. Excitation of the electrons from the HOMO to the LUMO in the case of aromatic compounds is not possible due to the high energy gap.

^{1}H-NMR data of some antiaromatic compounds

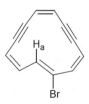

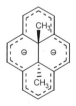

31   [12]-Annulene [11]   **32**   [12]-Annulene [12]   **33**   [16]-Annulene [13]

δ_{H_a} = 16.0 ppm δ_{ring} = 2.07–3.65 ppm δ_{ring} = –3.95 ppm

                                                                                           δ_{CH_3} = 21.0 ppm

A series of experimental observations confirms the theoretical prediction by the detection of paramagnetic ring current effects. The dianion **33** of *trans*-10b,10c-dimethyldihydropyrene provides a striking example. In the neutral compound **26** the signal of the methyl protons is found at $\delta = -4.25$ ppm. In the absence of the specific shielding effect of the ring current, one would have expected a signal at $\delta \approx 1.0$ ppm. The aromatic protons in **26** resonate at 7.95–8.67 ppm. The methyl protons in **33** resonate at 21.0 ppm and the aromatic ring protons at -3.95 ppm.

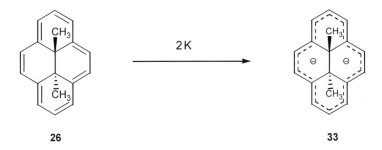

26   33

Dramatic shift differences between the spectra of the neutral compound **26** and **33** are observed. Compound **26** has 14 π-electrons and it is an aromatic compound. It shows a strong diamagnetic ring current. By reduction of compound **26** with metallic potassium, it is converted into the doubly charged dianion **33**, which has 16 π-electrons. This compound, **33**, generates a strong paramagnetic ring current as explained above. Therefore, we observe a change of approximately $\Delta\delta = 25$ ppm in the chemical shift of the methyl protons. Molecules that exhibit this behavior are called *paratropic*.

Chemical shifts of the substituted aromatic compounds: the effect of the substituents

Since all of the ring protons of benzene are equal, they resonate at 7.27 ppm as a singlet. By attachment of a substituent to the benzene ring, the electron density in the ring will either increase or decrease depending on the electronic nature of the substituents. In aromatic compounds the shielding is determined mainly by the mesomeric effect of the substituents. For example, electron-withdrawing substituents will decrease the electron density in the ring. As a result of this effect, the ring protons will be more deshielded and their resonances will appear at a lower field than benzene. However, the observed effect will not be of the same magnitude on the resonance frequencies of the ring protons. For example, let us analyze the acetophenone molecule. The acetyl group is an electron-withdrawing substituent and decreases the electron density in the ring.

   These resonance structures explain why the *ortho* and *para* positions of monosubstituted benzenes are more deshielded by electron-withdrawing substituents. Moreover,
the *ortho* protons are further deshielded because of the additional deshielding effect of the
carbonyl group (see carbonyl groups).
   Figure 26 shows the ^{1}H-NMR spectrum of acetophenone. By comparing this
^{1}H-NMR spectrum with those of benzene, it can be easily rationalized that all ring
protons of acetophenone are deshielded (resonate at a lower field). In the case of
nitrobenzene, all protons appear at a much lower field because of the strong electron-
withdrawing ability of the nitro group. Inspection of these NMR spectra (Figure 26)
indicates that the spectra consist of three different sets of signals that belong to
ortho-, *meta*-, and *para*-aromatic protons. These signals show further splitting. The
reason for this signal splitting will be explained in Chapter 4. The low field part of
these signals belongs to the *ortho* protons, which are most strongly affected. These
are followed by the *para* protons, while the *meta* protons remain almost unaffected.
It has been found experimentally that for multiple substitution the substituents
provide nearly constant additive contributions to the chemical shifts of the remaining
ring protons.
   Electron-releasing substituents ($-NH_2$, $-OH$, etc.) will increase the electron density,
and all protons will be more strongly shielded than those in benzene will. As a
consequence, the proton resonances will appear at a higher field. On this basis, it is easy
to determine the electronic nature of the substituents attached at the benzene ring, by
looking at the chemical shifts of the protons. It should always be taken into account that,
due to the simultaneous presence of an electron-withdrawing and an electron-releasing
substituent, these effects will cancel each other out.
   Figure 27 shows the ^{1}H-NMR spectra of benzene, anisol and aniline. The latter
two contain electron-releasing substituents. Oxygen and nitrogen are highly
electronegative and would be expected to inductively decrease the electron density
in the ring. In fact, these groups do have an electron-withdrawing inductive effect
($-I$ effect). However, the electron-releasing resonance effect ($+M$ effect) through
p-orbitals is stronger than the electron-withdrawing effect through σ-orbitals. Thus,

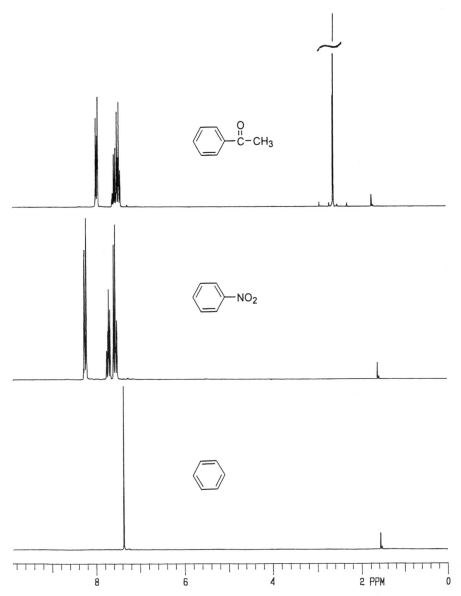

Figure 26   200 MHz ^{1}H-NMR spectra of benzene, nitrobenzene and acetyl benzene in CDCl$_3$.

these substituents have a net electron increasing effect. The resonance structures of anisols show that the electron density at the *ortho* and *para* positions is more increased. This does not mean that the electron density at the *meta* position is decreased. As mentioned above, the *meta* position remains almost unaffected by both kinds of substituents.

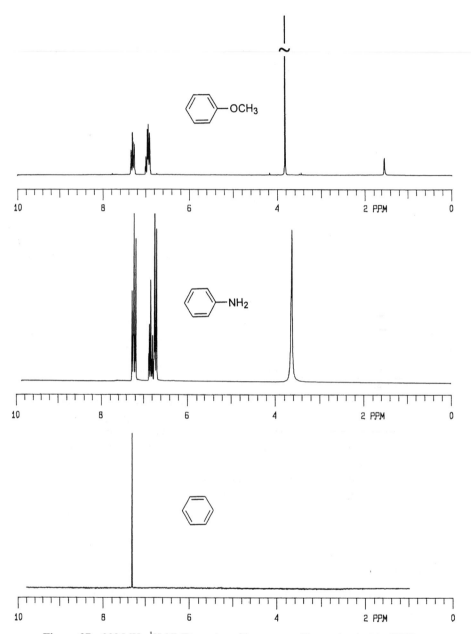

Figure 27   200 MHz ^{1}H-NMR spectra of benzene, aniline and anisol in CDCl$_3$.

^{1}H-NMR spectra of anisol and aniline also display three sets of signals. All of these signals are shifted to high field. The high-field part of these signals belongs to the *ortho* protons, which are most strongly affected. The *ortho* protons of aniline resonate at a higher field than those of anisols. This fact can be explained by the different electronegativities of oxygen and nitrogen. Since nitrogen is less electronegative than oxygen, it can release its nonbonding electrons more easily and increase the electron density of the ring much better than the oxygen atom. As a result, the ring protons of anilines resonate at a higher field.

Thus, we can draw the following conclusion: ^{1}H-NMR chemical shifts of the substituted benzenes largely depend on the mesomeric interaction between the substituent and benzene ring. These substituents have different effects on the chemical shifts of *ortho*, *meta* and *para* protons.

The influence of substituents on the resonance frequencies is to a first approximation additive. The systematic investigation of the resonance frequencies in substituted benzene has revealed empirical substituent constants that have proved to be useful for the prediction of ^{1}H resonance frequencies. These empirical substituent constants are tabulated for the various substituents (Table 3.7).

In cases where more than one substituent is attached at the benzene ring, the effect of the substituents on the chemical shift of the protons is additive. By using the substituent constant values, the resonance frequencies of the substituted benzene protons can be calculated with small deviations. Deviations from the additivity principle can be observed when more than two substituents are adjacent and cause some steric interactions.

We may demonstrate the applicability of these substituent constants by calculation of the proton resonances of *p*-nitroanisol. The ^{1}H-NMR spectrum of *p*-nitroanisol shown in Figure 28 displays two distinct proton resonances for the aromatic protons.

34

Table 3.7

Substituent constants (σ) for proton resonances in substituted benzenes [14]

Substituents	Substituent position		
	ortho	*meta*	*para*
−CH=CH−	0.16	0.00	−0.15
−C=N−	0.25	0.18	0.30
−CHO	0.55	0.19	0.28
−COCH$_3$	0.60	0.11	0.19
−COOCH$_3$	0.71	0.08	0.19
−COOH	0.75	0.14	0.25
−COCl	0.81	0.21	0.37
−NO$_2$	0.93	0.26	0.39
−OH	−0.48	−0.12	−0.48
−OCH$_3$	−0.49	−0.12	−0.44
−OCOCH$_3$	−0.21	−0.02	0.00
−F	−0.29	−0.02	−0.23
−Cl	−0.01	−0.06	−0.12
−Br	0.17	−0.11	−0.06
−I	0.38	−0.23	−0.01
−NH$_2$	−0.80	−0.25	−0.64
−N(CH$_3$)$_2$	−0.67	−0.18	−0.66

An electron-withdrawing and electron-releasing group are attached at the benzene ring. The chemical shift of the α-protons will be influenced by the nitro group and methoxyl group, which are located in the *ortho* and *meta* position with respect to the α-protons. From the data given in Table 3.7 we can obtain the substituent constants for these substituents and calculate the chemical shifts for the α-protons as well as for the

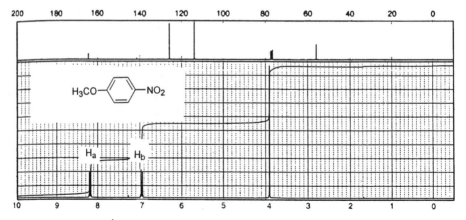

Figure 28   300 MHz ^{1}H-NMR spectrum of *p*-methoxy-nitrobenzene in CDCl$_3$. (Reprinted with permission of Aldrich Chemical Co., Inc. from C.J. Pouchert and J. Behnke, *The Aldrich Library of 13C and 1H FT-NMR Spectra*, 1992.)

β-protons:

$$\delta_{H_a} = 7.27 + NO_{2(ortho)} + OCH_{3(meta)}$$
$$\delta_{H_a} = 7.27 + 0.93 + (-0.12) = 8.08 \text{ ppm (calculated); } 8.10 \text{ ppm (experimental)}$$

$$\delta_{H_b} = 7.27 + NO_{2(meta)} + OCH_{3(ortho)}$$

$$\delta_{H_b} = 7.27 + 0.26 + (-0.49) = 7.04 \text{ ppm (calculated); } 6.90 \text{ ppm (experimental)}$$

7.27 ppm is the resonance frequency of benzene.

One can easily see that these results are in excellent agreement with the experimental data.

As a further example, we may calculate the resonance frequencies of a trisubstituted benzene and compare the values with the experimental ones. The 200 MHz ^{1}H-NMR spectrum of this compound is given in Figure 29 [15].

$$\delta_{H_a} = 7.27 + COOCH_{3(ortho)} + OH_{(meta)} + CHO_{(meta)}$$
$$\delta_{H_a} = 7.27 + 0.71 + (-0.12) + 0.19 = 8.05 \text{ ppm (calculated); } 8.01 \text{ ppm (experimental)}$$

$$\delta_{H_b} = 7.27 + COOCH_{3(meta)} + OH_{(para)} + CHO_{(ortho)}$$

$$\delta_{H_b} = 7.27 + 0.08 + (-0.48) + 0.55 = 7.42 \text{ ppm (calculated); } 7.43 \text{ ppm (experimental)}$$

$$\delta_{H_c} = 7.27 + COOCH_{3(meta)} + OH_{(ortho)} + CHO_{(ortho)}$$

$$\delta_{H_c} = 7.27 + 0.08 + (-0.48) + 0.55 = 7.42 \text{ ppm (calculated); } 7.40 \text{ ppm (experimental)}$$

In this case, the agreement between the experimental and calculated values is also

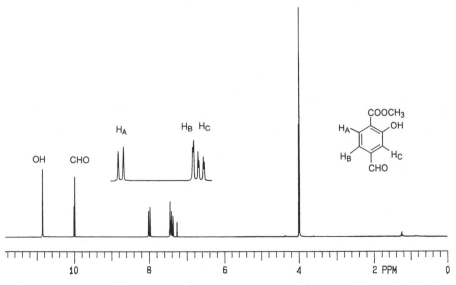

Figure 29   200 MHz ^{1}H-NMR spectrum of methyl 4-formyl-2-hydroxybenzoate in CDCl$_3$.

excellent. These examples demonstrate that the ^{1}H chemical shift values of substituted benzenes can be predicted using the additivity relationship.

The chemical shifts of some substituted aromatic compounds are given in Table 3.8.

Alkynes

We have already discussed how the benzene ring generates magnetic anisotropy in the presence of an external magnetic field and how this induced field affects the chemical shifts of the ring protons. As with the benzene ring, the multiple bonds (the C=C, C=O and N=O double bonds and the C≡C and C≡N triple bonds) possess particularly strong magnetic anisotropies when they are placed in a magnetic field. Let us begin with

Table 3.8

Chemical shifts of selected aromatic and heteroaromatic compounds

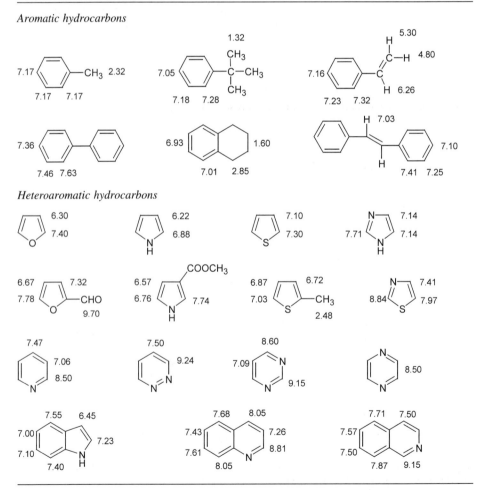

acetylene. Because of sp hybridization, acetylene is a linear molecule and the triple bond electrons are distributed cylindrically about the C–C axis. If an acetylene molecule is placed in a magnetic field, there are mainly two different alignments (parallel and perpendicular) with the external magnetic field. If the C–C axis is aligned with the applied magnetic field, the π-electrons of the triple bond can circulate at right angles to the applied field, thus inducing their own magnetic field, as we have seen in the case of the benzene ring (Figure 30).

This induced magnetic field opposes the applied magnetic field along the molecular axis, but reinforces the applied field outside the axis (Figure 30). Since acetylene protons are located along the molecular axis, they will be strongly shielded. Therefore, the chemical shift of acetylenic protons is found at high field ($\delta = 2.0-3.0$ ppm).

Chemical shifts of some acetylenic compounds:

$$HC\equiv CH, 1.80 \text{ ppm}; \quad H_2C=CH_2, 5.31 \text{ ppm}; \quad H_3C-CH_3, 0.88 \text{ ppm}$$

Acetylene carbons are sp-hybridized. Since the 's' ratio in the hybrid orbital is greater than the 's' ratio in the alkane and alkene C–H bonds, acetylenic protons are more acidic. In other words, the electron density around the acetylenic proton is decreased. Therefore, one would expect the acetylenic protons to resonate at a lower field than the olefinic protons. The unusual high-field resonance of acetylenic protons can be explained only by the formation of the above-mentioned magnetic anisotropy followed by strong shielding of acetylenic protons. Similar considerations apply to the nitrile group. The ^{1}H-NMR spectra of two acetylenic compounds are presented in Figure 31.

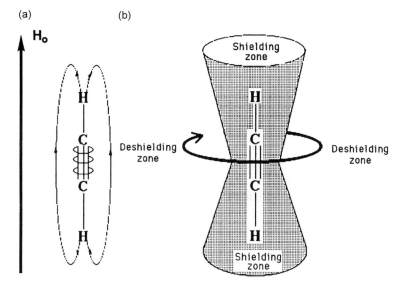

Figure 30   (a) Schematic representation of the magnetic anisotropic effect of the triple bond. (b) Shielding and deshielding zones around a triple bond.

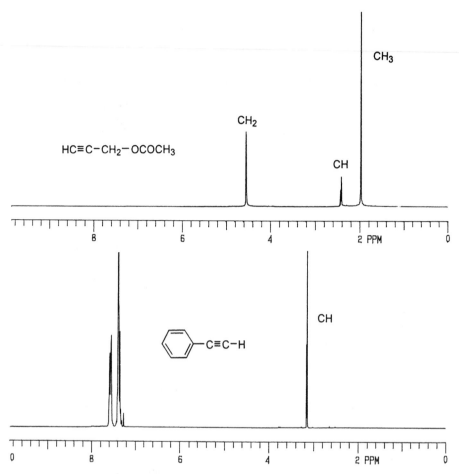

Figure 31   200 MHz ^{1}H-NMR spectra of 1-ethynylbenzene and propargyl acetate in CDCl$_3$.

Carbon–carbon and carbon–oxygen double bonds

The carbon–carbon and carbon–oxygen double bonds also generate magnetic anisotropy when placed in an external magnetic field. The circulation of the double bonds electrons generates a secondary magnetic field that accounts for the formation of the magnetic anisotropy, as we have already described, by the aromatic and acetylenic compounds. As a consequence of this induced magnetic field, the protons within the conical zones above and below the double bond will be shielded and the olefinic and aldehyde protons within the lateral zones will be deshielded (Figure 32). The chemical shifts of the olefinic and aldehyde protons will therefore move to low field, whereas the resonance frequencies of the protons situated above or below the double bond system will shift toward a higher field.

Proton chemical shifts are strongly affected by the distance r to the point dipole's center of the shielding and deshielding zones and are inversely proportional to the cube

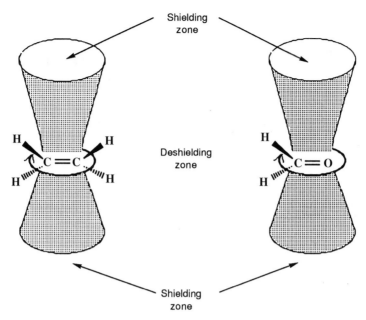

Figure 32   Schematic representation of the magnetic anisotropic effect of the –C=C– double bond and carbonyl group. Shielding and deshielding zones.

of the distance. Since an aldehyde proton is directly attached to the carbonyl carbon, it is close to the center of the deshielding zone. Furthermore, because of the electronegative effect of the carbonyl oxygen, the electron density at the carbonyl carbon atom is decreased, which will cause further deshielding. The particularly low field absorption of the aldehyde proton resonance ($\delta = 9.0–10.0$ ppm) is the result of the combined electronic and magnetic effects. The signal of aldehydic protons can be recognized easily from their characteristic position in that region (Figure 33).

α-Methylenic protons of aldehydes are also located in the deshielding zone generated by the carbonyl group. However, α-protons are far from the deshielding center compared with the aldehydic proton. Therefore, the effect of the carbonyl group on the chemical shift of the α-protons is less than the effect on the aldehyde proton. In a saturated aldehyde, the signal of the aldehydic proton appears at $\delta = 9.7$ ppm, while that of α-protons appear at approximately $\delta = 2.4$ ppm and the β-protons at $\delta = 1.7$ ppm.

$$
\begin{array}{cccc}
& \overset{\displaystyle O}{\overset{\displaystyle \|}{}} & & \\
H-C & -CH_2 & -CH_2 & -CH_2- \\
9.7 & 2.4 & 1.7 & 1.3
\end{array}
$$

α-Methylenic protons of esters and ketones are also located in the deshielding zone of the carbonyl group, as in the case of aldehydes, and their resonance frequencies are observed at $\delta = 2.0–2.5$ ppm.

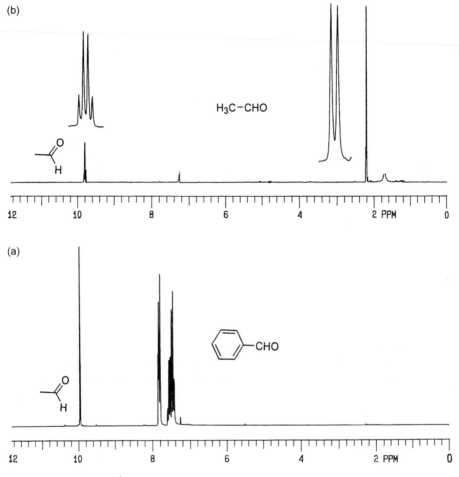

Figure 33   200 MHz ^{1}H-NMR spectra of (a) benzaldehyde and (b) acetaldehyde in CDCl$_3$.

The following examples demonstrate that the effect of the carbonyl groups on the chemical shift is additive. The resonance frequencies of $-CH_2-$ groups adjacent to two carbonyl groups can shift down to 4.0 ppm.

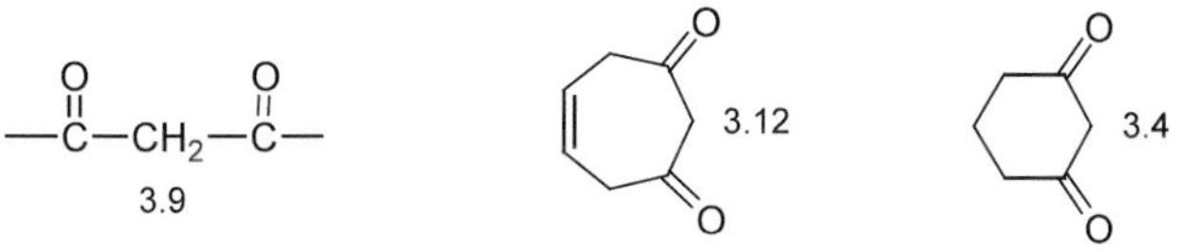

In the cyclic ketones, α-protons also resonate at low field.

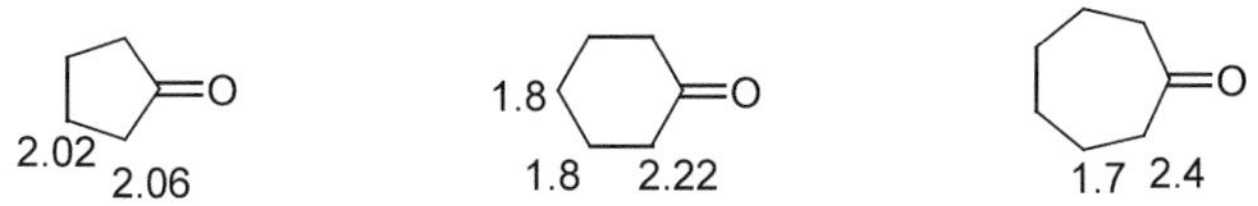

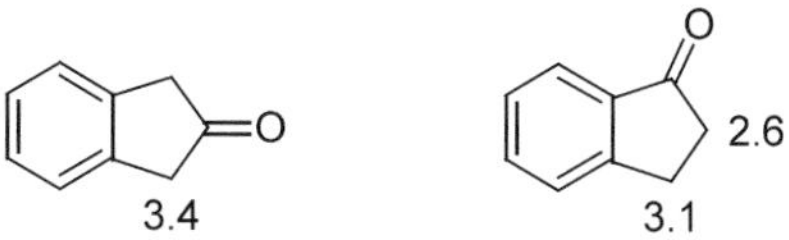

In esters, the resonances of the methylene protons can be observed at approximately 2.0 ppm. Between the resonance frequencies of methylene groups of esters and ketones, a difference of 0.2–0.3 ppm is observed where ketones resonate at a lower field. This difference arises from the conjugation of the carbonyl group with the nonbonding electrons of the ester oxygens that increase the electron density at the carbonyl carbon atom.

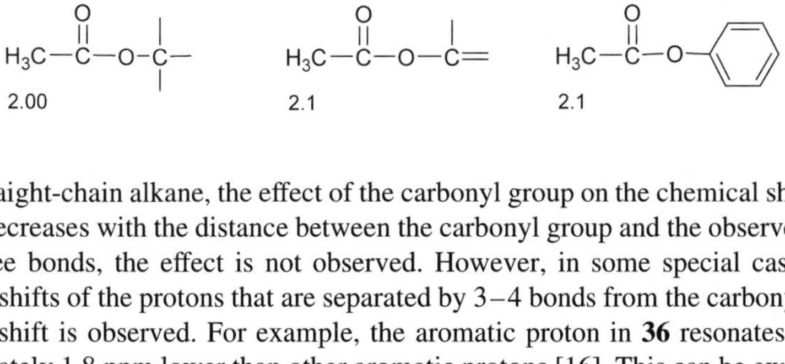

In a straight-chain alkane, the effect of the carbonyl group on the chemical shifts of the protons decreases with the distance between the carbonyl group and the observed proton. After three bonds, the effect is not observed. However, in some special cases, in the chemical shifts of the protons that are separated by 3–4 bonds from the carbonyl group a dramatic shift is observed. For example, the aromatic proton in **36** resonates at a field approximately 1.8 ppm lower than other aromatic protons [16]. This can be explained by the fact that this proton lies in the nodal region of the carbonyl group because of the geometry of the compound.

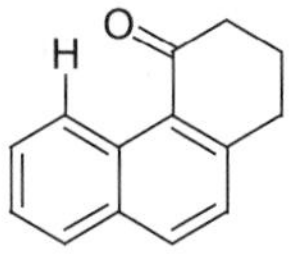

$\Delta\delta = 1.8$ Hz

36

Depending on the position of the carbonyl groups, the chemical shifts of the neighboring protons can be affected differently. For example, in compounds **37–39** [17] the resonance frequencies of the olefinic protons vary with the configuration of the double bond, i.e. with the position of the carbonyl group.

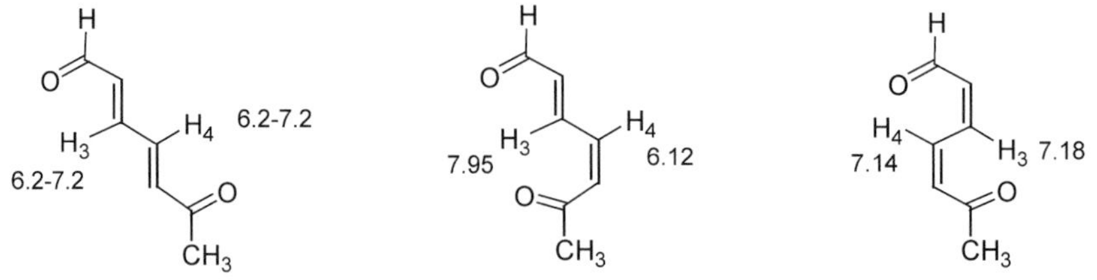

37 *trans–trans*   **38** *trans–cis*   **39** *cis–cis*

As we have seen above, in saturated ketones and aldehydes, the α-protons resonate at a lower field than β-protons. However, in α,β-unsaturated ketones and aldehydes the β-protons resonate at a lower field. The participation of resonance structures shown below is of special significance.

Because of the conjugation, the β-carbon of an α,β-unsaturated ketone bears a partial positive charge. As a result, the chemical shift is dominated by electronic effects and β-protons are strongly deshielded.

For example, in cyclopent-2-enone, the β-proton resonates at 7.71 ppm, whereas the α-proton absorbs at 6.61 ppm (Figure 34). This difference of 1.1 ppm is due to the conjugation. In cyclohexe-2-none, similar effects are observed. Comparison of the olefinic proton resonances of cyclopentenone and cyclohexenone reveals that the resonances of the five-membered ring are shifted around 1 ppm downfield. This can be explained by the better coplanar arrangement of the π-system, thereby decreasing the electron density at the β-carbon atom, in which is caused by an effective delocalization of the π-electrons. Chemical shifts of some selected carbonyl compounds are presented in Table 3.9.

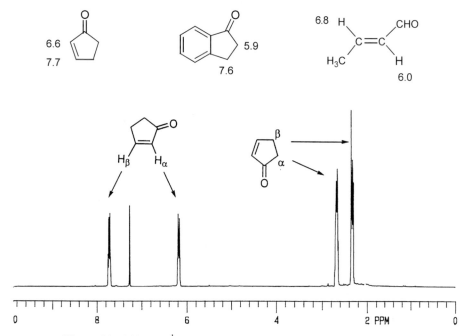

Figure 34   200 MHz ^{1}H-NMR spectrum of cylopent-2-enone in CDCl$_3$.

A similar argument can be adduced to rationalize the magnetic anisotropy that is formed upon placement of a carbon–carbon double bond in an external magnetic field. As one can see from Figure 32, the induced magnetic field creates a shielding cone above and below the double bond plane and a deshielding zone outside of this cone. Since the double bond protons are located in the double bond plane, they are deshielded. They absorb in a range of 5.0–6.5 ppm. Depending on the electronic nature and the number of substituents, resonances outside this range can be frequently encountered.

Terminal vinyl groups (–CH=CH$_2$) resonate at around 5.0–5.5 ppm. In the conjugated systems, protons beside the terminal protons resonate at lower field. For example, the central protons H$_2$ and H$_3$ in 1,3-butadiene absorb at 6.62 ppm, whereas the terminal protons H$_1$ and H$_4$ appear at 5.05–5.16 ppm.

This difference can be explained as follows. Since the H$_2$ and H$_3$ protons are simultaneously located in the deshielding zone of both –C=C– double bonds, their resonances are shifted to a lower field, whereas the terminal protons are under the influence of one double bond. Cycloheptatriene demonstrates these effects proficiently.

Table 3.9

^{1}H chemical shift of selected carbonyl compounds (δ in ppm)

γ-Proton resonance in cycloheptatriene is shifted to a lower field compared with the α- and β-protons (Figure 35).

Proton chemical shifts of some representative alkenes are shown in Table 3.10. Inspection of this table shows that the cyclopropene protons resonate at an unusually low field ($\delta = 7.06$ ppm). On the other hand, cyclopropane itself resonates at high field ($\delta = 0.22$ ppm). It is well known that the ring electrons in cyclopropane generate a diamagnetic anisotropy as we have seen with the $-C{=}C-$ double bond and aromatic ring. However, as a consequence of the different orientation of the C–H bonds (located in the

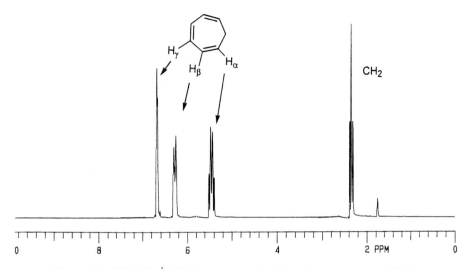

Figure 35   200 MHz ^{1}H-NMR spectrum of 1,3,5-cycloheptatriene in CDCl$_3$.

shielding zone) in comparison with those in benzene, shielding of the ring protons of cyclopropane results. On the other hand, the olefinic protons in cyclopropene are located in the ring plane, i.e. in the deshielding zone, which accounts for the low field shift of these protons.

Let us reiterate the fact that most of the olefinic protons resonate in the region of 5.0–6.5 ppm. By the attachment of a substituent at the double bond, the chemical shifts of the olefinic protons can be seen in a wider range of 4.0–8.0 ppm. Similarly, for aromatic compounds, substituent constants have been proposed in order to predict proton shifts of olefinic protons from the reference value of ethene (5.28 ppm).

By the attachment of a substituent to the ethylene molecule, the symmetry is destroyed, and there are three different protons with respect to the substituent. Substituent constant values (σ) of a substituent for *cis*-, *trans*-, and *geminal* protons are different.

Table 3.10

^{1}H chemical shift of selected unsaturated hydrocarbons (δ in ppm)

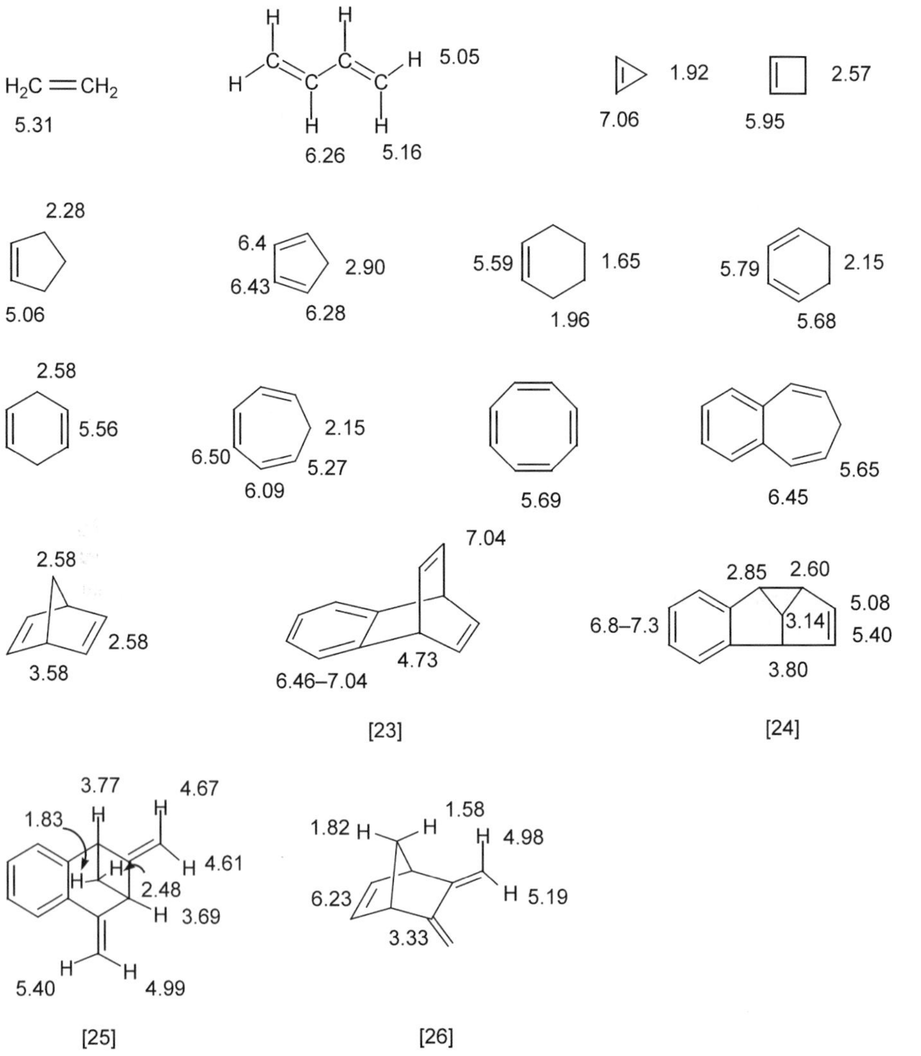

H$_2$C ═ CH$_2$
5.31

H$_2$C ═ CH─CH ═ CH$_2$
H 5.05
6.26   5.16

1.92
7.06

2.57
5.95

2.28
5.06

6.4
6.43
2.90
6.28

5.59
1.96
1.65

5.79
5.68
2.15

2.58
5.56

6.50
6.09
2.15
5.27

5.69

6.45
5.65

2.58
3.58
2.58

7.04
4.73
6.46–7.04

[23]

6.8–7.3
2.85 2.60
3.14
3.80
5.08
5.40

[24]

3.77   4.67
1.83
H 4.61
2.48
H 3.69
5.40   4.99

[25]

1.82 H
1.58
H 4.98
6.23
3.33
H 5.19

[26]

Table 3.11 shows the substituent constants of different substituents that are based on extensive experimental data. Mesomeric electron-withdrawing substituents ($-$M-effect) shift the resonance frequencies of all the double bond protons to lower field; the mesomeric electron-donating substituents have a different effect. For example, substituents such as $-$OR, $-$SR, or $-$NHR with inductive electron-withdrawing ($-$I effect) and mesomeric electron-donating ($+$M effect) abilities deshield (low-field

Table 3.11

Substituent constants (σ) for estimating ^{1}H chemical shifts in alkenes [27]

Substituent	$H_{geminal}$	H_{cis}	H_{trans}
$-$Alkyl	0.45	-0.22	-0.28
$-$Cycloalkyl	0.69	-0.25	-0.28
$-CH_2-O-$	0.64	-0.01	-0.02
$-CH_2-S-$	0.71	-0.13	-0.22
$-CH_2-$halide	0.70	0.11	-0.04
$-CH_2-C{=}O$	0.69	-0.08	-0.06
$-CH_2-C{\equiv}N$	0.69	-0.08	-0.06
$-CH_2-C_6H_5$	1.05	-0.29	-0.32
$-C{=}C-$ (isolated)	1.00	-0.09	-0.23
$-C{=}C-$ (conjugated)	1.24	0.02	-0.05
$-C{\equiv}N$	0.27	0.75	0.55
$-C{\equiv}C-$	0.47	0.38	0.12
$-C{=}O$ (isolated)	1.10	1.12	0.87
$-C{=}O$ (conjugated)	1.06	0.91	0.74
$-$COOH	0.97	1.41	0.71
$-$COOR (isolated)	0.80	1.18	0.55
$-$COOR (conjugated)	0.78	1.01	0.46
$-$CHO	1.02	0.95	1.17
$-$CO$-$N$-$	1.37	0.98	0.46
$-$COCl	1.11	1.46	1.01
$-$OR (aliphatic)	1.22	-1.07	-1.21
$-$OR (conjugated)	1.21	-0.60	-1.00
$-$OCOR	2.11	-0.35	-0.64
$-$Cl	1.08	0.18	0.13
$-$Br	1.07	0.45	0.55
$-$I	1.14	0.81	0.88
$-$NR$-$ (aliphatic)	0.80	-1.26	-1.21
$-$NR$-$ (conjugated)	1.17	-0.53	-0.99
$-$N$=$C$=$O	2.08	-0.57	-0.72
$-C_6H_5$	1.38	0.36	-0.07
$-$SR	1.11	-0.29	-0.13
$-SO_2-$	1.55	1.16	0.93

resonance) the α-proton by an inductive effect and shield (high-field resonance) the β-protons by resonance.

$$R{-}\overset{\frown}{\underline{O}}{-}CH{=}\overset{\frown}{CH_2} \quad \longleftrightarrow \quad R{-}\overset{\oplus}{\underline{O}}{=}CH{-}\overset{\ominus}{CH_2}$$

We have already discussed the effect of the mesomeric electron-withdrawing substituent on the chemical shifts of α- and β-protons on page 62. Since chemical shift constants are approximately additive, it is possible to calculate the carbon–carbon double bond proton shifts in di- and trisubstituted olefines with a small deviation.

For example, let us calculate the resonance frequencies of the protons in *trans*-crotone aldehyde by using the corresponding substituent constant values.

42

Chemical shifts can be calculated by the following equation:

$$\delta_{(H)} = 5.28 + \sum(\sigma) = 5.28 + \sigma_{geminal} + \sigma_{cis} + \sigma_{trans}$$

$\delta = 5.28$ is the chemical shift of the protons in ethylene and $\sum(\sigma)$ are the substituent increment values.

The substituent constants are given in Table 3.11. For the α-proton we have to use the *geminal* substituent constant value of the aldehyde group and the *cis* substituent constant value of the methyl group.

$$\delta_{(H_\alpha)} = 5.28 + \sigma_{geminal}(CHO) + \sigma_{cis}(CH_3)$$

$$\delta_{(H_\alpha)} = 5.28 + 1.02 + (-0.22) = 6.08 \text{ ppm (calculated);} \quad 6.06 \text{ ppm (experimental)}$$

$$\delta_{(H_\beta)} = 5.28 + \sigma_{cis}(CHO) + \sigma_{geminal}(CH_3)$$

$$\delta_{(H_\beta)} = 5.28 + 0.95 + 0.45 = 6.68 \text{ ppm (calculated);} \quad 6.72 \text{ ppm (experimental)}$$

The calculated values show the additivity of the increment values and they are in good agreement with the observed ones.

After analyzing the electronic effects on the chemical shifts of the olefinic protons in detail, let us examine the geometrical effects. If the protons of interest are situated above the double bond system, in the shielding cone, because of the geometrical arrangement, the resonance frequencies are shifted toward a higher field. We mostly encounter these kinds of arrangements in bicyclic systems.

43   **44**   **45**   **46**

In the chemical shifts of the bridge protons in compounds **43**–**46** dramatic high field shifts are observed because of their location in the shielding cone of the double bond. For example, the resonance of the *syn* proton in the tricyclic system **43** appears at high field (-0.42 ppm) on the right side of the TMS signal. The *anti* proton that is attached to the same carbon atom resonates in the expected region (1.42 ppm) [28, 29]. The observed chemical shift difference of 1.8 ppm arises from the shielding effect of the double bonds. By taking this chemical shift difference into account, one can distinguish between the possible isomers. For example, the tricyclic compounds **45** and **46** can be easily identified on the basis of the observed chemical shifts of the cyclopropane protons. The cyclopropyl proton of the *exo*-isomer (**45**) resonates at a higher field [30].

Magnetic anisotropy of carbon–carbon and carbon–hydrogen bonds and their effect on chemical shifts

Let us recall the chemical shifts of the saturated hydrocarbons.

$$CH_3$$
$$|$$
CH$_4$ H$_3$C—CH$_3$ H$_3$C—CH$_2$—CH$_3$ H$_3$C—CH—CH$_3$

0.23   0.86   1.23   1.77

Methane resonates at 0.23 ppm and ethane at 0.86 ppm. Replacement of one proton in ethane by a methyl group shifts the resonance of the methylene protons in *n*-propane to 1.23 ppm. Introduction of an additional methyl group causes a chemical shift of the tertiary proton down to 1.77 ppm. These examples show us that the more the number of the alkyl substituents is increased, the more resonances are shifted to low field. On the other hand, it is well known that the alkyl groups donate electrons by an inductive effect (+I) and increase the electron density around the adjacent carbon atom. Since the increased electron density will cause more shielding, the resonance frequency of a tertiary proton should shift to high field. Contrary to this expectation, tertiary protons resonate at lower field. At this point, one must suppose that factors other than electron density may influence the chemical shifts. We have already discussed the magnetic anisotropies of C=C, C=O double bonds, C≡C triple bonds and aromatic rings. Furthermore, we have shown how these magnetic anisotropies influence the chemical shift of protons. As with these functional groups, C–C single bonds also have magnetic anisotropy.

To understand the magnetic anisotropy of a C–C bond, let us first consider a diatomic molecule AB. If this molecule is placed in a magnetic field, the external magnetic field H_0 will induce secondary magnetic fields around A and B, respectively. Let us consider only the secondary magnetic field induced around A. This molecule AB can have different alignments. When the axis of AB is parallel to the applied field, the secondary field at B will oppose H_0, and B will be shielded (Figure 36a). However, in an arrangement of AB

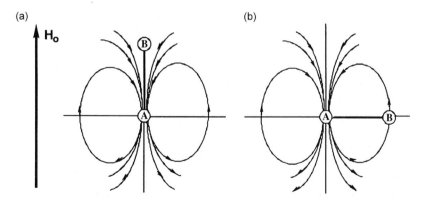

Figure 36 Different alignments of a diatomic molecule A–B in a magnetic field: (a) parallel alignment of A–B with the applied field; (b) perpendicular alignment with the applied field.

perpendicular to the applied field, the secondary magnetic field at B is parallel to H_0, and B will be deshielded (Figure 36b). In explanation, the AB molecules undergo rapid rotation and averaging takes place. The resulting net effect is then zero if the components of the susceptibility χ have the same values. However, if this is not the case, then A will possess magnetic anisotropy. The magnetic anisotropy along the C–C bond is shown in Figure 37. The deshielding zone is located contrary to the acetylene molecule along the molecular axis, whereas the lateral zones are shielding zones.

Since the protons in the saturated hydrocarbons are located in the deshielding zone of a C–C bond, those protons are deshielded. This can explain why the proton chemical shifts move to a lower field by increasing the number of substituents attached to the carbon bearing the resonating proton.

The shielding effect of a C–C single bond can also be found in cyclic systems. For example, when the ^{1}H-NMR spectrum of cyclohexane is recorded at room temperature, a single peak is observed. Because of the ring inversion process, axial protons in a one-chair form become equatorial protons in the ring-flipped chair form, and vice versa. Therefore, we observe a rapidly interconverting average spectrum. If the system is cooled (to $-70\,°C$), the ring inversion process will slow down so that the axial and equatorial protons can be observed separately as broad singlets in the ^{1}H-NMR spectrum.

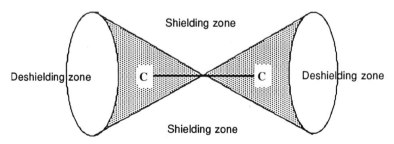

The chemical shift difference between the observed peaks is $\Delta\delta = 0.5$ ppm. The observation that an equatorial proton is consistently found at a lower field than the axial proton on the same carbon can be explained by the fact that the equatorial proton is located in the deshielding cone of the C–C bonds (Figure 38).

The chemical shift of a substituted methane can be calculated by means of substituent constants. The substituent constant values of different substituents are given in Table 3.12.

The chemical shift values of substituted methanes can be predicted with a small deviation using the additivity relationship given in the following equation:

$$\delta = 0.23 + \sum S(\sigma)$$

Figure 37 Schematic representation of the magnetic anisotropic effect of the C–C single bond. Shielding and deshielding zones.

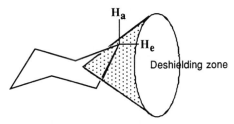

Figure 38   Schematic representation of the deshielding zone of the C–C single bond in cyclohexane.

The ^{1}H-NMR chemical shift values of some saturated hydrocarbons are given in Table 3.13.

van der Waals effects

A strong steric interaction between a proton and a neighboring group (it can also be a proton) will exist when these groups are in close proximity because of the molecular geometry. This steric interaction will arise from the touching and overlapping of van der Waals radii. The steric perturbation of the C–H bond involved causes the charge to drift towards the carbon atom. The reduced spherical symmetry of the electron distribution causes a paramagnetic contribution to the shielding constant. As a result, the chemical shift of the proton moves to low field, whereas the chemical shift of the

Table 3.12

Substituent constants (σ) for estimating ^{1}H chemical shifts
in methane [31]

Substituent	Substituent constant (σ)
$-CH_3$	0.47
$-CF_3$	1.14
$-C=C-$	1.32
$-C\equiv C-$	1.44
$-COOR$	1.55
$-COR$	1.70
$-CONR_2$	1.59
$-SR$	1.64
$-COPh$	1.84
$-C_6H_5$	1.85
$-I$	1.82
$-Br$	2.33
$-Cl$	2.53
$-OH$	2.56
$-OR$	2.36
$-OPh$	3.23
$-OCOPh$	3.13
$-OSO_2R$	3.13

Table 3.13

^{1}H chemical shift of selected saturated hydrocarbons (δ in ppm)

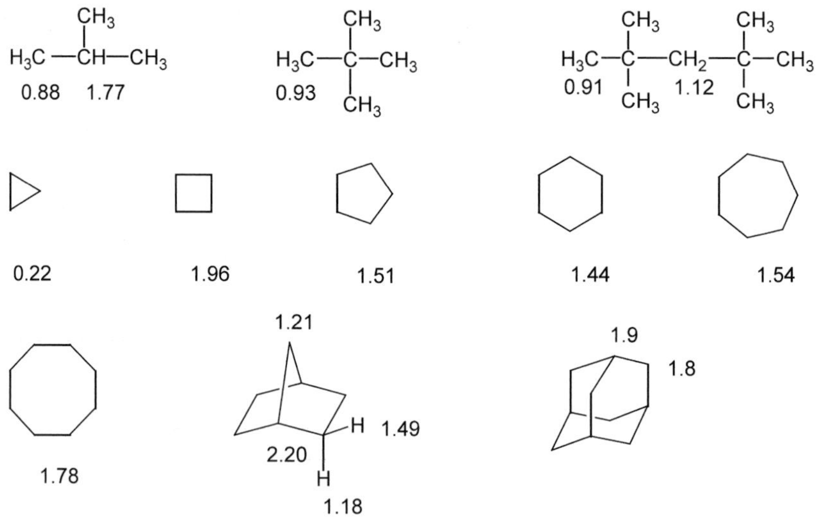

corresponding carbon moves to high field due to the increased electron density around the carbon atom.

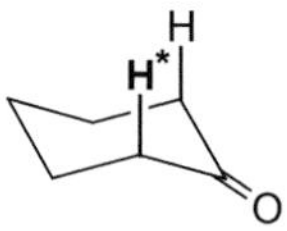

47 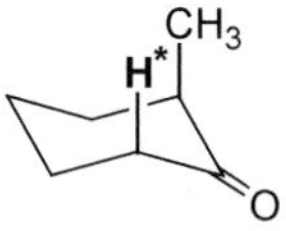　　　　　　　　**48**

For example, replacement of one of the axial protons in cyclohexanone **47** by a methyl group will shift the resonance of the other proton in **48** to low field. This shift cannot be explained by the inductive effect of the methyl group. This observed paramagnetic shift arises from the intramolecular steric interaction between the proton and the methyl group.

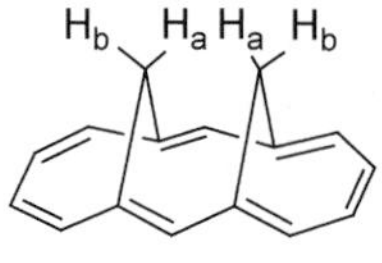

28 　　　　　　　　**49** 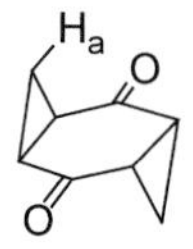　　　　　　　　**50**

$\delta H_a = 0.9$ ppm　　　　　　$\Delta \delta H_a = 0.9$ ppm
$\delta H_b = -1.2$ ppm

In the ^{1}H-NMR spectrum of the *syn*-bismethano[14]annulene (**28**), the bridge inner protons (H$_a$) resonate at 0.9 ppm, while the outer protons (H$_b$) appear at -1.13 ppm [32]. The observed chemical shift difference ($\Delta\delta = 2.08$ ppm) arises partly from the steric compression of the inner protons. Studies have revealed that van der Waals repulsion can cause a low-field shift of a maximum of 1 ppm. In this particular case, this chemical shift difference might also be attributed to the fact that the aromatic ring current in the outer rings is stronger than it is in the middle ring. The measured chemical shift difference ($\Delta\delta = 0.66$ ppm) of H$_a$ protons in the tricyclic diketones **49** and **50** is associated with the steric interaction of these protons.

In hydrocarbons **51** and **52**, the steric effect on inner cyclopropane protons is also observed. The chemical shift of the inner cyclopropane protons in **51** is shifted to a lower field [33].

By increasing the number of substituents exerting compression on a proton, the chemical shift will move to a lower field. This is demonstrated by the following ketones **53**, **54** and **55** [34]. The chemical shift of the *endo*-proton moves from 1.68 to 3.27 ppm, by increasing the number of bromine atoms.

The effect of concentration, solvent and temperature on the chemical shift

The changes in resonance frequency affected by the solvent in the case of proton resonances are usually small. Nowadays, a 1 mg sample in 0.5 ml solvent would be a strong solution for a medium to high field instrument. However, higher concentrations will increase the viscosity of the solution, which will affect the resolution obtainable, particularly in proton spectra. Furthermore, a concentration increase of the aromatic compound can affect the chemical shift.

Generally, temperature does not have a noticeable effect on chemical shifts. However, if there are some dynamic processes such as bond rotation, ring inversion and valence isomerization, temperature can have a dramatic effect on chemical shifts (see Chapter 8). For example, cycloheptatriene (**56**) is in equilibrium with its valence isomer, norcaradiene (**57**).

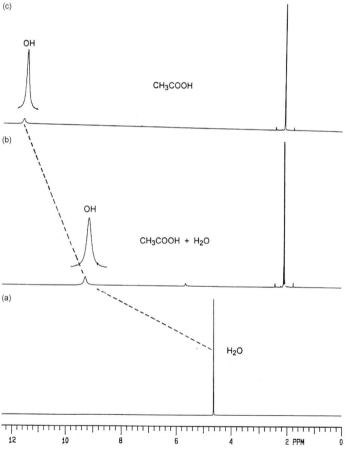

This equilibrium can be affected by a change in the temperature. For example, the H_a proton in cycloheptatriene (**56**) is bonded to a double bond. If the equilibrium shifts to the right, towards norcaradiene, this proton will change its position and will be a part of the cyclopropane ring. The chemical shift of this proton will change; in this particular case the resonance frequency of this proton will shift to a higher field. In general, the temperature will shift the equilibrium to the right or left, which will affect the position of the observed proton. This will be reflected in chemical shifts.

Figure 39   200 MHz ^{1}H-NMR spectra of (a) water, (b) water + acetic acid mixture and (c) acetic acid.

On the other hand, the chemical shift of protons attached to the electronegative atoms, such as oxygen, nitrogen and sulfur, strongly depends on the concentration and the temperature. Intermolecular hydrogen bonding explains why the shifts depend on concentration and temperature. Hydrogen bonding leads to significant shifts to low field although formally the electron density is increased through an interaction with the free electron pair of the acceptor atom. However, the electric dipole field of the hydrogen bond appears to affect deshielding. A decrease in concentration in a nonpolar solvent disrupts such hydrogen bonding, and the peak appears at high field. For example, the shift of the hydroxyl proton in neat ethanol is 5.28 ppm, but upon dilution in CCl_4, which breaks up the hydrogen bondings, the hydroxyl protons move upfield. Temperature has a similar effect.

Rapid exchangeability explains why the hydroxylic peak of alcohols is usually observed as a singlet. For example, acetic acid protons resonate at approximately 11.0 ppm, while water protons appear at 4.5 ppm. By mixing water and acetic acid, the hydroxyl protons resonate at 9.0 ppm (depending on the ratio of the components) as a broad singlet instead of emitting two separate signals, which shows the rapid exchangeability of these protons (Figure 39).

3.5 EXERCISES 1–30

Propose plausible structures for the following ^{1}H-NMR spectra. The molecular formulae of the compounds are given on the spectra. First determine the relative numbers of protons responsible for these peaks. All of these ^{1}H-NMR spectra consist of singlets. It is very important to note at this stage that the interpretation of ^{1}H-NMR spectra consisting of only singlets is much more difficult than that of spectra with split signals. One should always keep in mind the fact that the more peak splitting there is, the more information about the structures is provided. Since the readers may encounter some difficulties during the interpretation of these ^{1}H-NMR spectra, they should not be discouraged at this point. So far we have dealt only with the chemical shift and not with peak splitting. Therefore, these spectra are chosen. Peak splitting is dealt with in Chapter 4.

When the molecular formula of a compound is given, the double bond equivalency (the number of the double bonds and rings in the molecule) can be calculated using the following equation:

$$\text{Double bond equivalency} = \frac{(2a + 2) - b}{2} \quad [C_a H_b (O_c)]$$

When the compound contains halogen atoms, they will be considered H-atoms. For example, we calculate the double bond equivalency of benzene as follows:

$$\text{DBE (benzene)} = \frac{(2 \times 6 + 2) - 6}{2} = 4$$

The number 4 indicates the *total* number of double bonds (3) and rings (1). From the calculated double bond equivalency it is not possible to predict the exact number of double bonds and rings.

All the spectra (1–30) given in this section are reprinted with permission of Aldrich Chemical Co., Inc. from C.J. Pouchert and J. Behnke, *The Aldrich Library of 13C and 1H FT-NMR Spectra*, 1992.

1.

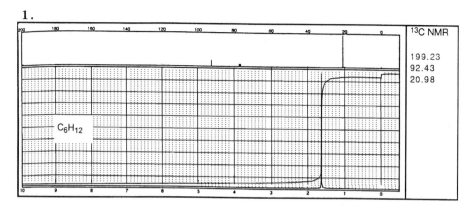

2.

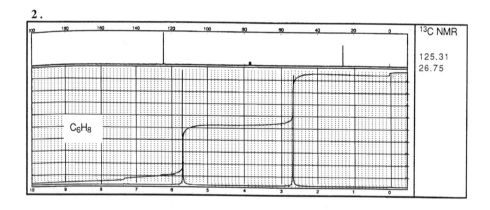

3.

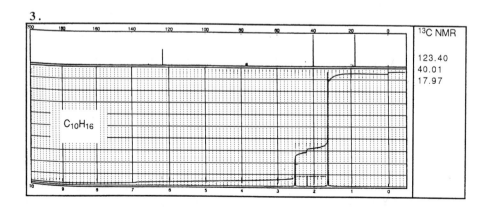

4.

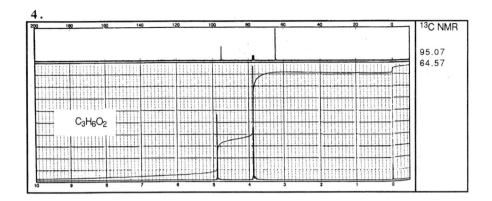

^{13}C NMR

95.07
64.57

$C_3H_6O_2$

5.

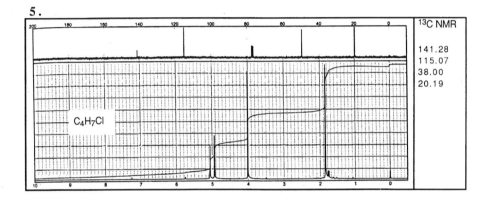

^{13}C NMR

141.28
115.07
38.00
20.19

C_4H_7Cl

6.

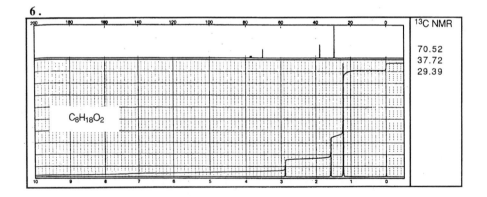

^{13}C NMR

70.52
37.72
29.39

$C_8H_{18}O_2$

7.

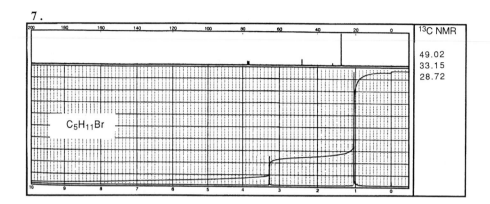

C₅H₁₁Br

¹³C NMR

49.02
33.15
28.72

8.

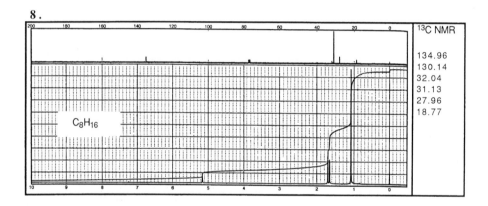

C₈H₁₆

¹³C NMR

134.96
130.14
32.04
31.13
27.96
18.77

9.

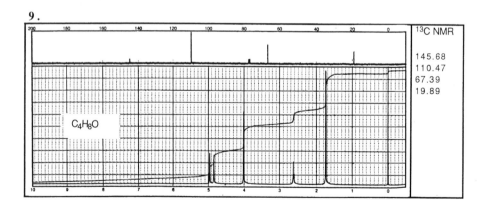

C₄H₈O

¹³C NMR

145.68
110.47
67.39
19.89

10.

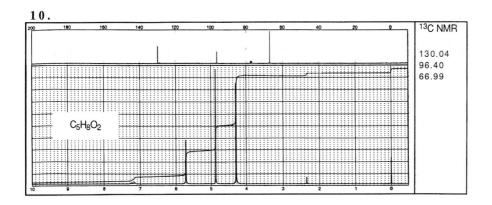

$C_5H_8O_2$

13C NMR

130.04
96.40
66.99

11.

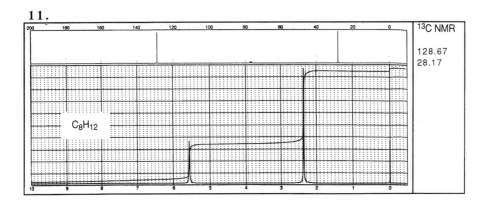

C_8H_{12}

13C NMR

128.67
28.17

12.

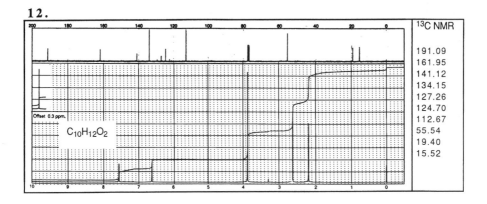

Offset 0.3 ppm.

$C_{10}H_{12}O_2$

13C NMR

191.09
161.95
141.12
134.15
127.26
124.70
112.67
55.54
19.40
15.52

13.

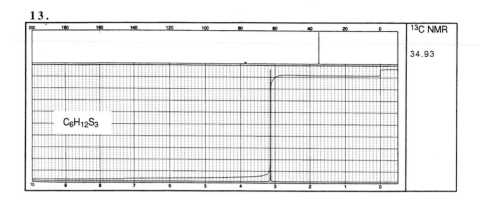

$C_6H_{12}S_3$

^{13}C NMR

34.93

14.

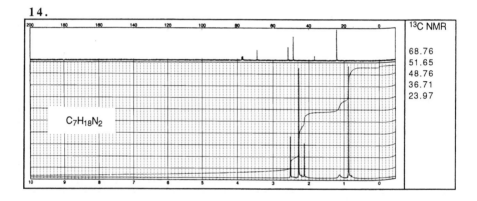

$C_7H_{18}N_2$

^{13}C NMR

68.76
51.65
48.76
36.71
23.97

15.

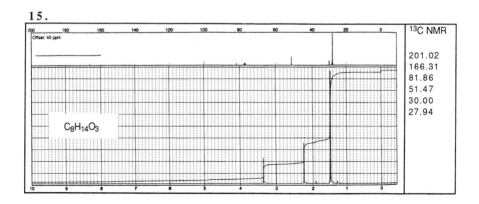

$C_8H_{14}O_3$

^{13}C NMR

201.02
166.31
81.86
51.47
30.00
27.94

16.

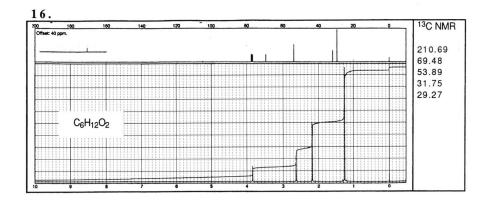

^{13}C NMR

210.69
69.48
53.89
31.75
29.27

$C_6H_{12}O_2$

17.

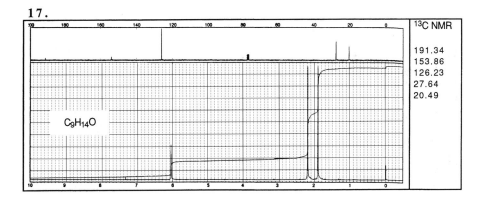

^{13}C NMR

191.34
153.86
126.23
27.64
20.49

$C_9H_{14}O$

18.

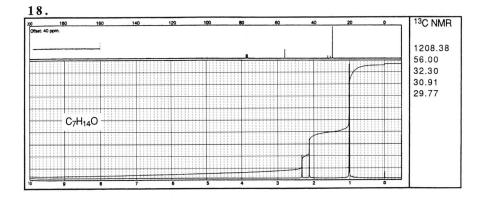

^{13}C NMR

1208.38
56.00
32.30
30.91
29.77

$C_7H_{14}O$

19.

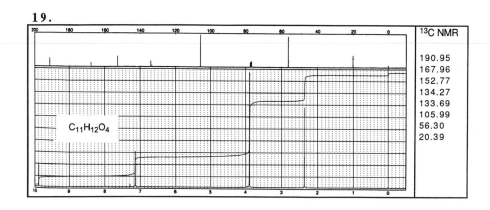

		^{13}C NMR
		190.95
		167.96
		152.77
		134.27
		133.69
		105.99
		56.30
		20.39

$C_{11}H_{12}O_4$

20.

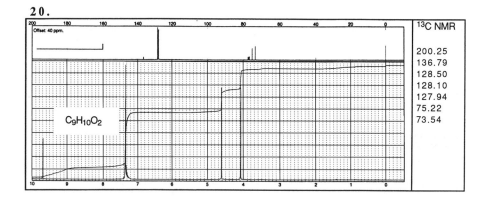

Offset: 40 ppm.

		^{13}C NMR
		200.25
		136.79
		128.50
		128.10
		127.94
		75.22
		73.54

$C_9H_{10}O_2$

21.

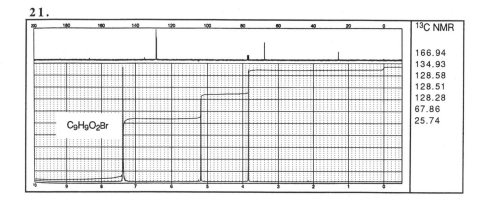

		^{13}C NMR
		166.94
		134.93
		128.58
		128.51
		128.28
		67.86
		25.74

$C_9H_9O_2Br$

22.

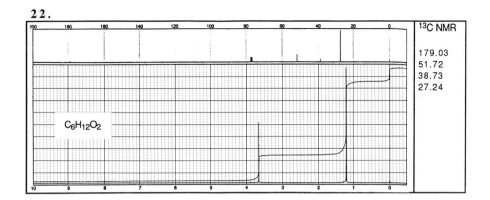

^{13}C NMR

179.03
51.72
38.73
27.24

$C_6H_{12}O_2$

23.

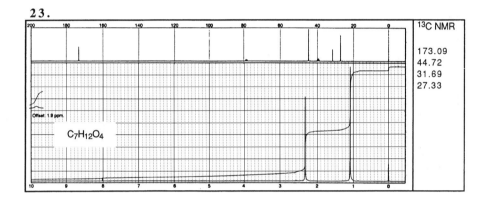

^{13}C NMR

173.09
44.72
31.69
27.33

Offset: 1.9 ppm.

$C_7H_{12}O_4$

24.

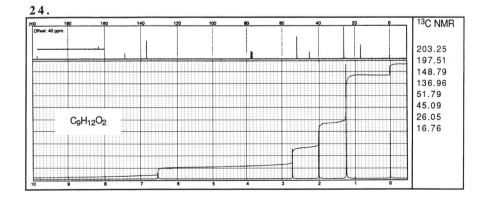

^{13}C NMR

203.25
197.51
148.79
136.96
51.79
45.09
26.05
16.76

Offset: 40 ppm.

$C_9H_{12}O_2$

25.

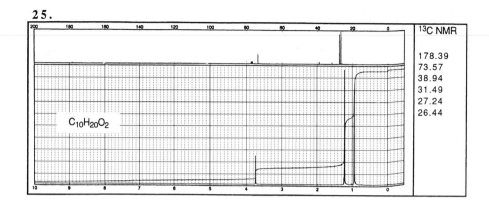

26.

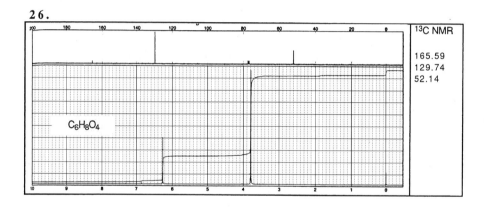

27.

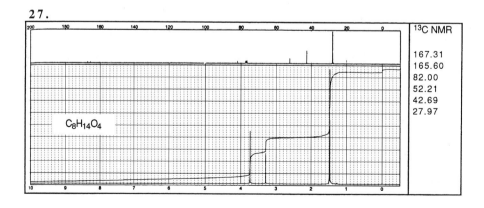

28.

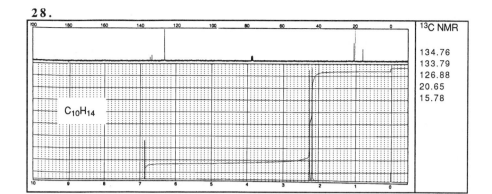

^{13}C NMR

134.76
133.79
126.88
20.65
15.78

$C_{10}H_{14}$

29.

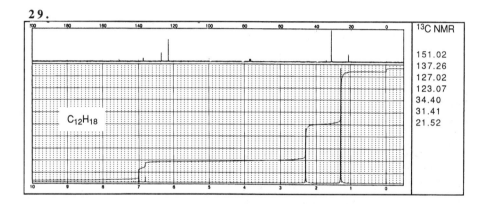

^{13}C NMR

151.02
137.26
127.02
123.07
34.40
31.41
21.52

$C_{12}H_{18}$

30.

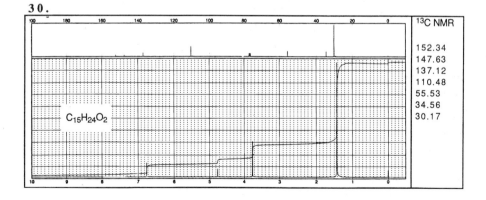

^{13}C NMR

152.34
147.63
137.12
110.48
55.53
34.56
30.17

$C_{15}H_{24}O_2$

– 4 –

Spin–Spin Splitting in ^{1}H-NMR Spectra

We have seen that differences in the chemical shift of protons are caused by the small local magnetic fields of the electrons that surround the various nuclei. Chemical shift is also affected by the magnetic anisotropy of different functional groups.

$$-\overset{|}{\underset{|}{C}}-\overset{|}{\underset{|}{C}}-$$
$$\quad H_a \quad H_b$$

In looking at these factors, we have neglected the point of how the alignment (parallel or antiparallel) of the magnetic moment of proton H_b in the external magnetic field can affect the chemical shift of the neighboring proton H_a. Different alignment of the magnetic moment of the proton H_b will cause fine splitting in the signal of proton H_a. As a consequence, the number of signals in the spectrum will increase and the appearance of the spectrum will be complex. At the same time, we will obtain more information from the spectrum for interpretation. As mentioned above, the interpretation of spectra consisting of single lines (*singlets*) is difficult. Singlets arise from the equal protons that do not have neighboring protons. Single line resonances can arise from methyl (CH_3), methylene (CH_2) and methine protons (CH). It is not easy to differentiate between these groups. Let us briefly review spin–spin splitting qualitatively with some examples.

4.1 EXPLANATION OF SPIN–SPIN SPLITTING

The ^{1}H-NMR spectra of ethyl acetate (**58**) and diethyl malonate (**59**) are shown in Figure 40.

$$H_3C-\overset{\overset{\displaystyle O}{\|}}{C}-O-CH_2-CH_3$$

58

$$H_2C\overset{\overset{\displaystyle O}{\|}}{\underset{\underset{\displaystyle O}{\|}}{\overset{C-O-CH_2-CH_3}{C-O-CH_2-CH_3}}}$$

59

Let us first analyze the ^{1}H-NMR spectrum of ethyl acetate (**58**) (Figure 40). As can be seen from the formula, ethyl acetate contains two different methyl groups, the chemical shifts of which, as expected, are different. One methyl group is attached directly to

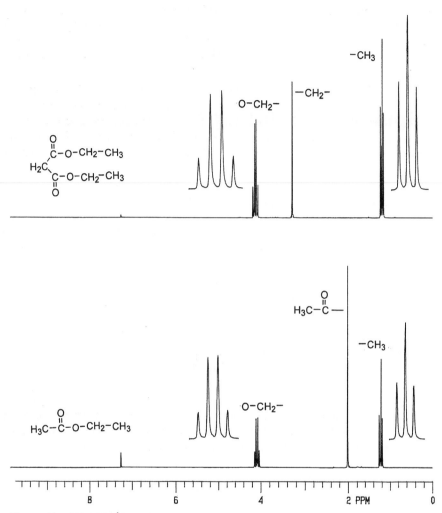

Figure 40   200 MHz ^{1}H-NMR spectra of ethyl acetate and diethyl malonate in CDCl$_3$.

the carbonyl group, and the other to the methylene (–CH$_2$–) group. However, the appearances of the signals belonging to these methyl groups are completely different. While the methyl group that is bonded to the carbonyl group resonates as a *singlet* at around 2.00 ppm, the other methyl group resonates at 1.3 ppm as a *triplet* (three peaks). The methyl group that is bonded to the carbonyl carbon does not have any neighboring protons. However, the other methyl group has two neighboring protons. We can conclude from this observation that the neighboring protons (–OCH$_2$–) are responsible for the splitting in the signal of the methyl group. Furthermore, methylenic protons resonate at 4.1 ppm as a *quartet* (four peaks). It is evident that the methyl protons are responsible for the splitting in the resonance signal of the –CH$_2$– protons.

In the ^{1}H-NMR spectrum of diethyl malonate (**59**) we observe similar signal splittings (Figure 40). The methylenic protons located between two carbonyl groups are equivalent and possess no neighboring protons. While their resonance signal appears as a sharp singlet at 3.2 ppm, the other methylenic protons next to the methyl group resonate as a *quartet*.

Analysis of these structures indicates that ethyl acetate has two different methyl groups. One of them resonates as a singlet, the other one as a triplet. In the case of diethyl malonate, there are two different methylenic protons, which resonate as a singlet and a quartet, respectively. These observations lead us to the following conclusion: a methyl group ($-CH_3$) can resonate as a singlet, doublet or triplet depending on the molecular structure. This splitting mode is also valid for methylene ($-CH_2-$) and methine protons ($-CH-$). *Generally, spin–spin splitting of single absorption peaks into multiplets (doublets, triplets, quartets, etc.) arises from different numbers of neighboring protons.* Conversely, the splitting in the resonance signals gives us exact information about the number of neighboring protons. This is a very important point in NMR spectroscopy in regard to structural analysis. This phenomenon, signal splitting, is called *spin–spin splitting*.

Let us now show how this splitting arises. For example, let us look at the ^{1}H-NMR spectrum of 4-methoxy-3-buten-2-one (**60**) in Figure 41.

$$H_3C-\overset{\overset{\displaystyle O}{\|}}{C}-\underset{\underset{\displaystyle H_a}{|}}{C}=\underset{\underset{\displaystyle H_b}{|}}{C}-OCH_3$$

60

There are two kinds of methyl groups, each of which resonates at different ppm values. Inspection of the olefinic protons shows their nonequivalency. However, these signals, in contrast to the other methyl resonances, are split into doublets.

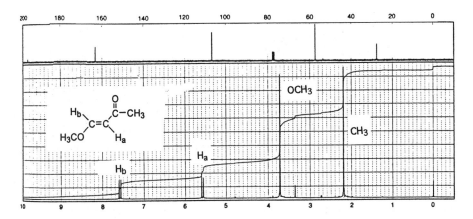

Figure 41   300 MHz ^{1}H-NMR spectrum of 4-methoxy-3-buten-2-one in CDCl$_3$. (Reprinted with permission of Aldrich Chemical Co., Inc. from C.J. Pouchert and J. Behnke, *The Aldrich Library of 13C and 1H FT-NMR Spectra*, 1992.)

Let us first analyze the H_a proton. If this proton is irradiated with electromagnetic radiation of the proper frequency, energy absorption occurs, and the lower energy state (parallel alignment) flips to the higher energy state (antiparallel alignment), which is called *resonance*. Now we address the question of how proton H_b behaves in the magnetic field while proton H_a resonates. Nucleus H_b possesses a magnetic moment (μ_a). According to the Boltzmann distribution law, proton H_b aligns either with or against the external magnetic field. If the neighboring proton H_b is aligned parallel to the external field, the total effective magnetic field at the proton H_a is slightly larger than it would be otherwise. Conversely, if proton H_b is aligned antiparallel to the external magnetic field, the effective field at proton H_a is slightly smaller than it would be otherwise. Thus, the applied field that is needed to bring proton H_a into resonance is slightly increased. In summary, proton H_a will be under the influence of an increased (H_1) and decreased (H_2) magnetic field depending on how the neighboring proton H_b is aligned. These magnetic fields are shown below:

$$H_1 = H_{total} + H_b \qquad (H_b \text{ parallel})$$

$$H_2 = H_{total} - H_b \qquad (H_b \text{ antiparallel})$$

where H_{total} is the magnetic field (including the effect of the neighboring groups) influencing proton H_a and H_b is the magnetic moment of H_b.

Since proton H_a will be under the influence of an increased and a decreased magnetic field, according to the resonance condition equation:

$$\nu = \gamma \frac{H_0}{2\pi} \qquad (13)$$

the H_a proton resonance will occur at two different values of the applied field, and proton H_a will have two different resonance frequencies. Therefore, the signal of proton H_a will split into a doublet. This phenomenon is called *spin–spin coupling*. We can also say that H_a couples with proton H_b and resonates as a doublet.

The reader should not assume that a nucleus is simultaneously under the influence of an increased and a decreased magnetic field. Approximately half of the H_a protons are under the influence of an increased magnetic field, and the other half of a decreased magnetic field. In the above-mentioned simple system, we have shown that proton H_a couples with the adjacent proton H_b and gives rise to a doublet. The question of how proton H_b resonates may be raised. Since proton H_a also has two different alignments in the external magnetic field, it influences the magnetic field around H_b in exactly the same way as described above. Consequently, proton H_b will also couple with the neighboring proton H_a and resonate as a doublet. Different orientations of the magnetic moments of a nucleus in an applied external magnetic field cause splitting in the resonance signals of the neighboring protons.

In the above-mentioned coupled system, protons H_a and H_b resonate as doublets. How can we determine the chemical shift? The consequence of the coupling with the neighboring proton is that the proton comes into resonance at two slightly different values of the applied field. One resonance is slightly above where it would be without coupling, and the other resonance is slightly below where it would be without coupling. Therefore, the real *chemical shifts* of the individual protons are located exactly in the center of the doublet lines. The distance between individual peaks in the frequency unit is called the *coupling constant* and is abbreviated as $^3J_{ab}$. The superscript on the left-hand side shows the number of bonds (double bonds and triple bonds are counted as single bonds) that separate the coupled protons. The subscript on the right-hand side indicates the coupled protons. *Note that a coupling constant is shared by both groups of coupled protons.* For example, proton H_a is coupled with proton H_b and appears as a doublet with $J = 5.0$ Hz. The H_b proton appears as a doublet with exactly the same coupling constant, $J = 5.0$ Hz. Figure 42 shows the analysis (determination of chemical shifts and coupling constant) of the doublet lines. Coupling constants are measured in hertz and fall in the range of 0–20 Hz. In some special cases, one may encounter coupling constants falling outside of this range. If the resonance frequencies of the doublet lines are given in ppm values, we cannot determine the coupling constant. In this case, we have to know the strength of the magnetic field of the NMR instrument. For example, on an instrument operating at 60 MHz, 1 ppm is equal to 60 Hz. However, on an instrument operating at 400 MHz, 1 ppm is equal to 400 Hz (see page 32. *Generally, spin–spin coupling constants are measured in hertz, and chemical shifts that are measured in ppm are not dependent on the magnitude of the applied field (the size of the NMR instrument). If a compound is measured on different NMR instruments, only the distance in hertz between the signals varies from one instrument to another.*

We have already discussed the electronic integration of the ^{1}H-NMR peak areas, which tells us the relative number of protons responsible for those peaks. Since both lines of a doublet resonance belong to proton H_a (or H_b), the total integration value of these lines has to be taken.

We have explained why the olefinic proton resonances in 4-methoxy-3-buten-2-one (**60**) (Figure 41) are split into doublets. In this system, both protons (H_a as well as H_b) have only one neighboring proton. To answer the question of what kind of splitting will occur in the case of an increased number of neighboring protons, we analyze a system where H_a has two equal neighboring protons H_b.

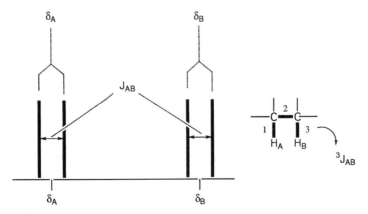

Figure 42 Spin–spin splitting of two coupled protons H_A and H_B. Representation of the chemical shifts (δ_A and δ_B) and the coupling constant $^3J_{AB}$.

$$Ha-C-C-Cl$$

61

For example, let us study the compound 1,1,2-trichloroethane (**61**). This compound has two different sets of protons. While proton H_a has two equal neighboring protons, these protons only have one neighboring proton. Let us first discuss the chemical shifts of these protons. Proton H_a resonates at a lower field (around 5.8 ppm) because two electronegative chlorine atoms are bonded to the same carbon atom (Figure 43). Inspection of the ^{1}H-NMR spectrum of **61** reveals that the H_a resonance at 5.8 ppm splits into a triplet; however, the two H_b protons resonate as a doublet. H_b protons will be aligned parallel and antiparallel to the external magnetic fields. However, the spins of these two equivalent protons can be aligned in three different combinations (Figure 44).

(1)   Both proton spins can be aligned antiparallel to the applied field (H_0).
(2)   One proton spin can be aligned parallel while the other is aligned antiparallel to the applied field (H_0). This combination has two possible variations.
(3)   Both proton spins can be aligned parallel to the applied field (H_0).

When two proton spins (H_b) are aligned antiparallel to the applied field, the total effective field at proton H_a is slightly reduced. In the second case, where one proton is aligned with the applied field and the other one against, these effects cancel each other out, so that the effective field at proton H_a is the same as the applied field H_0. If the neighboring protons are aligned parallel with the applied field, the total effective field at proton H_a is slightly increased. In summary, proton H_a will be under the influence of three

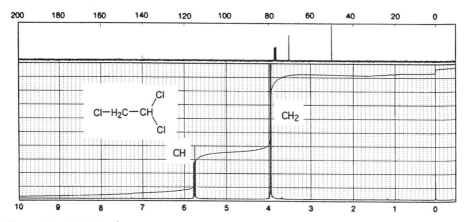

Figure 43   300 MHz ^{1}H-NMR spectrum of 1,1,2-trichloroethane in CDCl$_3$. (Reprinted with permission of Aldrich Chemical Co., Inc. from C.J. Pouchert and J. Behnke, *The Aldrich Library of 13C and 1H FT-NMR Spectra*, 1992.)

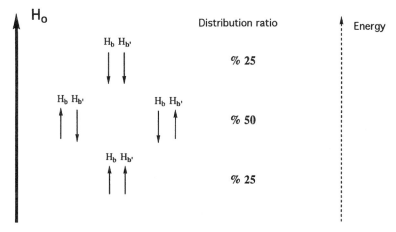

Figure 44 Energy levels for the three different spin states of the methylene protons that produce triplet and distribution ratio of the magnetic moments in a magnetic field.

different magnetic fields, as shown below:

$$H_1 = H_{\text{total}} - 2H_b \qquad (H_b \text{ antiparallel})$$

$$H_2 = H_{\text{total}} - H_b + H_b = H_{\text{total}} \qquad (H_b \text{ parallel and antiparallel})$$

$$H_3 = H_{\text{total}} + 2H_b \qquad (H_b \text{ parallel})$$

where H_{total} is the magnetic field (including the effect of the neighboring groups) influencing proton H_a and H_b is the magnetic moment of H_b.

Again, according to the resonance condition equation (eq. 13) the resonance of proton H_a occurs at three different values of the applied field. Proton H_a will have three different resonance frequencies. Therefore, the signal of proton H_a will split into triplets. After explanation of this triplet resonance, we look at the resonance signal of the equal H_b protons. From the ^{1}H-NMR spectrum (Figure 43), we see that these protons resonate as a doublet, since these protons have only a single neighboring proton (H_a).

Analysis of the triplet signal reveals that the intensities are not equal. It is evident that the intensity distribution is 1:2:1. From where do these different intensities arise? In Figure 44, one can observe that the distribution probabilities of the different spin combinations are 25, 50, and 25%. The populations of the three different energy levels are reflected in the signal intensities (1:2:1). We explained previously that the intensity of a signal depends directly on the number of nuclei populating an energy level. In summary, proton H_a 'sees' 25% of the H_b protons at the lower energy level (H_b protons are aligned parallel), 50% of the H_b protons at the middle energy level (one H_b proton is aligned parallel, the other antiparallel) and 25% of the H_b protons at the higher energy level (H_b protons are aligned antiparallel) (Figure 45).

The distances between the triplet lines must be equal and this value in hertz directly gives the coupling constant J_{ab} between the coupled protons. The same coupling has to be extracted from the distance between the doublet lines of proton H_b. In general, two groups of protons coupled to each other must have the same coupling constant J. The inner line of the triplet resonance gives the chemical shift of the corresponding proton. Since

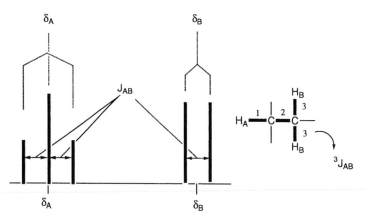

Figure 45 Spin–spin coupling between CH and CH$_2$ with different chemical shifts. Representation of the chemical shifts (δ_A and δ_B) and the coupling constant $^3J_{AB}$.

coupling is a reciprocal interaction between the spins of the protons, we can use this fact to predict which doublets, triplets or other split signals in a complex spectrum are related to each other. On the basis of this fact, the adjacent protons in a coupled system can be easily recognized.

When the number of neighboring protons is three, the splitting pattern and signal intensities of the lines will of course be different. For example, let us analyze 3-chloro-2-butanone (**62**).

62

In this compound, the H$_a$ proton has three equivalent protons on the adjacent carbon atom. Different alignments of these equivalent protons with the applied field may affect the magnetic field at proton H$_a$ in different ways. Let us indicate these equivalent protons with H$_{b'}$, H$_{b''}$ and H$_{b'''}$. These protons have eight possible orientations in the applied magnetic field when we differentiate between them. When three protons are aligned antiparallel to the applied field, the total effective field at proton H$_a$ is reduced. Conversely, three protons may be aligned parallel to the applied field. In this case, the total effective field at proton H$_a$ is increased. Finally, there are two other orientations: the two protons oppose the applied field and one reinforces, or one proton opposes and two reinforce. Since these protons are equivalent, we cannot distinguish which of the H$_b$ protons oppose and which reinforce the applied magnetic field.

Proton H$_a$ is under the influence of four different magnetic fields, of which two are slightly increased and the other two are decreased (Figure 46). Again, according to the resonance condition equation (eq. 13) proton H$_a$ will have four different resonance frequencies. Therefore, the signal of proton H$_a$ will split into four lines, which is called

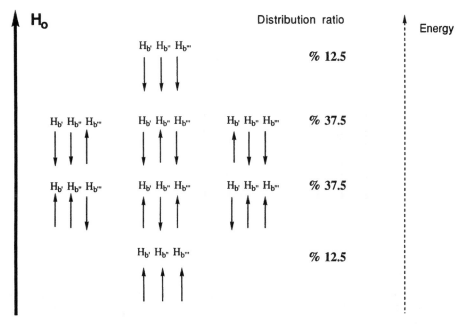

Figure 46 Energy levels for the four different spin states of the methyl group that produce quartet and distribution ratio of the magnetic moments in a magnetic field.

a *quartet* (Figure 47):

$$H_1 = H_{total} - 3H_b \qquad (H_b \text{ all antiparallel})$$

$$H_2 = H_{total} - H_b \qquad (H_b \text{ two antiparallel, one parallel})$$

$$H_3 = H_{total} + H_b \qquad (H_b \text{ two parallel, one antiparallel})$$

$$H_4 = H_{total} + 3H_b \qquad (H_b \text{ all parallel})$$

where H_{total} is the magnetic field (including the effect of the neighboring groups) influencing proton H_a and H_b is the magnetic moment of H_b.

Since the probability of second (two parallel, one antiparallel) and third arrangements (one parallel, two antiparallel) is three times that of either of the other two arrangements (all parallel or all antiparallel), the center peaks of the quartet are three times as intense. The H_b protons in **62** will couple with proton H_a and resonate as a doublet (Figure 48).

The distances between the quartet lines are equal and give the coupling constant J_{ab}. Since the coupling constant is shared by both groups of coupled nuclei, the same coupling J_{ab} can also be extracted from the doublet line, which arises from H_a resonance. The center of the quartet lines exactly gives the chemical shift of the resonating proton.

As a final example, we will briefly discuss the splitting pattern of an ethyl group, which is encountered very frequently in organic compounds. For example, diethyl ether has two sets of equivalent protons ($-CH_2-$ and $-CH_3$). Let us now apply the basic splitting concept, which we have already seen, to the ethyl group. We can predict that the methylene group will split into a quartet due to the three neighboring protons, and the methyl protons into a triplet. Inspection of the ^{1}H-NMR spectrum of diethyl ether

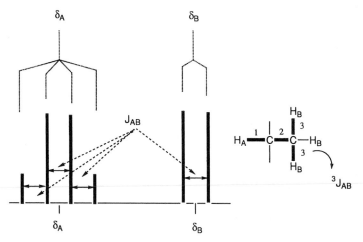

Figure 47 Spin–spin coupling between CH and CH_3 with different chemical shifts. Representation of the chemical shifts (δ_A and δ_B) and the coupling constant $^3J_{AB}$.

(Figure 49) supports our prediction. The quartet resonance is shifted downfield due to the electronegativity of the oxygen atom, which is directly bonded to the methylene carbon atom.

All of the above-mentioned examples show that resonance signals are split by the neighboring protons and the splitting pattern depends on the number of adjacent protons. All of these observations can be summarized by the following general rules:

(1) Chemically equivalent protons do not exhibit spin–spin splitting. The equivalent protons may be on the same or on different carbon atoms, but their signals do not split.
(2) Resonance lines are split by neighboring protons and the multiplicity of the split is determined by the number of protons. The signal of a proton that has n equivalent

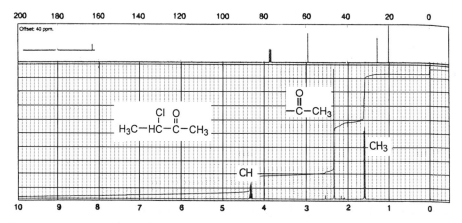

Figure 48 300 MHz ¹H-NMR spectrum of 2-chloro-2-butanone in $CDCl_3$. (Reprinted with permission of Aldrich Chemical Co., Inc. from C.J. Pouchert and J. Behnke, *The Aldrich Library of 13C and 1H FT-NMR Spectra*, 1992.)

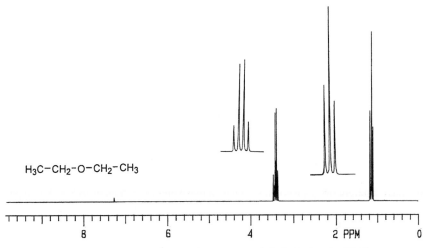

H₃C–CH₂–O–CH₂–CH₃

Figure 49 200 MHz ^{1}H-NMR spectrum of diethyl ether in CDCl₃.

neighboring protons is split into a multiplet of $n + 1$ peaks. The general formula, which covers all nuclei, is $2nI + 1$, I being the spin quantum number of the coupled element.

(3) The relative intensities of the peaks and multiplicity can be directly determined from Pascal's triangle.

(4) The distances between the split lines are equal and correspond to the coupling constant between the coupled nuclei.

With the help of Pascal's triangle the multiplicity and relative intensities may be easily obtained (Figure 50). We will demonstrate the applicability of Pascal's triangle with some selected examples. Let us consider 1-bromopropane (**63**).

Number of equivalent adjacent protons	Relative intensities observed							Type of signal
0				1				**singlet**
1			1		1			**doublet**
2			1	2	1			**triplet**
3		1	3	3	1			**quartet**
4	1	4	6	4	1			**quintet**
5	1	5	10	10	5	1		**sextet**
6	1	6	15	20	15	6	1	**septet**

Figure 50 Pascal's triangle. Relative intensities and types of signals.

$$H_3C-CH_2-CH_2-Br$$
$$\quad\ 3\quad\ \ 2\quad\ \ \ 1$$

63

Here, there are three sets of equivalent protons (Figure 51). We have no problem deciding what kind of signal splitting to expect from the protons of the $-CH_3$ group and the methylene protons adjacent to the bromine atom. Since the methyl protons are coupled to only two protons of the central $-CH_2-$ group, they resonate as a triplet. The protons of the CH_2-Br group are similarly coupled to only two protons of the central $-CH_2-$ group and their resonance also appears as a triplet. As for the protons of the C_2 methylene group, they will couple to two CH_2-Br protons, as well as to three CH_3 protons. The central methylene protons have 5 neighboring protons, which lead to 6 peaks ($n = 5$ leads to $5 + 1 = 6$ peaks). We should observe a sextet with the intensity distribution of 1:5:10:10:5:1, which can be extracted from Pascal's triangle. The 200 MHz ^{1}H-NMR spectrum of **63** is in full agreement with our expectation. The central methylene protons resonate at 1.9 ppm as a sextet.

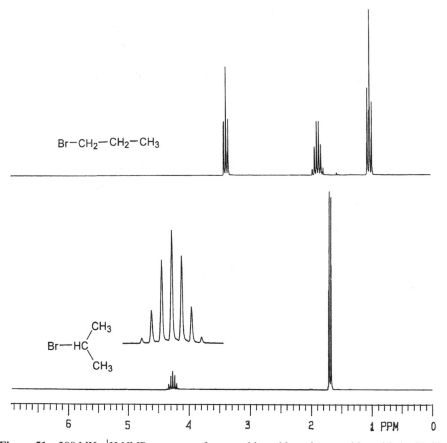

Figure 51   200 MHz ^{1}H-NMR spectrum of *n*-propyl bromide and *i*-propyl bromide in CDCl$_3$.

The ^{1}H-NMR spectrum of 2-bromopropane further illustrates the preceding rules and the applicability of Pascal's triangle. The 2-bromopropane ^{1}H-NMR spectrum shows two sets of equal signals resonating at 1.7 and 4.3 ppm, respectively.

$$H_3C$$
$$\diagdown$$
$$CH—Br$$
$$\diagup$$
$$H_3C$$

64

The six equivalent methyl protons appear as a doublet due to the signal splitting by the adjacent single –CHBr– proton. The methine proton –CHBr– has six equivalent neighboring protons on the two methyl groups and couples to these protons. Its signal appears at 4.3 ppm as a septet with an intensity distribution of 1:6:15:20:15:6:1, as expected from Pascal's triangle.

So far we have called the protons bonded to the adjacent carbon atoms (vicinal protons) *neighboring protons*. Attention should be paid to the fact that in all the examples discussed, three bonds separate the coupled protons. Protons that are farther than two carbon atoms apart do not usually couple.

splitting observed   splitting not usually observed

Let us consider 1-bromopropane (**63**) again. We have discussed the couplings between the protons bonded to C_1, C_2, and C_3 carbon atoms in this compound. We have not discussed any spin–spin coupling between the protons bonded to C_1 and C_3 carbon atoms. These protons are separated by four σ-bonds. Coupling is ordinarily not important beyond three bonds. In some special cases, coupling can be observed even between protons separated by four or five σ-bonds. Couplings of this kind (long-range couplings) are observed mostly in small rings or bridged systems and in the systems where the coupled protons are separated by π-bonds (see Section 4.3.3). Generally speaking, the magnitude of the spin–spin coupling between protons decreases as the number of bonds between the coupled protons increases. In saturated acyclic systems, coupling through more than three bonds is not observed. Therefore, there is no coupling between the H_1 and H_3 protons in **63**.

Now that we have shown that the splitting pattern and intensity distribution can be determined by Pascal's triangle, let us turn our attention to some general rules.

(1) If a proton H_a couples to different sets of protons H_b and H_c, splitting pattern and intensity distribution can be determined by Pascal's triangle only in cases where the coupling constant J_{ab} is equal to J_{ac} ($J_{ab} = J_{ac}$).

(2) So far as the chemical shift difference ($\Delta\delta$) between the coupled protons in hertz is much larger than the coupling constant:

$$\frac{\delta_{H_a} - \delta_{H_b}}{J_{ab}} \geq 10 \tag{25}$$

we can apply Pascal's triangle to determine splitting pattern and intensity distribution.

All the ^{1}H-NMR spectra that meet the above criteria are called *first-order spectra*. If the ratio of the chemical shift difference of the coupled protons over the coupling constant J_{ab} is smaller than 10, then we cannot apply the general rules to determine multiplicity and intensity distribution. The spectra become much more complex. The number of peaks increases and the intensities are not those that might be expected from the *first-order spectra*. These spectra are called *second-order spectra*, which we will discuss in Chapter 6.

At this point let us briefly discuss the effect of the magnitude of the applied field on the spectra. Scientists working with an NMR spectrometer always prefer an NMR instrument with a higher magnetic field. We will use an example to explain why they are interested in buying such instruments (400, 500, 600, 700, 800, 900 MHz machines are available). Let us consider a spin system consisting of the vicinal protons H_a and H_b. We assume that these protons resonate at 2.5 and 2.0 ppm with a coupling constant of $J_{ab} = 10$ Hz.

$$
\begin{array}{ccc}
-\overset{|}{\underset{|}{C}}-\overset{|}{\underset{|}{C}}- & & \delta_a = 2.0 \text{ ppm} \\
\quad H_a \quad H_b & & \delta_b = 2.5 \text{ ppm} \\
& & J_{ab} = 10 \text{ Hz}
\end{array}
$$

The reader should remember that the chemical shifts of NMR absorptions given in ppm or δ values and spin–spin couplings J are constant, regardless of the operating frequency of the spectrometer. Imagine that we measure the compound having the protons H_a and H_b on two different NMR instruments (60 and 400 MHz). Of course, the chemical shifts of the protons H_a and H_b and the spin–spin coupling constant between these protons will not change. However, the chemical shift difference between the resonance frequencies of the protons H_a and H_b in hertz will change. On a 60 MHz instrument 1 ppm is equal to 60 Hz, and on a 400 MHz instrument 1 ppm is equal to 400 Hz. The chemical shift difference between the resonance frequencies of H_a and H_b is 0.5 ppm in both cases. Therefore, the two signals are only 30 Hz apart at 60 MHz (0.5 ppm) and 200 Hz apart at 400 MHz (still 0.5 ppm) (Figure 52). When we determine the ratio of the chemical shift difference of the coupled protons over the coupling constant J_{ab}, we obtain two different values (eq. 25):

$$
\frac{\delta_{H_a} - \delta_{H_b}}{J_{ab}} = \frac{30}{10} = 3 \qquad \text{for a 60 MHz instrument}
$$

$$
\frac{\delta_{H_a} - \delta_{H_b}}{J_{ab}} = \frac{200}{10} = 20 \qquad \text{for a 400 MHz instrument}
$$

Since the ratio $\Delta\delta/J$ on a 400 MHz instrument is equal to 20 (greater than 10), we then speak of a *first-order spectrum*. In this case, we can apply the first-order rules concerning the multiplicity and intensity distribution. However, on a 60 MHz instrument the ratio $\Delta\delta/J$ is smaller than 10. More lines are observed in the spectrum and the intensity distribution in the lines of the signals is dramatically affected.

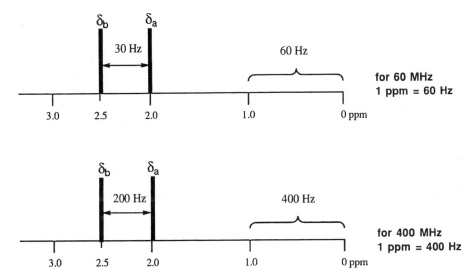

Figure 52   Representation of chemical shift difference (in hertz) of two different protons in different magnetic fields.

We have discussed in detail how the multiplicity and the intensity distribution can be determined by Pascal's triangle. We have to place some restrictions on these rules. These rules (see Figure 50) can be applied only to elements with a spin quantum number of $I = 1/2$. In the simplest case, if a proton is directly attached to a ^{13}C atom (the spin quantum number of the ^{13}C atom is also 1/2), the proton signal is split into a doublet by the ^{13}C atom. We will discuss spin–spin couplings of this kind in the ^{13}C-NMR chapter.

For nuclei that have a spin of greater than 1/2, the splitting patterns are completely different. Since in many mechanistic experiments, protons are replaced by deuterium atoms in order to follow the mode of the reaction, it is worth it to discuss a brief explanation of the spin–spin coupling between a proton and deuterium atoms. When the spin of any nucleus that couples to the proton has a spin other than 1/2, the signal splitting changes. Deuterium has a spin of $I = 1$ and $m = 3$, which means that deuterium atoms have three different alignments in an applied field. The deuterium atom will affect the magnetic field of proton H in three different ways. Finally, the proton resonance signal is split into a triplet by the neighboring deuterium. This triplet is different from those we have seen thus far. The intensity distribution in the triplet lines arising from the coupling of a proton to two neighboring protons is 1:2:1. However, the intensity distribution in triplet lines arising from H/D coupling is 1:1:1 (Figure 53).

The proton–deuterium couplings can be seen in the NMR spectra of the commonly used deuterated solvents such as acetone, dimethylsulfoxide and methanol. These solvents contain impurities of some degree of a partly deuterated solvent molecule as shown below.

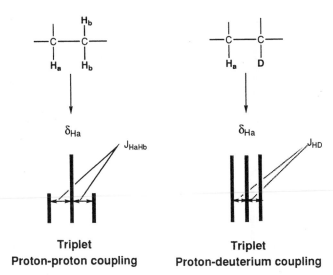

Triplet
Proton-proton coupling

Triplet
Proton-deuterium coupling

Figure 53 Appearance of a triplet arising by the coupling of a proton (a) with two equal protons and (b) with deuterium.

In partly deuteurated acetone (CD_3COCHD_2), the proton couples to two neighboring deuterium atoms and gives rise to a quintet (Figure 54). The relative intensities (1:4:6:4:1) may be easily obtained from Pascal's triangle. The general formula for multiplicity, which covers all nuclei, is given below, I being the spin number:

$$k = 2nI + 1 \qquad (26)$$

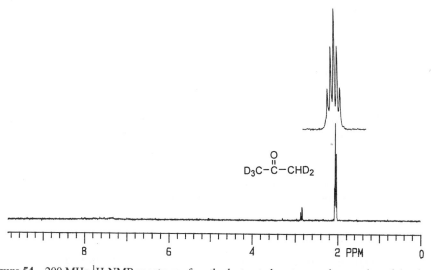

$$D_3C-\overset{\overset{\displaystyle O}{\|}}{C}-CHD_2$$

Figure 54 200 MHz ^{1}H-NMR spectrum of partly deuterated acetone and expansion of the signal.

where k is the number of lines, n is the number of coupled nuclei, and I is the spin number of the coupled nucleus.

This equation explains the quintet splitting of the proton signal in partly deuterated acetone.

4.2   SPIN–SPIN COUPLING MECHANISM [35]

Before considering the factors that affect the spin–spin coupling constants let us discuss briefly the mechanism of the spin–spin coupling interaction. How does spin–spin coupling arise?

Although there are three different mechanisms that contribute to spin–spin coupling, the *Fermi contact term* mechanism is the dominant one for all couplings involving hydrogen. According to this mechanism, the spin–spin coupling between two nuclei A and B arises from the interaction of nuclear spins A and B, which involves the bonding electrons connecting these nuclei. A and B can be either of the same isotope or of different isotopes attached by one bond. We can use a greatly simplified model to understand the coupling mechanism, taking as an example the $^1H–^{13}C$ molecule. Two atoms are bonded by a σ-bond having two electrons. The coupling mechanism for the $^1H–^{13}C$ coupling can be visualized as follows (Figure 55).

The magnetic moment of the hydrogen atom ^1H can be either parallel or antiparallel to the applied field. We begin first with the ground state orientation of the hydrogen atom ^1H, which has a parallel orientation. The energetically preferred state is that in which the nuclear magnetic moment of the hydrogen atom ^1H and the magnetic moment of the nearest bonding electron are in an antiparallel configuration. According to the *Pauli principle*, the two electrons in the bond between the nuclei must have their spins paired. Therefore, the magnetic moment of the second electron adjacent to ^{13}C will have an orientation parallel to the applied field, which will force the nuclear magnetic moment of the carbon atom ^{13}C to align itself antiparallel to the

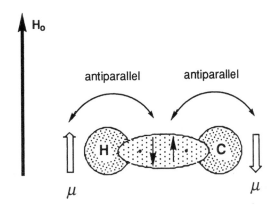

Figure 55   Schematic representation of the spin–spin interaction (proton and carbon) through the bonding electrons in a C–H bond.

applied field. In this way each nucleus responds to the orientation of the other. The antiparallel orientation of H will force carbon atom ^{13}C to align itself in parallel orientation to the applied field in the same manner. Consequently, the magnetic field around the resonating nucleus will be affected by the orientation of the adjacent nucleus. In this particular case, the nuclear magnetic moment of one nucleus forces the nuclear magnetic moment of the other nucleus (through the interaction with the bonding electrons) to be in the opposite direction, in a stable state. In this case, the *coupling constant is defined as having a positive sign.* All ^{13}C–^{1}H coupling constants have positive signs.

As for the second case, let us discuss coupling through two bonds, *geminal coupling.* For example, we again analyze the orientation of the nuclear magnetic moments of the nuclei and the magnetic moments of electrons in the structure H_a–C–H_b. Again, we start with the ground state orientation of the hydrogen atom H_a, which has parallel orientation (Figure 56). In the energetically preferred state, the nuclear magnetic moment of the hydrogen atom H_a and the magnetic moment of the nearest bonding electron will be in antiparallel configuration. The magnetic moment of the second electron adjacent to the carbon atom will have an orientation parallel to the applied field. According to the *Hund's rule,* the energetically preferred state is that in which the electron spins of the two bonding electrons near the carbon atom are parallel. Therefore, the second electron near the hydrogen atom will have an antiparallel orientation, which will force the nuclear magnetic moment of the hydrogen atom H_b to align itself parallel to the applied field. The antiparallel orientation of H_a will force the other hydrogen atom H_b to align itself antiparallel to the applied field. Therefore, *the geminal coupling constant is defined as having a negative sign.*

The next coupling is *vicinal spin–spin coupling,* which can always be encountered in NMR spectra and provide very useful information for the determination of the chemical structures. By applying the same models discussed above, we see the nuclear magnetic moment of one nucleus forces the nuclear magnetic moment of the coupled nucleus (over three bonds) to orient itself in the opposite direction (Figure 57). Therefore, most vicinal couplings have a positive sign.

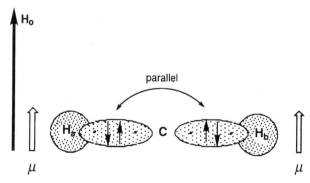

Figure 56 Schematic representation of the geminal spin–spin interaction (proton and proton) through the bonding electrons in a CH$_2$ group.

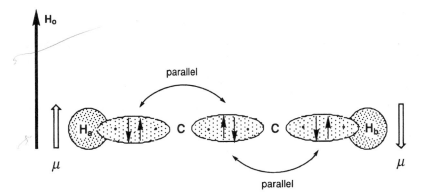

Figure 57   Schematic representation of the vicinal spin–spin interaction (proton and proton) through the bonding electrons in an HC–CH group.

4.3   FACTORS INFLUENCING SPIN–SPIN COUPLING CONSTANTS

The spin–spin couplings between the protons are always given in hertz. Coupling constants provide very useful information about chemical structures. Before starting to discuss the factors influencing coupling constants, it is worthwhile to qualitatively state which kind of information can be gathered after the accurate measurement of the coupling constants.

For example, a disubstituted C=C double bond can have three different isomers (*trans*, *cis*, *geminal*). The measured coupling constant between the protons bonded to the C=C double bond can determine the exact position of the substituents. Imagine a benzene ring having more than one substituent. The number of isomers depends on the number of substituents. The exact location of these substituents can be easily determined by measuring the coupling constants between the protons, which are attached to the benzene ring. During a chemical reaction, configurational isomers such as *exo/endo* or *syn/anti* can be formed. Again, coupling constants can distinguish in most cases between these formed isomers. Through the analysis of an unknown compound, the geminal coupling constants can provide some idea about the size of the rings. In addition, the dynamic processes existing in the compound can be studied with the aid of the coupling constants. Even the dihedral angles between the coupled nuclei and the bond distances can be determined approximately by the coupling constants. In summary, coupling constants provide us with information about the constitution, configuration and conformation of the molecule. In the following chapter, the importance of the coupling constant will be made much clearer.

Before discussing the factors influencing coupling constants, we will use a classification indicating the number of bonds between the coupled nuclei. As we discussed before, coupling constants between nuclei are denoted by J. The number of bonds between the coupled nuclei is indicated by a superscript n, which is given on the left-hand side of J. The coupling constant between nuclei is denoted by a subscript and is given on the right-hand side of J:

$$^{n}J_{ab}, \quad ^{n}J_{12}$$

The number of bonds between the coupled nuclei:

(1) $n = 1$. In ^{1}H-NMR spectroscopy, only one molecule exists for this special case, which is the hydrogen molecule. The coupling between the hydrogen atoms is $^1J = 280$ Hz and is of only theoretical interest.

(2) $n = 2$. When the number of bonds between the coupled protons is two, these protons are bonded to the same carbon atom and the coupling is called *geminal coupling* ^2J. Geminal coupling can be observed between the protons of a $-CH_2-$ group, provided that they are not chemically equivalent.

Geminal coupling

$^2J_{ab}$

(3) $n = 3$. The number of bonds between the coupled protons is three. The commonest and most useful coupling encountered in ^{1}H-NMR spectroscopy is *vicinal coupling*, which is denoted by ^3J. This coupling is observed between adjacent protons. The bond between the carbon atoms can be a single as well as a double bond. A double bond is counted as a single bond. A few selected examples of vicinal coupling are illustrated below.

Vicinal coupling

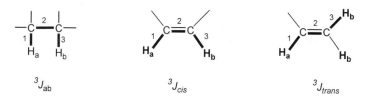

$^3J_{ab}$ $^3J_{cis}$ $^3J_{trans}$

(4) $n = 4$. The number of bonds between the coupled protons is four. When the number of bonds between the coupled protons is greater than three, these couplings are called *long-range couplings*. Long-range couplings are observed in cyclic saturated compounds, in particular when the ring system is strained and has a bicyclic structure. Allylic coupling and *meta* coupling are also typical examples of long-range couplings through four bonds. Some examples of long-range couplings over four bonds are presented below.

long-range coupling

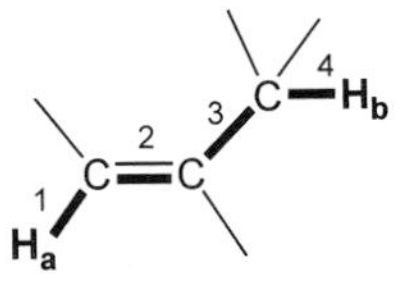

$^4J_{allylic}$ 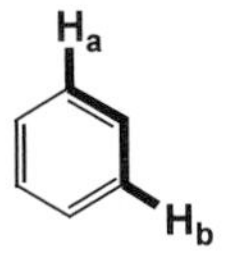 $^4J_{meta}$ 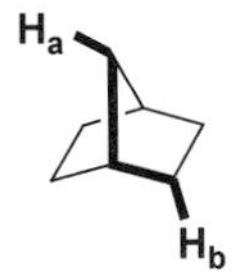 $^4J_{ab}$

(5) $n = 5$. Couplings that arise from the interaction of protons separated by five bonds are also called *long-range couplings*. These couplings are observed in aromatic systems and homoallylic systems.

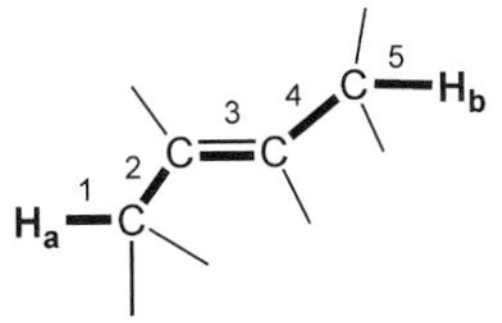

$^5J_{ab}$  $^5J_{para}$

   Now that we have classified coupling constants according to the number of bonds between the coupled protons, let us now discuss the factors influencing coupling constants. As we have discussed previously, the magnitude of the spin–spin coupling of the *first-order spectra* can be determined by measuring the distance between the lines in hertz. However, while the sign of the coupling constant cannot be directly determined, special experiments allow us to determine the sign of the coupling constants. It should be noted that the signs of all ^3J and most ^5J values are positive while those of ^2J and ^4J values can be either positive or negative. To interpret the couplings in connection with the structure, it is not necessary to know their sign.

4.3.1   Geminal spin–spin coupling ^2J [36]

Geminal spin–spin coupling constants ^2J range from -20 to $+40$ Hz. There are four structural features that affect the geminal spin–spin coupling constants:

(1)   dependence on the hybridization of the carbon atom;
(2)   dependence on the bond angle ϕ;
(3)   the effect of the substituents;
(4)   the effect of the neighboring π-substituents.

Dependence on the hybridization of the carbon atom

The geminal spin–spin coupling constant strongly depends on the hybridization of the carbon atom bearing the coupled protons. For example, in the methane molecule, which is sp^3 hybridized, the geminal coupling constant is $^2J = -12.5$ Hz, whereas the geminal coupling constant in ethene, which is sp^2 hybridized, is $^2J = +2.5$. Spin–spin coupling is not observed for protons that are chemically equivalent. The measurement of the coupling constants between the equivalent protons is explained in a later section. This observed difference between the coupling constants in methane and ethene also depends on factors other than hybridization. It is well known that the hybridization in cyclopropane is between sp^3 and sp^2. The coupling constant of -4.5 Hz for cyclopropane lies between the values of methane and ethylene.

increased s character

$^2J_{HH} = -12.5$ Hz $^2J_{HH} = -4.5$ Hz $^2J_{HH} = +2.5$ Hz

Dependence on bond angle ϕ

An increase in the angle (HCH) ϕ causes very important changes in the geminal proton coupling ^2J values. In cyclic systems, the bond angle depends on the strain of the ring. Therefore, the measured geminal coupling constants in the cyclic systems is a measure of the ring strain, in other words, of the ring size. In strain-free or less-strained systems such as cyclohexane and cyclopentane, the bond angle (HCH) ϕ is close to the angle in methane ($\phi = 109°$). Therefore, the geminal coupling observed in five- and six-membered rings is close to that of methane.

$^2J_{HH} = -12.5$ Hz -13.0 Hz -14.0 Hz -10.0 Hz -4.0 Hz

When the ring size is decreased, the bond angle is increased. The greater the bond angle ϕ between the coupled protons is, the more positive (less negative) the geminal coupling ^2J. Since geminal couplings are generally negative, an increase in the coupling decreases the absolute value of the coupling constants.

In some particular cases, it is possible to predict the size of the ring by looking at geminal coupling constants. For example, compounds **68** and **69** with different ring sizes are formed after the dehydrobromination reaction of the corresponding dibromides **65** and **66** [28]. The question of whether or not the formed norcaradiene structures **68** and **69** are in equilibrium with their valence isomer, cycloheptatriene, can be easily answered by measuring the geminal coupling constants between the bridge methylene protons.

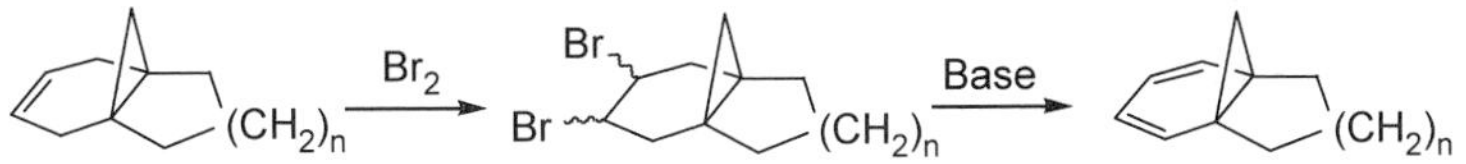

65 $n = 1$
66 $n = 2$

67

68 $n = 1$
69 $n = 2$

The measured geminal coupling constants between the bridge protons in **68** and **69** are $^2J = 4.0$ and 9.0 Hz, respectively [37]. The coupling constant of $^2J = 4.0$ in **68** indicates that it is a cyclopropane coupling. The compound has a norcaradiene structure and the equilibrium **68** ⇌ **70** is completely shifted towards the norcaradiene structure (**68**). The geminal coupling, $^2J = 9.0$ in **69** shows that the formed norcaradiene **69** is in equilibrium with its valence isomer cycloheptatriene, where the equilibrium is shifted towards cycloheptatriene structure **71**.

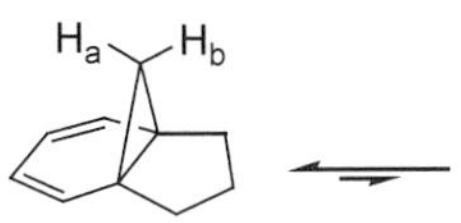

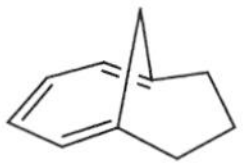

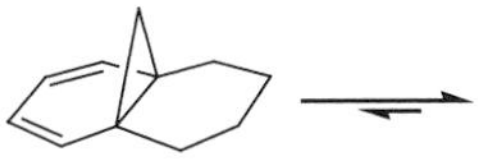

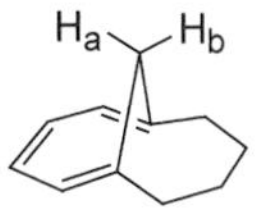

68

70

69

71

$^2J_{ab} = 4.0$ Hz

$^2J_{ab} = 9.0$ Hz

This experiment demonstrates that the existence of an equilibrium in a dynamic system can be easily determined with the aid of the coupling constants.

Further examples of geminal coupling constants determined from different ring size are illustrated below. All of these geminal couplings that are measured in systems that do not contain electronegative or electropositive substituents give us some idea about the size of the ring.

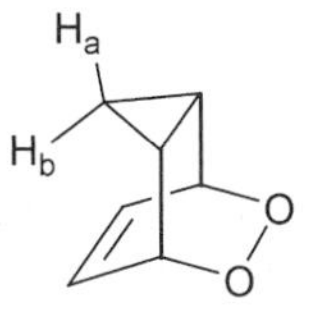

$^2J_{ab} = 5.9$ Hz [38]

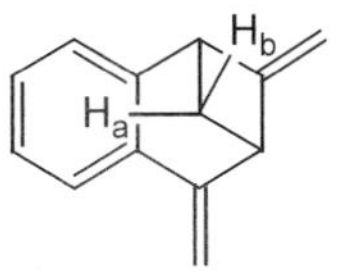

$^2J_{ab} = 8.0$ Hz [25]

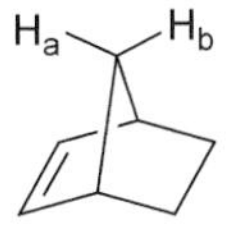

$^2J_{ab} = 8.5$–9.5 Hz

$^2J_{ab}$ = 10.7 Hz $^2J_{ab}$ = 14.2 Hz [39]

The effect of substituents

The geminal coupling constants are subject to the influence of both α- and β-substituents. The electronegative substituent directly bonded to the methylene group causes a positive change in the coupling constants, so that the absolute value of the coupling decreases. The influence of electronegative substituents is clearly observed in saturated systems. For example, the geminal coupling constant in methane ($^2J = -12.4$ Hz) is increased by the introduction of substituents such as $-OH$ (CH$_3$OH, $^2J = -10.8$ Hz) and F (CH$_3$F, $^2J = -9.6$ Hz). Some characteristic data are presented in Table 4.1. The effect of the substituents on the coupling constants is additive.

Table 4.1

Geminal H–H coupling constants ^2J in some selected compounds [35, 40]

Compound	Coupling constant ^2J (Hz)
CH$_4$	− 12.4
CH$_3$CCl$_3$	− 13.0
CH$_3$Cl	− 10.8
CH$_3$OH	− 10.8
CH$_3$F	− 9.6
Br–**CH$_2$**–CH$_2$–OH	− 10.4
Br–CH$_2$–**CH$_2$**–OH	− 10.4
Br–CH$_2$–**CH$_2$**–CN	− 17.5
H$_2$C=**CH$_2$**	+2.5
HFC=**CH$_2$**	− 3.2
LiHC=**CH$_2$**	+7.1
	− 8.3
	0.0
	− 5.0

The electronegative effect of the substituent on the geminal coupling can also be observed in unsaturated systems. In ethene the geminal coupling is $+2.5$ Hz. However, formaldehyde has a coupling constant of $+41.0$ Hz, which is the maximum value observed in organic compounds. This can be explained by the strong electronegativity of the oxygen atom, as well as the back donation of the oxygen π-orbitals in the CH_2 plane, which further increases the inductive effect.

$$H_2C=C\begin{smallmatrix}H\\H\end{smallmatrix} \qquad\qquad O=C\begin{smallmatrix}H\\H\end{smallmatrix}$$

$$^2J_{HH} = 2.5 \text{ Hz} \qquad\qquad\qquad ^2J_{HH} = 41.0 \text{ Hz}$$

The effect of the β-substituents on the geminal coupling constants is observed in vinyl systems. In the case of ethylene derivatives, an electronegative substituent at the β-position causes a negative contribution to the geminal coupling. Conversely, an electropositive substituent such as lithium increases the geminal coupling (Table 4.1).

The effects of neighboring π-substituents

One of the most important effects influencing geminal coupling constants is the number of neighboring π-bonds directly attached to the methylene carbon atom. This effect of the π-bonds has to be considered separately in two different systems:

(1) acyclic systems;
(2) cyclic systems.

The effects of the neighboring π-bonds on geminal coupling constants depend strongly on the stereochemistry of the system. In the acyclic system, the rapid rotation of the substituent will not result in optimal orientation. There is an almost constant change of -2.0 Hz for each neighboring π-bond in a freely rotating fragment. The measured value is an average.

$$\begin{array}{c}H\\ \diagdown\\ C\\ \diagup \quad \diagdown\\ H \qquad C=X\end{array} \qquad\qquad X = C, O, N$$

For example, the geminal coupling constant in methane is $^2J = -12.4$ Hz. When two protons in methane are replaced by π-bonds, the geminal coupling constant decreases, and the absolute magnitude increases. In toluene, where a proton in methane is replaced by a π-bond, the methyl protons have a geminal coupling constant of -14.5 Hz (Table 4.2). This value is only 2 Hz more negative than that of methane (-12.4 Hz). In acetonitrile, the magnitude of the geminal coupling constant changes from -12.4 Hz in methane to -16.2 Hz (the nitrile group contains two double bonds). In the case of malonitrile, the expected change is around 8 Hz because of the four neighboring π-bonds.

Table 4.2

Dependence of geminal H–H coupling constants ^2J on the adjacent π-bonds

Compound	^2J (Hz)	Number of π-bonds
CH$_4$	− 12.4	0
⬡—CHH$_2$	− 14.5	1
N≡C–CHH$_2$	− 16.2	2
N≡C–CH$_2$–C≡N	− 20.3	4

The measured value is − 20.3 Hz, which is in agreement with the expectation. These values indicate that the effects of the neighboring π-bonds on geminal coupling constants are additive.

In cyclic systems, the methylene group has mostly a fixed conformation. The effect of the neighboring π-bond on the geminal coupling constant is a function of the angle ϕ between the π-orbital and the C–H bond. This dependence is clearly shown in Figures 58 and 59.

This curve shows just the changes in geminal coupling constants, and not the absolute values. The largest effect is observed when the angle ϕ between the π-orbital and C–H bond is 30°. When the angle ϕ is 90°, geminal coupling constants are not affected. This angle plays an important role in the determination of the structures. Starting from the measured geminal coupling constants, one can easily estimate the angle, which will provide important information about the configuration and conformation of the molecule. The examples given below clearly demonstrate the dependence of geminal coupling constants on this angle ϕ.

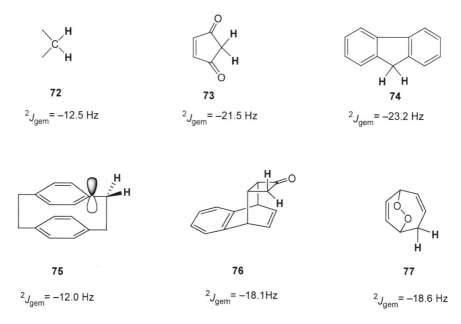

72

$^2J_{gem}$ = −12.5 Hz

73

$^2J_{gem}$ = −21.5 Hz

74

$^2J_{gem}$ = −23.2 Hz

75

$^2J_{gem}$ = −12.0 Hz

76

$^2J_{gem}$ = −18.1 Hz

77

$^2J_{gem}$ = −18.6 Hz

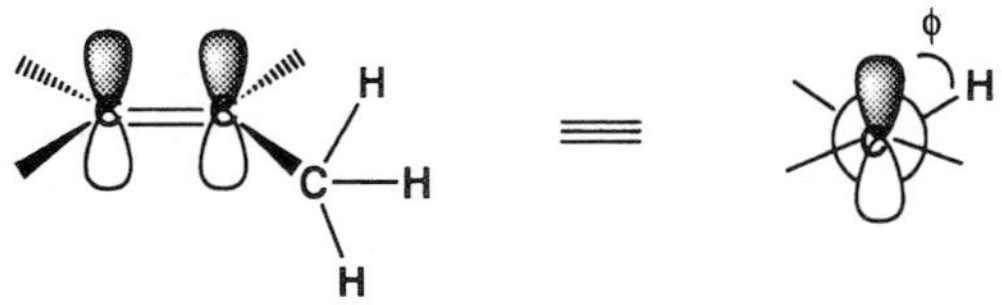

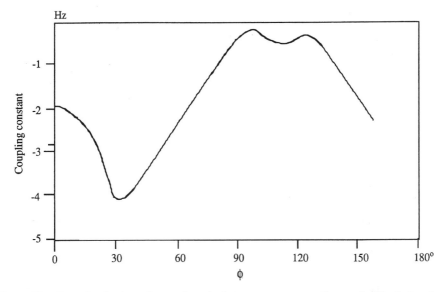

Figure 58 Schematic representation of the angle between the π-orbital and C–H bond for geminal coupling.

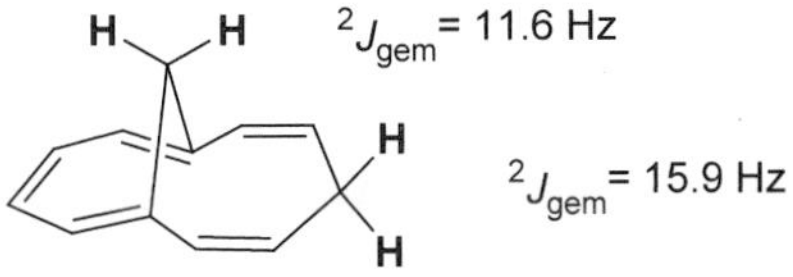

78

The geminal coupling constants in cyclopentadienone **73** ($^2J = -21.5$ Hz) and fluorene **74** ($^2J = -23.2$ Hz) indicate the favorable orientation of the π-system and the CH$_2$ group [41]. In contrast, in metacyclophane **75**, the measured coupling constant $^2J = -12.0$ Hz is very close to that of methane [42]. The π-system of the benzene ring does not contribute to the coupling. The perpendicular orientation of the π-system to the plane of the methylene group can explain this observation. In the other reported examples **76–78** the large effects of a carbonyl group and C=C double bonds on the geminal coupling constants are shown [38, 43, 44].

Figure 59 Curve for the dependence of geminal proton–proton coupling on the dihedral angle ϕ.

Measurement of geminal coupling constants between equivalent protons

When the geminal protons have different chemical shifts (in other words, when they are nonequivalent), they couple and split each other's signals. If there are no additional protons to further coupling, their resonances appear as doublets. Analysis of the doublet by way of the measurement of the distance between the doublet lines directly gives the corresponding geminal coupling constant. When the geminal protons are equivalent, the coupling cannot be directly observed because they resonate as a singlet. For example, CH_2Cl_2 resonates as a singlet; therefore, the geminal coupling constant cannot be directly determined.

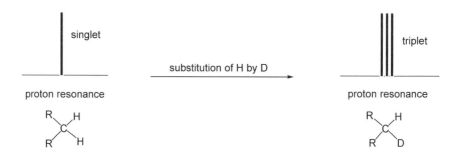

Since the bridge methylene protons in **71** are located in different chemical environments, they resonate in different regions as two distinct doublets. The distance between the doublet lines gives the coupling constant. The bridge protons in [10]annulene **20** are equal and resonate as a singlet. In this case, the coupling cannot be determined. The question arises of how to determine the geminal coupling constants between equal protons. To determine the geminal coupling constant, we have to distort the symmetry in the molecule in some way. In many cases the geminal couplings are obtained from the deuterium derivatives. When we replace one of these geminal protons in **20** by deuterium, we obtain compound **79**. In this case, the proton couples to deuterium (spin quantum number of deuterium is $I_D = 1$) and gives rise to a triplet consisting of three equally intense lines.

singlet	substitution of H by D	triplet
proton resonance		proton resonance

The distances between the lines of the triplet directly give the geminal coupling constant J_{HD} between the proton and deuterium (H–C–D). The H–H and H–D coupling constants are proportional to the gyromagnetic ratios:

$$J_{HH} = 6.55 J_{HD} \tag{27}$$

Eq. 27 can be used to calculate the geminal coupling constant J_{HH}.

For this process, one of the equal methylenic protons has to be replaced by a deuterium atom, which is not an easy process. The deuterated compound has to be synthesized.

4.3.2 Vicinal spin–spin coupling ^3J

The most useful and informative coupling encountered in NMR spectra is vicinal spin–spin coupling. In vicinal coupling, the number of bonds between the coupled protons is three.

$$^3J_{HH} = \text{Vicinal coupling constant}$$

Let us first consider the coupling in olefinic compounds. There are two different vicinal couplings.

(a) $^3J_{cis}$ vicinal coupling constant

(b) $^3J_{trans}$ vicinal coupling constant

In ethylene and ethylene derivatives, these couplings are different and the coupling between *cis* protons is smaller than that between *trans* protons. In the acyclic system, when the conformation is not fixed, we do not differentiate between *cis* and *trans* couplings. Because of the free-rotation about the C–C bond, we observe an average vicinal coupling in the acyclic systems.

However, in saturated cyclic systems, different couplings between vicinal protons are always observed. For example, in cyclohexane, the chair conformation is preferred. Therefore, we observe three different couplings: axial–axial, axial–equatorial, and equatorial–equatorial.

After this short introduction, we may look at the factors influencing vicinal coupling constants $^3J_{HH}$. Vicinal proton–proton couplings have been very thoroughly studied. The magnitude of $^3J_{HH}$ is always positive and mainly depends upon four factors:

(1) the torsional or dihedral angle ϕ (HCCH);
(2) the valence angle θ (HCC);
(3) the bond length (R_{CC});
(4) the electronegative effect of the substituents.

Dependence on the dihedral angle

The angle between the planes formed by H_1CC and H_2CC is called the *dihedral angle ϕ*. This angle can be visualized by an end-on view of the bond between the vicinal carbon atoms (see Figure 60).

   Vicinal proton–proton coupling depends primarily on the dihedral angle ϕ. The relationship between the dihedral angle and vicinal coupling was first predicted by Karplus and Conroy, and is given by the following equation [45]:

$$^3J = 4.22 - 0.5 \cos \phi + 4.5 \cos^2 \phi \tag{28}$$

The calculated coupling constants for the angles $\phi = 0$, 90, and 180° are given below:

$$^3J = 8.22 \text{ for } \phi = 0°, \qquad ^3J = -0.33 \text{ for } \phi = 90°, \qquad ^3J = 9.22 \text{ for } \phi = 180°$$

   It is very important to note that the vicinal coupling for $\phi = 90°$ is almost zero. On the other hand, the vicinal coupling constants vary from 8 to 10 Hz when the dihedral angle is $\phi = 0°$ and when it is $\phi = 180°$. In the case of $\phi = 0°$, the coupled protons have *cis* configuration (*synperiplanar*), and in the case of $\phi = 180°$, they have *trans* configuration (*antiperiplanar*). Generally, *trans* coupling is larger than *cis* coupling.

   The ^{1}H-NMR spectrum of the dibromo benzonorbornene derivative **80** is given in Figure 61. One of the methylene protons H_a resonates at 2.46 ppm as a doublet of triplets due to the coupling to the geminal proton H_b and vicinal protons H_c and H_d [46]. However, the other methylenic proton H_b resonates at 2.26 ppm as a doublet of doublets in spite of the fact that this proton has the same neighboring protons as H_a. These different splitting modes indicate that the methylenic proton H_b has coupling to only one of the vicinal protons (H_c and H_d). Inspection of the model shows that the dihedral angle between H_b and H_d is close to 90°. Therefore, a coupling between these protons is not observed.

   Extensive experimental studies have shown that the relationship between dihedral angle and vicinal coupling does not always obey eq. 28. There are also other factors influencing vicinal couplings. Experiments indicate that there are some deviations up to 2 Hz from the calculated coupling values obtained from eq. 28. The lower curve of

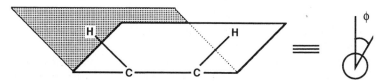

Figure 60   Schematic representation of the dihedral angle ϕ formed between two protons bonded to adjacent carbon atoms.

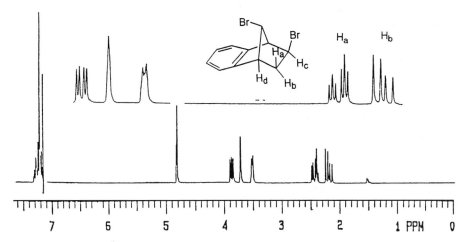

Figure 61   200 MHz ^{1}H-NMR spectrum of dibromo compound (**80**) in CDCl$_3$ and expansions of the signals.

Figure 62 corresponds approximately to the theoretical Karplus–Conroy curve. The shaded area shows the range of the vicinal couplings obtained experimentally. Taking these deviations into account, the Karplus equation has been described by the following relation:

$$^3J = 7.0 - 0.5 \cos \phi + 4.5 \cos^2 \phi \tag{29}$$

The dependence of vicinal couplings on the dihedral angle provides very useful information concerning the configuration and conformation of the molecules.

In cyclohexane, there are three types of vicinal couplings: These couplings can be visualized by an end-on view of the bond between the vicinal carbon atoms.

axial–axial vicinal coupling		$^3J_{aa}$
axial–equatorial vicinal coupling		$^3J_{ae}$
equatorial–equatorial vicinal coupling		$^3J_{ee}$

In the *axial–axial* orientation (*trans* configuration), the dihedral angle between the coupled protons is approximately 180°. Therefore, the maximum coupling in cyclohexane is the *axial–axial* coupling of 8.0–13.0 Hz. The experimentally determined values are in a range of 8.0–10.0 Hz. Newman projection shows that the protons in the *equatorial–equatorial* and *axial–equatorial* orientation have a dihedral angle of about 60°. The values found experimentally are about 2.0–4.0 Hz. This is an important

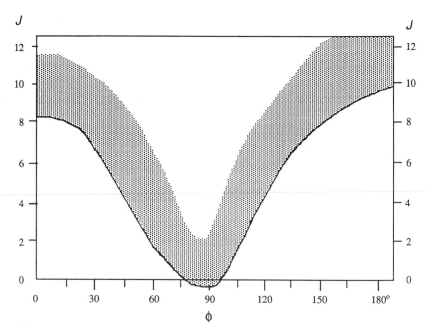

Figure 62 The Karplus–Conroy curve: Range of observed vicinal proton–proton coupling constants for different values of the dihedral angle ϕ.

criterion in the conformational as well as configurational analysis of cyclohexane derivatives.

Typical values

$^3J_{aa} = 8.0–13.0$ Hz	$8.0–10.0$ Hz
$^3J_{ae} = 2.0–5.0$ Hz	$2.0–4.0$ Hz
$^3J_{ae} = 2.0–5.0$ Hz	$2.0–4.0$ Hz

In cyclic systems smaller than cyclohexane, we encounter only two types of couplings: *cis* vicinal and *trans* vicinal.

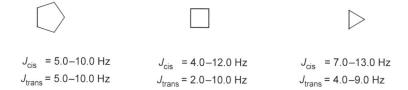

J_{cis} = 5.0–10.0 Hz	J_{cis} = 4.0–12.0 Hz	J_{cis} = 7.0–13.0 Hz
J_{trans} = 5.0–10.0 Hz	J_{trans} = 2.0–10.0 Hz	J_{trans} = 4.0–9.0 Hz

The above-mentioned values vary over a wide range depending on the electronic nature of the substituents. According to the Karplus–Conroy curve, *trans* coupling is always greater than *cis* coupling. However, in cyclopentane, *cis* and *trans* couplings have almost the same values, whereas in cyclopropane *cis* coupling is always greater than *trans* coupling. In the beginning there appears to be an anomaly in the cyclopropane ring. The dihedral angle for the *trans* protons is approximately 130°, and that for the *cis* protons is 0°. Because of the geometry, the dihedral angle between the *trans* protons in cyclopropane ring can never be 180°. Inspection of the Karplus–Conroy curve (Figure 62) clearly shows that the coupling for the dihedral angle of 0° is greater than that of 130°.

The Karplus–Conroy curve can also be applied to vicinal coupling in olefines. Since C=C double bonds have a planar structure, the dihedral angle between the *cis* protons is 0° and the dihedral angle between the *trans* protons is 180°. Consequently, *trans* coupling ($^3J_{trans}$ = 12.0–18.0 Hz) in olefines is always greater than *cis* coupling ($^3J_{cis}$ = 7.0–10.0 Hz). As these ranges are clearly separated, the configurational assignment of the double bond can be done clearly.

J_{cis} = 7.0–10.0 Hz J_{trans} = 12.0–18.0 Hz

Dependence on the valence angle θ

Another important factor influencing vicinal spin–spin coupling is the valence angle (CCH) θ.

The dependence of vicinal coupling on the valence angle is clearly seen in monocyclic olefines. The *cis* coupling increasing from 2.0 Hz in cyclopropene to the normal value of 10.5 Hz in cyclohexene, in which the internal angle is relatively unstrained.

0.5 – 2.0 Hz   2.5 – 4.0 Hz   5.1 – 7.0 Hz   8.8 – 10.5 Hz

8.0 Hz   9.0 – 12.0 Hz   10.0 – 13.0 Hz

Because of the constant dihedral angle of 0° in cyclic olefines and the absence of substituent effects, the dependence of vicinal couplings can be attributed solely to valence angles. Generally, when the valence angle increases (decreased ring size), the coupling decreases. A distinction between different sizes of rings can be made by measuring vicinal coupling constants. With the spin–spin coupling values given above, small rings in particular can be easily distinguished from the other rings. The vicinal coupling of benzene ($^3J = 8.0$ Hz) is smaller than that of cyclohexene. Besides the valence angle, the π-bond order (bond length) also influences vicinal coupling. This is the subject of the next chapter.

In acyclic systems, the valence angle will influence the vicinal spin–spin system. Since the dihedral angle also changes the vicinal coupling constants, it is difficult to solely observe the effect of the valence angle in acyclic systems.

Dependence on the bond length (π-bond order)

Another important parameter influencing vicinal spin–spin coupling is the distance between the carbon atoms bearing protons. Coupling decreases with increased bond length. Since C=C double bonds are shorter than C–C single bonds, the vicinal couplings observed between double bond protons are larger than those of saturated systems. There is a relationship between the C–C bond length and the vicinal coupling constants [48]. The bond length decreases with increased π-bond order. In ethylene the vicinal *cis* coupling is $^3J = 11.5$ Hz, whereas in benzene it is $^3J = 8.0$ Hz. Since the π-electrons in benzene are delocalized, the average electron density between two carbon atoms is less than that of ethylene. Therefore, the bond length in ethylene is shorter (1.33 Å) than that in benzene (1.38 Å). Consequently, vicinal coupling decreases with increased bond length.

$^3J_{cis}$ = 11.5 Hz $^3J_{cis}$ = 8.0 Hz

π-bond order = 1.0   π-bond order = 0.67

The measured vicinal coupling constants in aromatic compounds give us very useful information about the physical properties (electron distribution) of the compounds. For example, the vicinal couplings observed in naphthalene (**19**) and biphenylene (**81**) reveal the interesting electronic nature of these aromatic compounds. For these compounds two different vicinal (*ortho*) coupling constants are found ($^3J_{12}$ and $^3J_{23}$). In both compounds these couplings have different values. In naphthalene the vicinal coupling $^3J_{12} = 8.3$ Hz is greater than the vicinal coupling $^3J_{23} = 6.85$ Hz [48]. However, these coupling values are reversed in biphenylene [49].

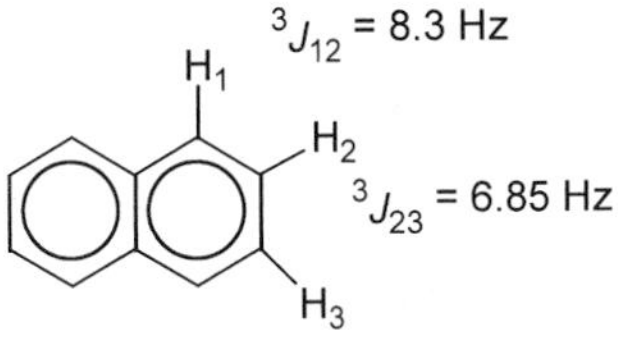

$^3J_{12}$ = 8.3 Hz

H_1

H_2

$^3J_{23}$ = 6.85 Hz

H_3

19

$^3J_{12}$ = 6.8 Hz

H_1

H_2

$^3J_{23}$ = 8.24 Hz

H_3

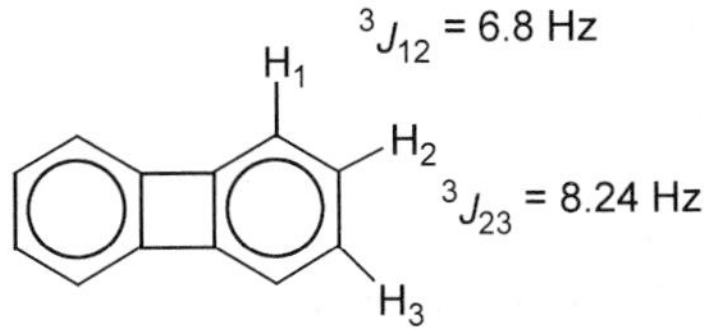

81

The difference between these vicinal (*ortho*) couplings arises from the different bond lengths, which can be understood in terms of partial bond localization. The fact that the vicinal coupling $^3J_{12}$ in naphthalene is larger than the vicinal coupling $^3J_{23}$ indicates that the corresponding C_1–C_2 bond is shorter than the C_2–C_3 bond. This can be explained with the partial localization of the π-electrons between carbon atoms C_1–C_2. The resonance structure **19A** is the dominant one.

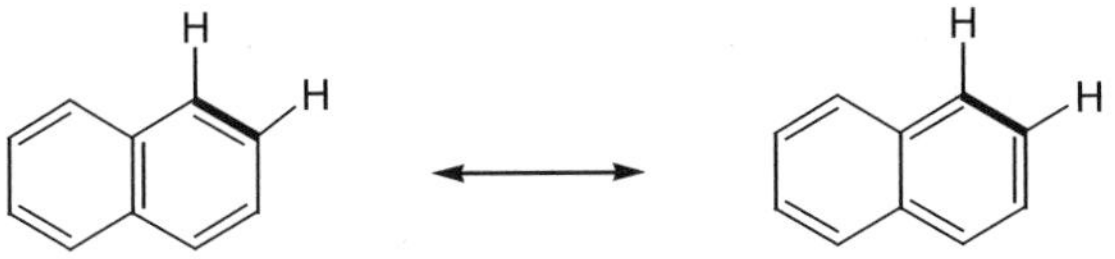

19A

19B

However, in biphenylene the vicinal coupling $^3J_{23}$ is larger, indicating bond localization between the carbon atoms C_2–C_3. The dominating resonance structure for biphenylene is **81B**. The structure **81A** is not the preferred one, mainly because of the cyclobutadiene structure formed.

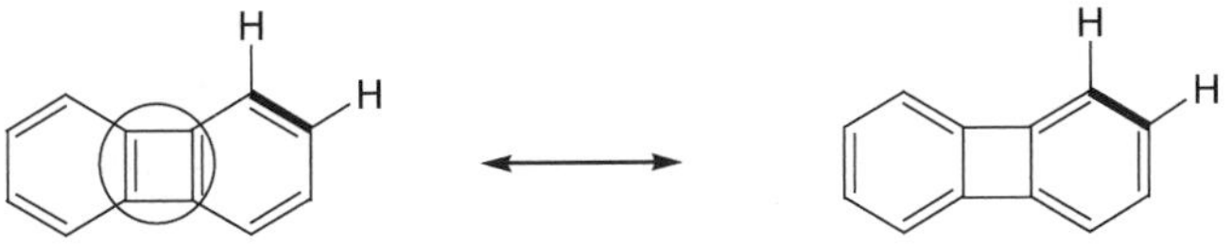

81A

Cyclobutadiene

81B

Since cyclobutadiene is an antiaromatic compound, biphenylene tries to escape this antiaromatic resonance structure. With the help of the vicinal coupling constants it is possible to determine bond localizations. The bond length can be determined since it correlates with the vicinal coupling constant.

56   **57**

Cycloheptatriene **56** is in a dynamic equilibrium with its valence isomer, norcaradiene **57** [50]. To which side this equilibrium is shifted can be determined by measuring the vicinal coupling constant between the protons H_1 and H_2. When the equilibrium is shifted toward the cycloheptatriene structure, then the protons H_1 and H_2 are bonded to the C=C double bond. The coupling will be large. However, when the equilibrium is shifted to the side of norcaradiene, protons H_1 and H_2 are connected through a C–C single bond, which will decrease the coupling constant. In the event that equilibrium is in the middle, an average coupling will be observed. However, in this particular case, the position of the equilibrium can be easily determined observing the chemical shifts of proton H_1. H_1 in cycloheptatriene is attached directly to the double bond, whereas in norcaradiene it is attached to an sp^3 hybridized carbon atom. There will be a large difference in chemical shift.

Substituent effects

In saturated systems, the substituents change the vicinal coupling constants depending on their electronic nature. Electronegative substituents decrease the vicinal coupling constants. The relationship between the vicinal coupling constant and the electronegativity of the substituent can be described by the following empirical equation:

$$^3J_{vic} = 8.0 - 0.80 \sum (E_X - E_H) \tag{30}$$

where E_X is the electronegativity of substituent and E_H is the electronegativity of hydrogen.

In this equation, the vicinal coupling constant measured in ethane is taken as the basis. For multiple substitution, the substituent effects are additive. For substituted ethanes of type X_1X_2CH–CHX_3X_4, the coupling constant $^3J_{HH}$ can be approximated by adding the coupling constant of ethane $^3J_{HH} = 8.0$ Hz and electronegativity differences $(E_X - E_H)$. For example, using eq. 30 we can calculate the vicinal coupling constant in trichloroethane.

$$E_{Cl} - E_H = 3.15 - 2.20 = 0.95$$

$$^3J_{vic} = 8.0 - 0.80(3 \times 0.95) = 5.72 \text{ Hz}$$

This calculated value of $^3J_3 = 5.72$ is in good agreement with the observed value of $^3J_3 = 6.0$ Hz. Generally, when the electronegativity of the substituents is increased, the vicinal coupling constant is decreased to the same extent.

In cyclic systems, the effect of a substituent on the vicinal coupling constant depends on the position of the substituent. Let us look at the coupling constant between the axial–equatorial protons in cyclohexane in the presence of an electronegative substituent attached to one of the carbon atoms. The maximum effect of a substituent on the vicinal coupling constant is observed when the substituent is in a planar *trans* orientation with one of the coupled protons, as shown below.

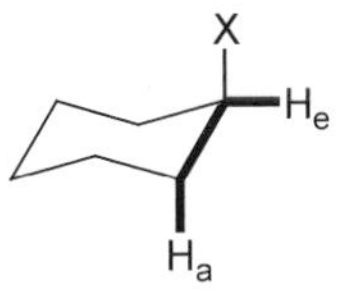

$$^3J_{ae} = 2.5 \text{ Hz} \qquad\qquad ^3J_{ae} = 5.5 \text{ Hz}$$

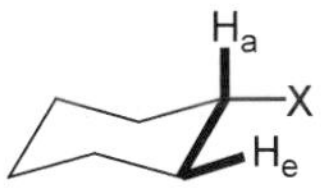

$$X = OH, OAc$$

The effect of the electronegative substituent on the vicinal coupling constant in olefines is also demonstrated. Again, the effect of the substituent on the vicinal coupling constants depends on the position of the substituent attached to one of the double bond carbons. The following equations show this dependency:

$$^3J_{trans} = 19.0 - 3.2(E_X - E_H), \qquad ^3J_{cis} = 11.7 - 0.41(E_X - E_H)$$

For example, the replacement of one proton in ethylene by an OR group decreases the vicinal *cis* and *trans* couplings. Although the substituents have a considerable influence on olefinic coupling constants, it is always true that for the same substituent, J_{trans} is larger than J_{cis}.

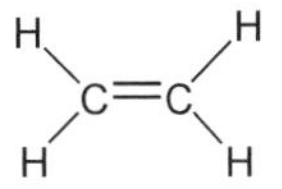

$$^3J_{cis} = 11.7 \text{ Hz} \qquad\qquad ^3J_{trans} = 19.0 \text{ Hz}$$

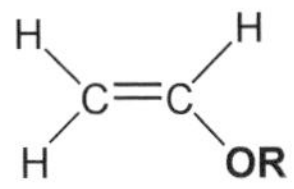

$$^3J_{cis} = 6.7 \text{ Hz} \qquad\qquad ^3J_{trans} = 14.2 \text{ Hz}$$

In heteroaromatic compounds, the vicinal coupling constants depend on the electronegativity of the heteroatom. Some interesting values are illustrated below.

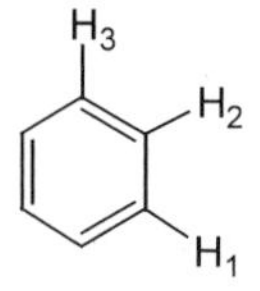

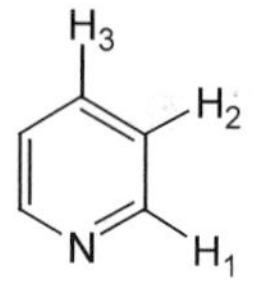

	82	83	84

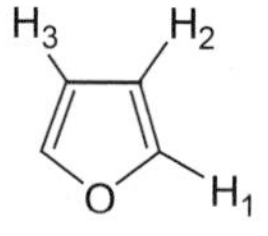

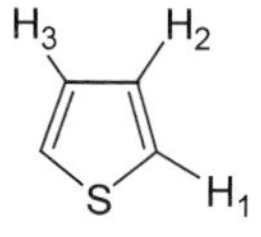

$^3J_{12}$= 8.0 Hz $^3J_{12}$= 4.88 Hz $^3J_{12}$= 1.75 Hz $^3J_{12}$= 5.0 Hz

$^3J_{12}$= 7.67 Hz $^3J_{12}$ = 3.30 Hz $^3J_{12}$= 3.5 Hz

In spite of the fact that the π-bond order between C_1–C_2 in pyridine (**82**) is greater (shorter bond) than that of C_2–C_3 (longer bond), the vicinal coupling $^3J_{12}$ is smaller than $^3J_{23}$. This is attributed to the electronegative effect of the nitrogen atom. In heterocycles such as furan (**83**) and thiophene (**84**), the difference between the coupling constants is much clearer. The effect of the substituents is less in the case of thiophene due to the smaller electronegative effect of the sulfur atom. Some representative vicinal H–H coupling constants are given in Table 4.3.

4.3.3   Long-range coupling constants

Proton–proton coupling beyond two bonds (^2J) and three bonds (^3J) is also observed very frequently in organic compounds. These couplings (over four or more bonds) are called long-range coupling [51]. We have already mentioned that the coupling constants generally decrease when the number of bonds between the coupled protons increases. Appreciable long-range coupling may occur in alkenes, alkynes, and aromatic systems, which gives very important information about the substitution pattern. In saturated systems, couplings through four or more bonds are usually not observed. However, in strained cyclic systems, they can be particularly large. We will discuss long-range couplings under four different categories:

(a)   allylic coupling;
(b)   homoallylic coupling;
(c)   proton–proton coupling in benzene derivatives;
(d)   coupling in cyclic systems.

Allylic coupling ^4J

A frequently observed long-range coupling is allylic coupling, which arises from the σ and π contribution to the proton–proton coupling (Figure 63).

Allylic proton–proton couplings range from 0.0 to 3.0 Hz. They are strongly dependent on the dihedral angle ϕ, which is defined as the angle between the planes formed by HC_2C_3 and $C_1C_2C_3$ (see Figure 64).

Table 4.3

Vicinal H–H coupling constants ^3J in some selected compounds [35, 40]

Compound	^3J (Hz)
CH_3-CH_3	8.0
CH_3-CH_2Li	8.4
$CH_3-CH_2-CH_3$	7.3
CH_3-CH_2OH	7.0
$CH_3-CH(OH)_2$	5.3
CH_3-CHO	2.9
$-CH-CHO$	0.0–3.0
$HC{=}CH-CHO$	5.0–8.0
$-CH_3-CH{=}C$	6.4
$C{=}CH-CH{=}C$	9.0–13.0

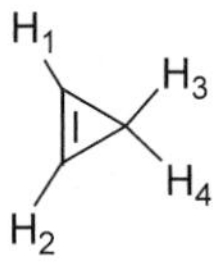

J_{12}, 1.3; J_{23}, 1.8

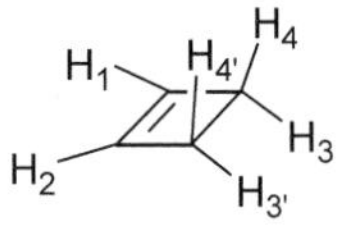

J_{12}, 2.9; J_{14}, 1.0; $J_{44'}$, 4.7; $J_{34'}$, 1.8

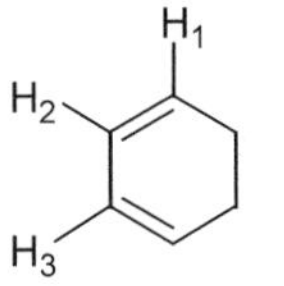

J_{12}, 9.6; J_{23}, 5.8

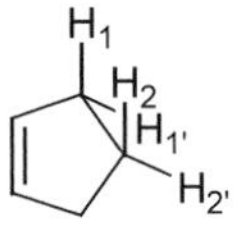

$J_{12(cis)}$, 9.6; $J_{12'(trans)}$, 5.8

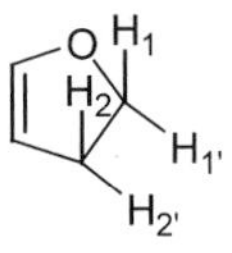

$J_{12(cis)}$, 10.7; $J_{12'(trans)}$, 8.3

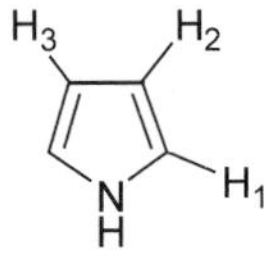

J_{12}, 2.6; J_{23}, 3.7

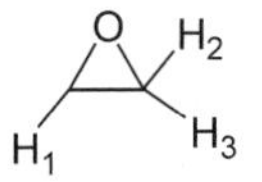

J_{12}, 2.5; J_{13}, 4.0; J_{23}, 6.0

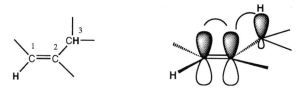

Figure 63 Schematic representation of the ideal conformation desired for maximum allylic coupling.

Since the $\sigma-\pi$ interaction is involved in the allylic coupling, there is a certain amount of overlapping of the π-orbital of the double bond and the C–H bonding orbital (Figure 63). Consequently, a maximum π-contribution to the allylic coupling is observed when the dihedral angle ϕ is 90°. At the same time, for conformation with $\phi = 0°$ or $\phi = 180°$, i.e. when the coupled CH proton is in the plane of the double bond, the contribution of the $\sigma-\pi$ interaction will be zero. The observed allylic coupling will then have a minimum value. Therefore, allylic coupling plays an important role in determining the configuration and the conformation in certain systems. Since a double bond can have *cis* and *trans* configuration, we should also distinguish between *cis* and *trans* allylic coupling. In acyclic olefines, *trans* allylic coupling is slightly larger than *cis* coupling. The corresponding allylic coupling constants are given below.

cis-allylic coupling

$^4J_{cis}$ = 1.3 Hz

trans-allylic coupling

$^4J_{cis}$ = 1.8 Hz

The relationship between the dihedral angle and the allylic coupling can be seen better in cyclic systems because of the rigid geometries. Table 4.4 shows the allylic coupling constant and the corresponding dihedral angles from some examples.

Ring size also has an effect on the magnitude of allylic coupling constants. The absolute values of allylic couplings decrease as the ring size decreases. In particular, the allylic coupling in the four-membered ring is close to zero in spite of the suitable arrangement of the coupled protons.

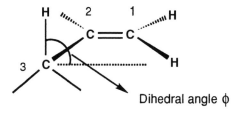

Dihedral angle ϕ

Figure 64 Schematic representation of the dihedral angle for the allylic coupling.

Table 4.4

Allylic H–H coupling constants ^4J (Hz) in some selected compounds

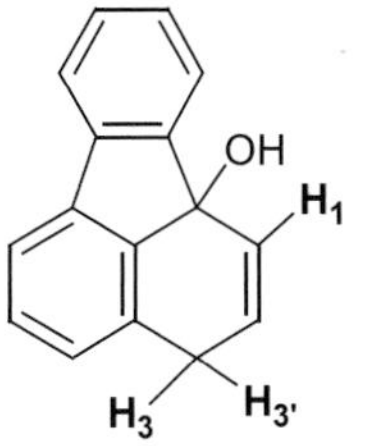

[52]

$\phi_{13} = 155^\circ$ $\phi_{13}' = 85^\circ$
$^4J_{13} = 0.8$ Hz $^4J_{13}' = 2.6$ Hz

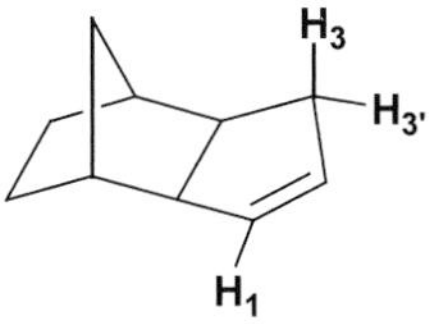

[53]

$\phi_{13} = 120^\circ$ $\phi_{13}' = 120^\circ$
$^4J_{13} = 2.0$ Hz $^4J_{13}' = 2.2$ Hz

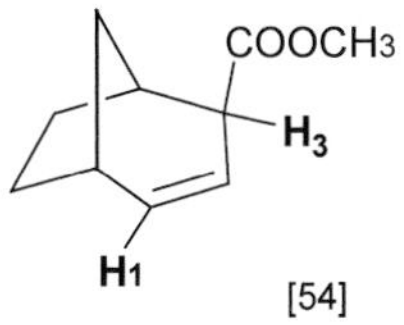

[54]

$\phi_{13} = 125^\circ$
$^4J_{13} = 2.3$ Hz

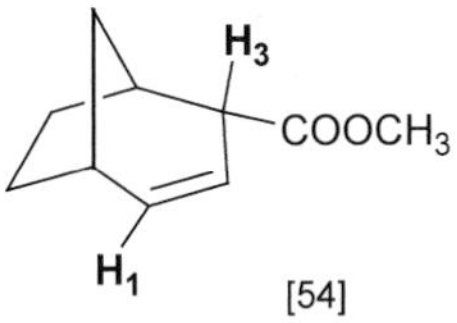

[54]

$\phi_{13} = 115^\circ$
$^4J_{13} = 2.6$ Hz

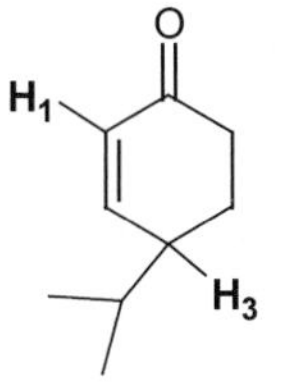

[55]

$\phi_{13} = 100^\circ$
$^4J_{13} = 1.3$ Hz

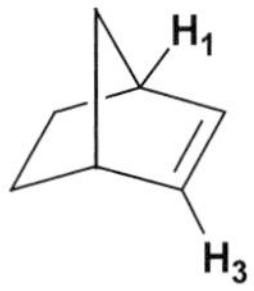

[56]

$^4J_{13} = 0.7$ Hz

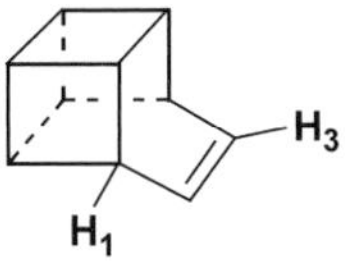

[57]

$^4J_{13} = 1.0$ Hz

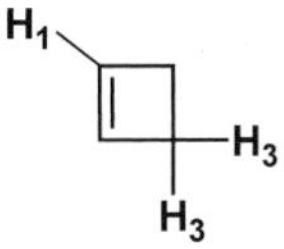

[58]

$^4J_{13} = -0.35$ Hz

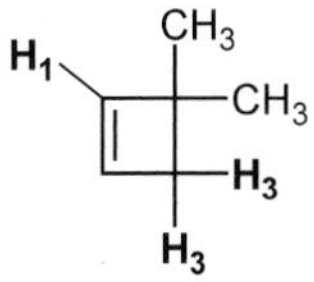

[59]

$^4J_{13} = 0.0$ Hz

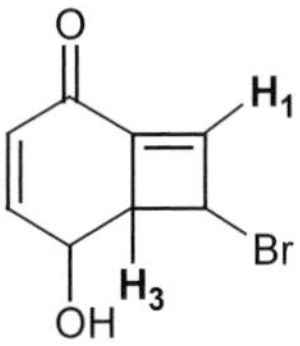

[60]

$^4J_{13} = 0.0$ Hz

Homoallylic coupling ^5J

Coupling through five bonds in allylic systems, called homoallylic coupling, is shown below.

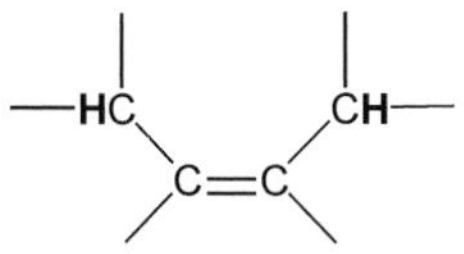

homo-allylic coupling ^5J

These couplings strongly depend on the conformation of the molecule and generally have values of 0.0–2.0 Hz. Since homoallylic coupling also operates through the σ–π interaction, the dihedral angle also plays an important role here. In homoallylic systems there are two different dihedral angles. When both of these angles are $\phi = 90°$ the coupling reaches a maximum value. These couplings are observed mostly in cyclic systems due to the rigid conformation of the molecule (Table 4.5).

Analysis of the ^{1}H-NMR spectrum of butadiene reveals all possible couplings, including long-range ones.

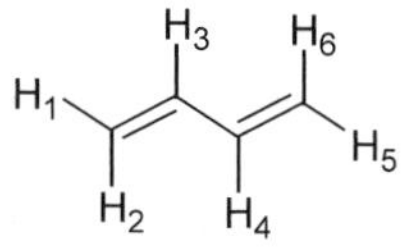

$^2J_{12}$= 1.8 Hz

$^3J_{13}$= 10.2 Hz $^3J_{23}$= 17.1 Hz $^3J_{24}$= 10.4 Hz

$^4J_{14}$= 0.9 Hz $^3J_{24}$= 0.8 Hz

$^5J_{16}$= 1.3 Hz $^5J_{15}$= 0.6 Hz $^5J_{25}$= 0.6 Hz

Table 4.5

Long-range H–H coupling constants (Hz) in some unsaturated compounds

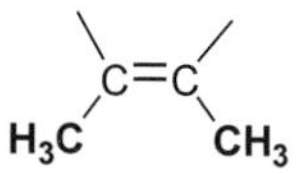

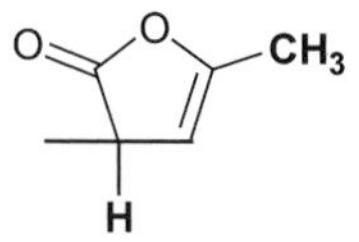

^5J = 1.2 – 1.8 Hz ^5J = 2.7 Hz ^5J = 2.7 Hz

$H_3C-C\equiv C-H$ $H_3C-C\equiv C-C\equiv C-CH_3$

^4J = 2.9 Hz ^6J = 1.3 Hz

Long-range couplings in aromatic systems

Beyond the vicinal coupling in aromatic systems ^3J, there are a number of other long-range couplings observed in aromatic systems. On the basis of long-range couplings, we can easily determine the arrangement of the substituents. Since the protons attached to an aromatic ring are located on a plane, the number of bonds separating the coupled protons determines the size of the coupling constants. Conformational factors do not have an influence on the coupling constant. For benzene two different long-range couplings are found: $^4J_{meta} = 1.0\text{--}3.0$ Hz and $^5J_{para} = 0.0\text{--}1.0$ Hz.

$$^3J_{12} = \ ^3J_{ortho} = 6.0\text{--}10.0 \text{ Hz}$$

$$^4J_{13} = \ ^4J_{meta} = 1.0\text{--}3.0 \text{ Hz}$$

$$^5J_{15} = \ ^5J_{para} = 0.0\text{--}1.0 \text{ Hz}$$

The long-range coupling constants depend on the electronegativity of the substituents attached to the benzene ring. Table 4.6 shows the long-range coupling constants in selected substituted benzene derivatives.

In aromatic compounds the geometric arrangement of the coupled protons has an influence on the long-range couplings. In particular, when the C–H and C–C bonds have a planar zigzag orientation, long-range couplings are observed. For naphthalene two different long-range couplings through five bonds are found: $^5J_{15} = 1.0$ Hz and $^5J_{14} = 0.5$ Hz (Table 4.7). The first arises from the favorable arrangement of the protons. In naphthalene, even a coupling through six bonds is observed. Generally, these couplings are transmitted through the σ-bonds.

Table 4.6

meta and *para* H–H coupling constants in monosubstituted benzene derivatives

	Coupling constants (Hz)		
	$^4J_{24}$	$^4J_{35}$	$^5J_{25}$
X=H	1.37	1.37	0.66
CH$_3$	1.54	1.29	0.60
COOCH$_3$	1.35	1.31	0.63
Br	1.12	1.78	0.46
OCH$_3$	1.03	1.76	0.44
N(CH$_3$)$_2$	1.01	1.76	0.43

Table 4.7

Long-range H–H coupling constants (Hz) in some aromatic compounds

zigzag configuration

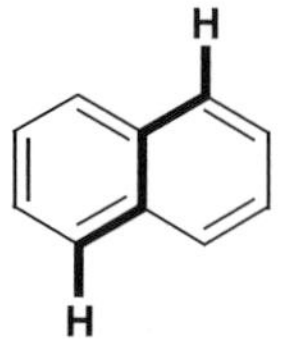

$^5J_{15} = 1.0$ Hz

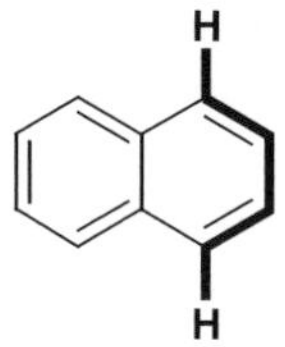

$^5J_{14} = 0.5$ Hz

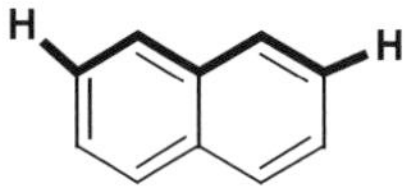

$^6J_{27} = 1.0$ Hz

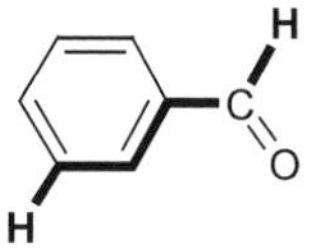

$^5J = 1.0$ Hz

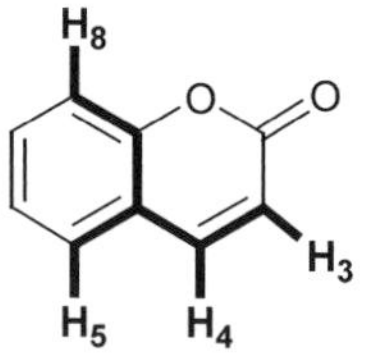

$^5J_{48} = 0.65$ Hz

$^5J_{45} = 0.34$ Hz

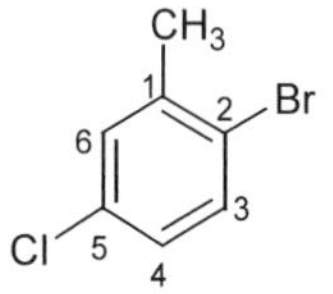

$^4J_{6CH3} = 0.63$ Hz

$^5J_{3CH3} = 0.40$ Hz

$^6J_{4CH3} = 0.58$ Hz

In benzaldehyde, it has been shown that only the *meta* proton couples with the aldehyde proton (over five bonds), and not the *ortho* protons (over four bonds), which is explained by the zigzag arrangement of the protons. Similar couplings are also observed in olefinic compounds.

Long-range couplings in saturated cyclic systems

We have already mentioned that in saturated systems couplings through more than three bonds are usually not observed. However, in some bicyclic strained compounds, long-range coupling constants can reach values of the same order as vicinal coupling constants. In those compounds the coupled protons exist in a zigzag arrangement. In the case of a coupling through four bonds, the 'W' or 'M' arrangement occurs. Both letters indicate the orientation of the coupled protons.

Between the equatorial–equatorial protons in cyclohexane, long-range couplings are observed that are attributed to the W or M conformation of the σ bonds between the coupled protons. In certain strained systems, for example, [2.1.1]bicyclohexane and

[1.1.1]bicyclopentane, the coupled protons are in a sterically fixed configuration as shown in Table 4.8. The corresponding long-range couplings are high: $^4J = 7.0$–8.0 Hz in bicyclohexane and $^4J = 10.0$ Hz in bicyclopentane.

For the assignment of correct stereochemistry in substituted bicyclic systems, these couplings can be very useful. For example, the long-range coupling constant was very helpful in assignment of the correct stereochemistry of the bromine atoms in **85**.

85

The ^{1}H-NMR spectrum of **85** (Figure 65) shows that the bridge methylenic protons resonate as two sets of signals [61]. Since the H_a proton is located in the shielding area of

Table 4.8

Long-range H–H coupling constants (Hz) in some saturated bicyclic compounds

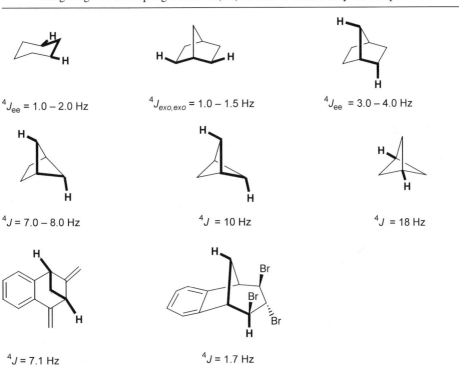

$^4J_{ee} = 1.0 - 2.0$ Hz $^4J_{exo,exo} = 1.0 - 1.5$ Hz $^4J_{ee} = 3.0 - 4.0$ Hz

$^4J = 7.0 - 8.0$ Hz $^4J = 10$ Hz $^4J = 18$ Hz

$^4J = 7.1$ Hz $^4J = 1.7$ Hz

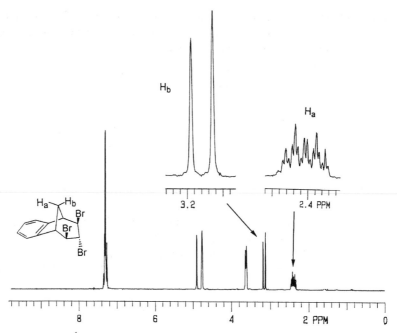

Figure 65   200 MHz ^{1}H-NMR spectrum of tribromo compound (**85**) in CDCl$_3$ and expansions of the signals.

the benzene ring, it resonates at a higher field (2.4 ppm) than the H$_b$ proton, which appears at 3.16 ppm as a doublet. The doublet splitting arises from the coupling with the geminal proton H$_a$. The fact that the H$_b$ proton does not couple with the bridgehead protons H$_c$ is attributed to the dihedral angle, which is nearly 90°. At the same time, proton H$_a$ couples with the geminal proton H$_b$ and bridgehead protons H$_c$. Furthermore, proton H$_a$ has long-range coupling to the H$_d$ protons. This can be clearly seen by the further triplet splitting of the signals. This long-range coupling arises from the zigzag orientation of protons H$_a$ and H$_d$. The zigzag orientation of protons H$_b$ and H$_d$ is impossible because of the rigid geometry. Consequently, there is no long-range coupling between these protons. The fact that proton H$_a$ has long-range coupling to H$_d$ protons clearly indicates the *exo* configuration of the bromine atoms. In the case of the *endo* configuration we should not observe any long-range coupling.

4.3.4   The effect of temperature

Under normal conditions, the spin–spin coupling constants given in hertz are constant, regardless of temperature. However, with dynamic processes such as valence isomerization, bond rotation and ring inversion, the structure or the conformation of the compound will change due to the temperature variations; consequently, the coupling will also change. In short, temperature change does not have any direct effect on coupling. For example, cycloheptatriene (**56**) is in equilibrium with its valence isomer norcaradiene (**57**). Let us consider the coupling between protons H_1 and H_2.

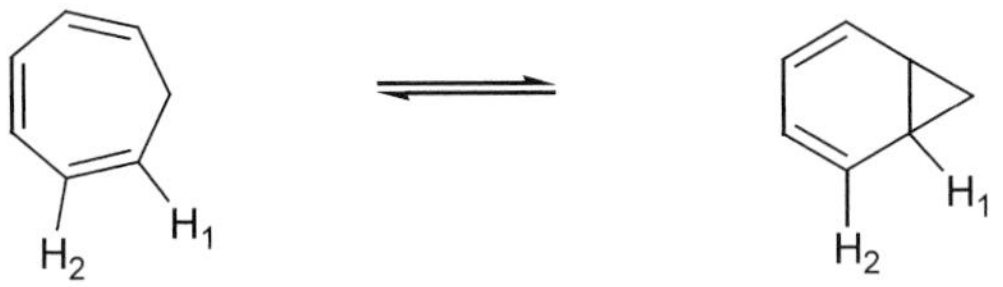

It is well known that this equilibrium shifts towards norcaradiene when the temperature is decreased. When the equilibrium shifts to the side of norcaradiene (**56**), the double bond between protons H_1 and H_2 changes to a single bond in **57**. Of course, the spin–spin coupling will change. This change with temperature is attributable to the valence isomerization, and not directly to the change of temperature.

Cyclohexane rapidly undergoes a ring-inversion process. In *trans*-1,2-disubstituted cyclohexane, two substituents are on opposite sides of the ring, and the compound can exist in either of the two chair conformations shown below.

axial–axial   equatorial–equatorial

This equilibrium is sensitive to variations in temperature. We have already seen that the largest vicinal coupling is observed when the coupled protons are located in the *axial–axial* position. Variations in temperature will affect the equilibrium, i.e. the position of the protons (*axial–axial* or *equatorial–equatorial*) will in turn influence the coupling constant.

It has long been known that spin–spin coupling constants change with solvent polarity. However, this change is so small that it is considered to be negligible.

– 5 –

Spin–Spin Splitting to Different Protons

5.1 GENERAL RULES

In spin–spin coupling we discussed the coupling of protons with their neighboring protons and we examined the splitting pattern. The ^{1}H-NMR signal of a ^{1}H nucleus with n equivalent neighboring adjacent protons splits into $n + 1$ peaks. Furthermore, we showed that the relative intensities of the signals may be easily obtained from Pascal's triangle. So far, we have restricted our attention to signal splitting arising from interactions of only one set of equivalent protons. What kind of patterns should we expect from compounds in which more than two sets of equivalent protons are interacting? Unfortunately, the formula $(n + 1)$ cannot be applied to these systems. If a proton is coupled to more than one neighboring proton with different coupling constants, the formula will not work and we will need more accurate approaches to determine the splitting modes of the signal. Let us consider the resonance signal of the protons labeled in bold in the following compounds.

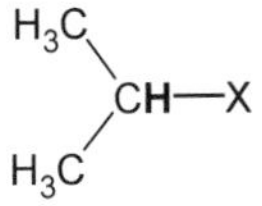

86

87

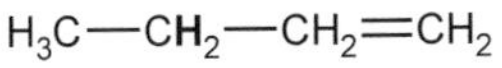

88

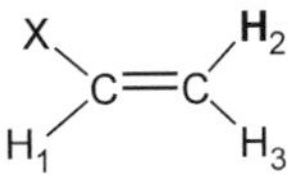

89

In the i-propyl derivative **86**, the tertiary proton has two neighboring methyl groups. Because of the equivalent methyl protons, the tertiary proton resonates as a septet. In the n-propyl derivative **87**, there are three sets of protons. We have no problem in deciding what kind of signal splitting to expect from the protons of the $-CH_3$ and $-CH_2X$ groups. These groups will resonate as triplets due to the coupling to the protons of the central $-CH_2-$ group. What about the protons of the central $-CH_2-$ group? The $-CH_3$ protons and $-CH_2X$ protons are not equivalent. Due to the free rotation of

135

C–C bonds, the coupling of the central $-CH_2-$ group to $-CH_3$ protons and $-CH_2X$ protons is nearly the same. Therefore, we use formula (26) and Pascal's triangle as mentioned above in order to determine the splitting pattern. Because the central $-CH_2-$ protons have five neighboring protons, these methylenic protons will resonate as a sextet (see Figure 5). However, in 1-butene (**88**) the coupling mode is different. The aliphatic $-CH_2-$ protons have $-CH_3$ and olefinic CH=C proton as neighbors. Since the coupling constants of aliphatic $-CH_2-$ protons with these groups are significantly different, the appearance of the ^1H-NMR spectrum will also be different. In this case, we cannot use formula (26) and Pascal's triangle to determine the splitting. In monosubstituted ethylene compound **89** we are faced with the same problem. While proton H_2 has *trans*-vicinal coupling (16–18 Hz) to proton H_1, it has *geminal* coupling (0–2 Hz) to proton H_3. As can be seen, there is a large difference between these couplings. The signal splitting will be different.

At this point the question arises of how the splitting patterns occur in the spectrum when a proton couples to two different protons with different coupling constants. In a case such as that, if the first-order spectra rules are valid, we do the following. For example, let us take the signal of proton H_2 in monosubstituted ethylene **89**. Firstly, the signal of proton H_2 will be split into a doublet by the neighboring proton H_1 (by *trans* coupling) and then each line of the doublet will be further split into doublets by the neighboring proton H_3 (by *geminal* coupling). Finally, we will observe a signal group for the resonance of proton H_2 consisting of four lines, which is called a *doublet of doublets* (Figure 66). The analysis of these lines as shown in Figure 66 will provide the coupling constants $^3J_{trans}$ and $^2J_{geminal}$ as well as the chemical shift of H_2.

Let us now investigate the coupling of a proton to two sets of different protons: one set has two equivalent protons and the second set has only one proton. For example, we may consider a disubstituted olefin having a $-CH_2-$ group as one of the substituents (Figure 67).

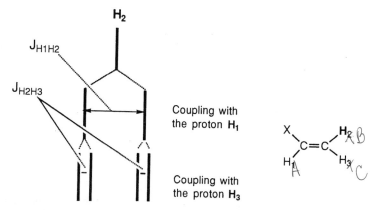

Figure 66   Doublet of doublets: Coupling of a proton (or set of equal protons) with two different protons with different coupling constants.

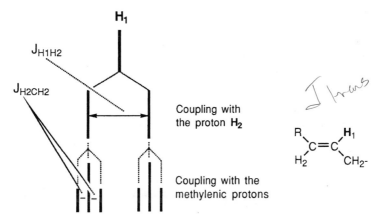

Figure 67 Doublet of triplets: Coupling of a proton (or set of equal protons) with two equal protons and one different proton with different coupling constants.

$$H_1 \underset{-CH_2}{\overset{}{\diagdown}} C = C \underset{H_2}{\overset{X}{\diagup}}$$

Here, we may analyze the signal of proton H_1. The proton will couple with olefinic proton H_2 and with aliphatic $-CH_2-$ protons. If the coupling constants of H_1-H_2 and H_1-CH_2 couplings were identical, the resultant signal would appear as a quartet. Since these couplings have quite different values, then proton H_1 will couple with proton H_2 and give rise to a doublet. Furthermore, the doublet lines will be further split into triplets by $-CH_2-$ protons. Proton H_1, therefore, will have a signal group of six lines, which is called a *doublet of triplets* (Figure 67). The corresponding coupling constants $^3J_{H_1H_2}$ and $^3J_{H_1CH_2}$ can be extracted from the spectrum. The center of the signal group is the chemical shift of the resonating proton.

Now let us consider a $-CH_3$ group instead of a $-CH_2-$ group in the latter molecule (Figure 68). In this case, if the coupling constants $^3J_{H_1H_2}$ and $^3J_{H_1CH_3}$ have quite different values, then proton H_1 will be split into a doublet by neighboring proton H_2 and each line of the doublet will be further split into quartets by adjacent methyl protons. This *doublet of quartets* will therefore have eight lines.

A proton may have three sets of different neighboring protons. If the coupling constants have different values, then a more complicated signal group appears. For example, in a $-CH_2-$ substituted ethylene group, proton H_1 has four vicinal protons.

$$H_1 \underset{-CH_2}{\overset{}{\diagdown}} C = C \underset{H_3}{\overset{H_2}{\diagup}}$$

These are H_2, H_3, and $-CH_2-$ protons. Note that two of these, methylene protons, are equivalent. Before attempting to draw the spectrum, let us assume $^3J_{13} > {}^3J_{12} > {}^3J_{1CH_2}$. Proton H_1 will be split into a doublet by proton H_3 (Figure 69). In fact, there is no splitting

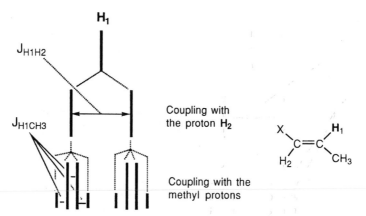

Figure 68   Doublet of quartets: Coupling of a proton (or set of equal protons) with three equal protons and one different proton with different coupling constants.

order (we will talk about this in the next example). The doublet will be split again by proton H_2 into two doublets. Consequently, the doublet of doublets lines formed will be further split into four separate triplets by two methylenic protons. The resultant peak is then a doublet of doublets of triplets, in short ddt.

All the splitting patterns we have shown are only valid for first-order NMR spectra. If the chemical shift difference $\Delta\delta$ between the coupled protons is small, then the spectra will be in the range of second-order spectra and their appearance will be very complex. To analyze complex NMR spectra, computer-aided methods are used.

The appearance of the signal groups will not always be as we have shown above. If the coupling constants differ from one another, then we encounter in most cases spectra with the expected splitting patterns as shown in Figure 69. If the difference between

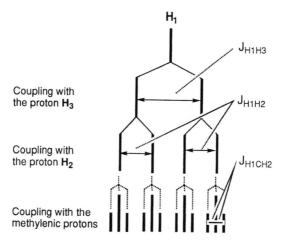

Figure 69   Doublet of doublets of quartets: Coupling of a proton (or set of equal protons) with three equal protons and two different protons with different coupling constants.

the coupling constants is small or two coupling constants have the same value, the spectra will be completely different from the one that is expected. In the above example, we assume that coupling $^3J_{12} = {}^3J_{1CH_2}$. To determine the signal splitting of proton H_1, we follow the exact same method as described above. We split the signal of proton H_1 first into a doublet by proton H_3 and then into a doublet of doublets by proton H_2. In the triplet splitting of these lines we have to ensure that we split the lines into triplets by exactly the same magnitude ($^3J_{12} = {}^3J_{1CH_2}$) as we did in the second doublet splitting with proton H_2. It is clear that some of the peaks usually appear on top of other peaks. Due to the overlapping of some peaks the signal group will appear as a doublet of quartets (Figure 70).

The chemical shifts of the protons are not important in determining the splitting patterns of the signals. *The only determining factor is the magnitude of the coupling constants.* In the above-mentioned example, we assumed that two of the couplings are equal, $^3J_{12} = {}^3J_{1CH_2}$. If this condition is met, proton H_1 will couple with protons H_2 and $-CH_2-$ with the same coupling. The chemical shifts of these protons (one is an olefinic proton and the others are aliphatic protons) are not very important, and proton H_1 will behave as it would in coupling to three equal protons and will resonate as a quartet. The signal obtained is called a *doublet of quartets.*

Let us analyze a system that should resonate as a doublet of doublets of quartets. We assume that the magnitudes of coupling constants are different and $^3J_1 > {}^3J_2 > {}^3J_3$. The expected signal of the corresponding proton will appear as a doublet of doublets of

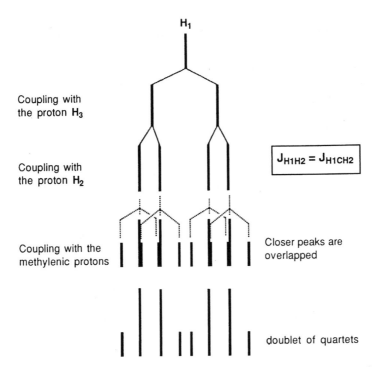

Figure 70 Change of a doublet of doublets of triplets into doublet of quartets in case where two of the coupling constants are equal.

quartets (Figure 71a). When two of these couplings are equal and the condition $(^3J_1 > {}^3J_2 = {}^3J_3$ and $J_2 = \frac{1}{2}J_1)$ is met, the spectrum appearance will be different. Instead of the 16 lines (doublet of doublets of quartets), we will observe a septet due to the overlapping of some lines (Figure 71b).

Hence, the analysis of spectra with the naked eye has to be done carefully. The creative imagination and the experience of the student play a very important role in analyzing these kinds of spectra. In this context, we should note that this experience cannot be gained in a very short time. The more a researcher deals with different NMR spectra, the more experience he or she obtains. For example, in chess, intelligence is of little use without experience: one cannot expect to play at one's best the first time even if one is a genius. In reality, experience is as crucial as intelligence. Similarly, to be a good NMR analyst one must not only obtain enough knowledge, but also apply it to real spectra.

Earlier we examined the splitting of a proton with different protons using several examples. In these examples we did not actually follow a sequence while we showed a doublet of triplets (e.g. Figure 67) and we first drew the doublet splitting and then triplets. In fact, the answer to the question of whether we can obtain the same spectrum if the inverse splitting order is used is most definitely yes. Let us see it in a quantitative drawing of the similar case of Figure 67 using two different sequential orders. Let us follow both of the splitting orders. In the first spectrum (Figure 72a), proton H_1 splits into a doublet by coupling with H_2 and then the doublet lines give two triplets by coupling with methylene protons. In the second (Figure 72b), inversely, proton H_1 forms a triplet by the interaction

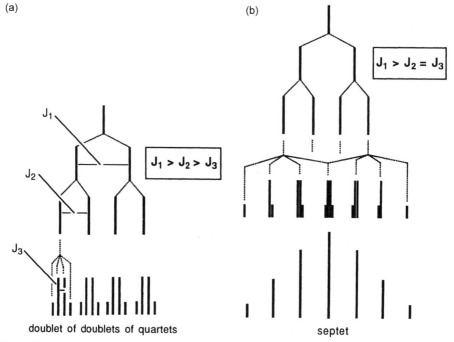

(a)   (b)

$J_1 > J_2 = J_3$

J_1

J_2

$J_1 > J_2 > J_3$

J_3

doublet of doublets of quartets   septet

Figure 71 Change of a doublet of doublets of quartets depending on the size of the coupling constants.

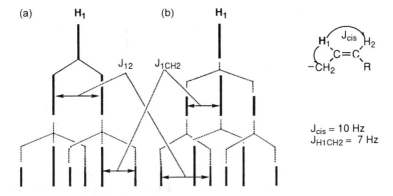

Figure 72 Drawing of a doublet of triplets by different orders: (a) doublet of triplets on the left side; (b) triplet of doublets on the right side. In both cases, the same spectrum is obtained.

with methylenic protons and then triplet lines split into three doublets with the coupling of proton H_2. Note that the results of the two sets of splitting orders are the same. Therefore, we can conclude that the order of splitting is of no importance. However, to avoid confusion that may arise from the drawing of the splitting patterns, the NMR language is standardized to show the least splitting first, followed by the greater ones.

5.2 EXAMPLES OF COUPLING WITH DIFFERENT PROTONS

After theoretically showing the coupling patterns of a proton coupled to two or more nonequivalent kinds of protons, we may analyze the ^{1}H-NMR spectra of some compounds. The ^{1}H-NMR spectrum of bromobenzocycloheptatriene **90** is illustrated in Figure 73.

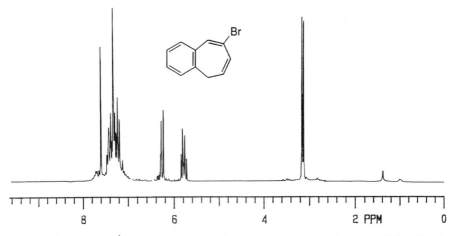

Figure 73 200 MHz ^{1}H-NMR spectrum of benzocycloheptatriene derivative (**90**) in CDCl$_3$.

5.2.1   The ¹H-NMR spectrum of bromobenzocycloheptatriene

90

The spectrum of **90**, according to the chemical structure, consists of three sets of signals appearing in the aliphatic, olefinic and aromatic regions. The high-field doublet is due to the splitting of the aliphatic protons' $-CH_2-$ signal by the neighboring olefinic proton H_1. The four aromatic proton signals appear as a multiplet at δ 7.0–7.4. The resonances of olefinic protons are different. The vinylic proton at C_4 is next to the aromatic ring and is therefore shifted downfield from the normal vinylic region. This proton is located in the deshielding region of the aromatic ring. Furthermore, the bromine at C_3 also shifts the resonance frequency of this proton downfield. Consequently, vinylic proton H_4 resonates as a singlet at δ 7.6 ppm due to the absence of any neighboring proton. Vinylic proton H_2 resonates at δ 6.27. Since it has one neighboring proton at C_1, its signal is split into a doublet with $J_{12} = 10.0$ Hz. The signal of vinylic proton H_1 appears at δ 5.8 and exhibits an interesting absorption pattern.

It is coupled to nonequivalent vinylic proton H_2 and aliphatic protons $-CH_2-$ to give a doublet of triplets. The analysis of this signal group is given in Figure 74. Two vicinal coupling constants of $^3J_{12} = 10.0$ Hz and $^3J_{17} = 7.1$ Hz are extracted from

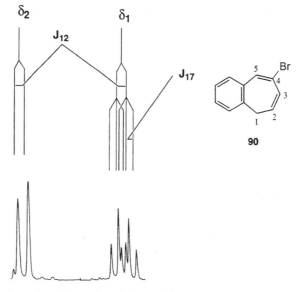

Figure 74   Expansion of the olefinic proton resonances of **90**.

this signal group. Doublet splitting arises from the coupling with vinyl proton H_2. The interesting feature of this signal group is that the triplet signals cross, which arises only from the size of the coupling constants. Since *two groups of protons that are coupled to each other must have the same coupling constants*, J, the vicinal coupling $^3J_{12}$ can be extracted from the doublet lines of H_2 as well as from the doublet of triplets lines of H_1. Starting from this point we can determine which signal groups are connected to one another in a spectrum having many split signals.

5.2.2 The ^{1}H-NMR spectrum of furfural

The ^{1}H-NMR spectrum of furfural (**91**) is more complex but can nevertheless be interpreted in a straightforward way (Figure 75a). The aldehyde proton signal appears in the normal low-field position at δ 9.67. The ring protons appear due to the asymmetry as three groups of signals in a region of 6.61–7.71 ppm. Let us first assign the resonances on the basis of the observed chemical shifts.

91

$^3J_{45} = 1.8$Hz

$^3J_{34} = 3.6$Hz

$^3J_{45} = 1.75$Hz

$^3J_{34} = 3.3$Hz

If an oxygen atom is directly attached to a C=C double bond, the resonance signal of the α-proton is shifted to low field (due to inductive effect) and the resonance signal of the β-proton to high field (due to the mesomeric effect). On this basis we can expect protons H_3 and H_4 to resonate at high field and proton H_5 at low field. On the other hand, if a carbonyl group is conjugated with a double bond, the resonance signal of the β-proton is shifted to low field. Therefore, proton H_3 will be under the influence of two opposite effects. We assume that they will partly cancel each other out and proton H_3 will resonate at the normal position. After these assumptions we can predict the following order of chemical shifts:

$$\delta_{H_5} > \delta_{H_3} > \delta_{H_4}$$

However, the correct order of resonances can be determined by analyzing the signal splitting. Let us start with proton H_4, which has two vicinal neighboring protons H_3 and H_5. Proton H_4 will couple to these protons and resonate as a doublet of doublets. In furan there are two different vicinal coupling constants: $^3J_{34} = 3.3$ Hz and $^3J_{45} = 1.75$ Hz. The signal group at high field appears as a doublet of doublets with coupling constants of $J = 1.8$ and 3.6 Hz. These couplings are in agreement with those observed in furan. Therefore, the signal group at high field can be assigned to proton H_4. Another doublet of doublets is observed at low field (7.3 ppm). The analysis reveals the following coupling

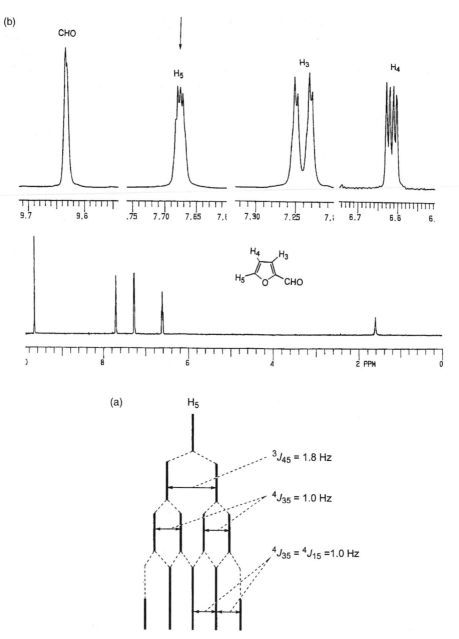

Figure 75 (a) 200 MHz ^{1}H-NMR spectrum of furfural in CDCl$_3$ and expansions of the signals. (b) Drawing of H$_5$ resonance signal of furfural.

constants: $J = 1.0$ and 3.6 Hz. The coupling $J = 3.6$ Hz has also been determined by the analysis of the signal group resonating at high field. This coupling constant arises from the interaction of protons H_3 and H_4. On this basis, the low-field resonance signal is assigned to proton H_3. The second coupling $J = 1.0$ Hz observed in the signal of H_3 arises from the long-range coupling with proton H_5. This value is also in good agreement with that of furan $(^4J_{35} = 0.88$ Hz$)$.

The remaining signal group at δ 7.72 belongs to proton H_5. We have already determined by analysis of the other signal groups that there are couplings between H_3–H_5 and H_4–H_5. We should find the same couplings in the resonance signal of proton H_5. In this case, proton H_5 should resonate at least as a doublet of doublets. However, the signal of this proton has a complex structure (Figure 75b). Careful analysis shows that proton H_5 resonates as a doublet of doublets of doublets where some peaks overlap. It is likely that proton H_5 has long-range coupling to the aldehyde proton. Otherwise, the complex structure of the signal cannot be explained. A doublet splitting in the resonance signal of aldehyde strongly supports our assumption. Vinylic proton H_5 can resonate as a doublet of doublets of doublets only when all of the couplings have quite different values. Analysis of the signal, which is a quintet, indicates that the long-range coupling constants $^4J_{35}$ and $^4J_{15}$ have approximately the same value, which leads to signal overlapping.

5.2.3   The ^{1}H-NMR spectrum of 2-butenal (crotonaldehyde)

The 200 MHz ^{1}H-NMR spectrum of crotonaldehyde is given in Figure 76. As expected from the structure, we see four groups of signals in the spectrum. In spite of the simple molecular structure of crotonaldehyde, its ^{1}H-NMR spectrum is complex.

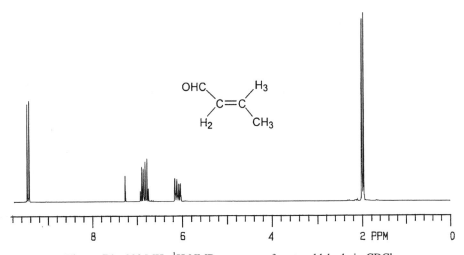

Figure 76   200 MHz ^{1}H-NMR spectrum of crotonaldehyde in CDCl$_3$.

$$\underset{2}{\overset{1}{OHC}}\diagdown \underset{H_2}{\overset{}{C}} = \underset{3}{\overset{}{C}}\diagup \overset{H_3}{\underset{CH_3}{}}$$

The low-field doublet at δ 9.40 belongs to the aldehyde proton. Since the aldehyde proton has only one neighboring proton at C_2, its signal is split into a doublet with $J = 8.0$ Hz. Careful analysis of the doublet lines does not reveal the existence of any long-range couplings. The high-field doublet at δ 2.05 arises from the methyl protons.

The expanded spectrum indicates that the doublet lines are further split into doublets (Figure 77). The methyl protons' signal is split first into a doublet by the adjacent vinylic proton H_3 and then each line of the doublet is further split into doublets by the other vinylic proton H_2 through long-range coupling. The extracted coupling constants are $J = 7.0$ and 1.8 Hz. The large coupling is the vicinal coupling between proton H_3 and methyl protons. The small coupling is the allylic coupling between proton H_2 and methyl protons. The question arises of how can it be determined whether this observed

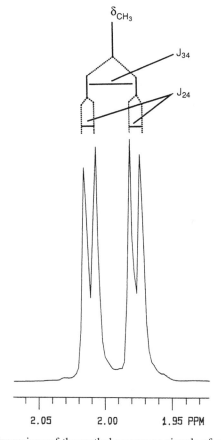

Figure 77   Expansions of the methyl resonance signals of crotonaldehyde.

long-range coupling arises from the coupling between the methyl protons and vinylic proton H_2 or the methyl protons and the aldehyde proton. This question is very easily answered. The aldehyde proton resonates as a doublet and does not exhibit any long-range coupling with the methyl protons. Therefore, we can exclude any coupling between the methyl protons and the aldehyde proton.

Olefinic protons resonate as two distinct multiplets. Due to the conjugation of the double bond with the carbonyl group, it is expected that the resonance of proton H_3 will appear at low field. For the correct signal assignment we have to analyze the splitting patterns of these signals. In Figure 78, the expanded spectrum of the olefinic protons is illustrated. Vinylic proton H_3 has vicinal coupling to vinylic proton H_2 and to the three methyl protons. The resonance signal of proton H_3 is split into a doublet and the doublet lines are further split into quartets. In the spectrum, two quartets are observed between 6.4 and 7.0 ppm. Analysis of this doublet of quartets reveals two different coupling constants of $J = 15.7$ and 7.0 Hz. The large coupling $J = 15.7$ Hz is assigned to the *trans* coupling, which clearly establishes the configuration of the double bond. The smaller coupling arises from the coupling with the methyl protons. This coupling was also determined from the signal group of the methyl resonance. Analysis of this signal group provides evidence that this group belongs to proton H_3. The high-field resonance shows a more complex structure. Inspection of the lines at high field shows 16 lines: a doublet of doublets of quartets. During the analysis of the resonance signals of the aldehyde proton and proton H_3, we determined that they couple to proton H_2. Furthermore, we noted long-range coupling between the methyl protons and H_2 protons. We must also be able to observe all of these couplings in the signal of proton H_2. Consequently, the vinylic proton H_2 couples to these protons and resonates as a doublet of doublets of quartets.

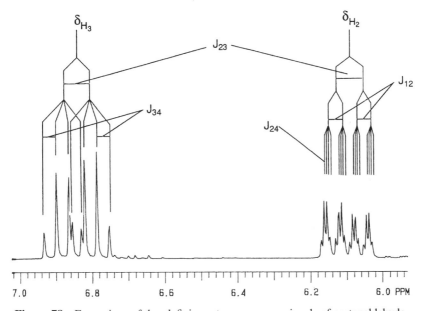

Figure 78 Expansions of the olefinic proton resonance signals of crotonaldehyde.

The extracted coupling constants are $J = 15.7$, 8.0 and 1.8 Hz. All of these couplings have exactly the same values as the coupling constants obtained after analysis of the other signals.

Finally, all of the coupling parameters obtained after analysis of the different peak groups are completely in agreement with the structure of 2-butenal and the *trans* configuration of the double bond.

5.3   EXERCISES 31–60

Propose structures for compounds whose ^{1}H-NMR spectra are given below. The corresponding molecular formulae are given above on the right-hand side of the spectra. Use the following procedures: first measure the area under each NMR absorption peak and determine the relative number of protons responsible for each peak. After this, determine the exact number of protons by using the molecular formula. All of these examples contain peak splitting. As you will notice, peak splitting gives us more information about the structure. The solution of these spectra will be easier than that of spectra having only singlets.

All the spectra (31–60) given in this section are reprinted with permission of Aldrich Chemical Co., Inc. from C.J. Pouchert and J. Behnke, *The Aldrich Library of 13C and 1H FT-NMR Spectra*, 1992.

31.

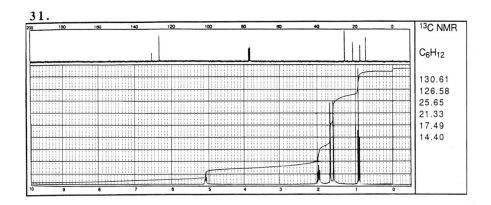

^{13}C NMR

C_6H_{12}

130.61
126.58
25.65
21.33
17.49
14.40

32.

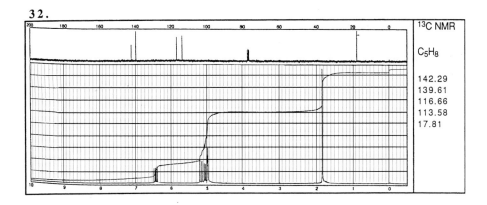

^{13}C NMR

C_5H_8

142.29
139.61
116.66
113.58
17.81

33.

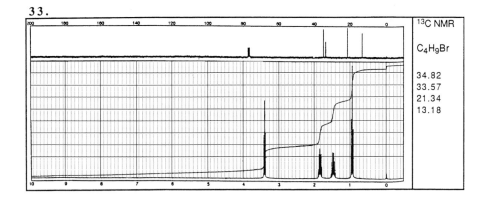

^{13}C NMR

C_4H_9Br

34.82
33.57
21.34
13.18

34.

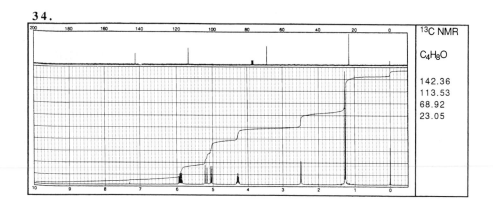

^{13}C NMR

C_4H_8O

142.36
113.53
68.92
23.05

35.

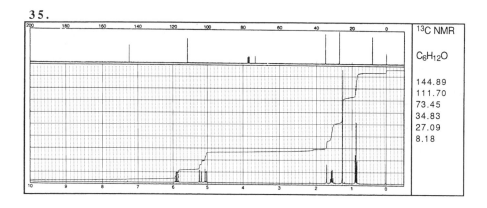

^{13}C NMR

$C_6H_{12}O$

144.89
111.70
73.45
34.83
27.09
8.18

36.

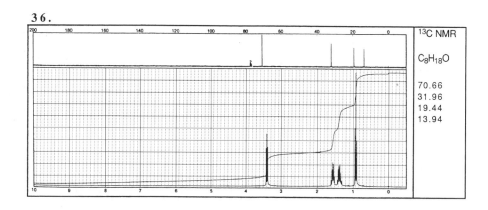

^{13}C NMR

$C_8H_{18}O$

70.66
31.96
19.44
13.94

37.

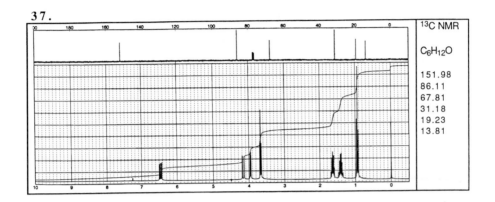

^{13}C NMR

$C_6H_{12}O$

151.98
86.11
67.81
31.18
19.23
13.81

38.

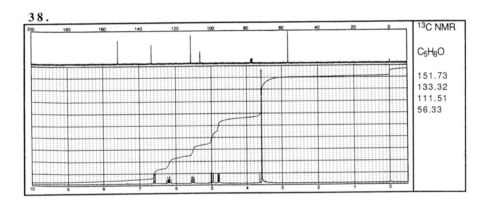

^{13}C NMR

C_5H_8O

151.73
133.32
111.51
56.33

39.

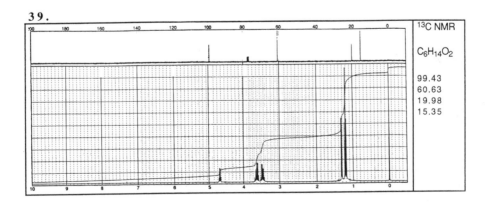

^{13}C NMR

$C_6H_{14}O_2$

99.43
60.63
19.98
15.35

40.

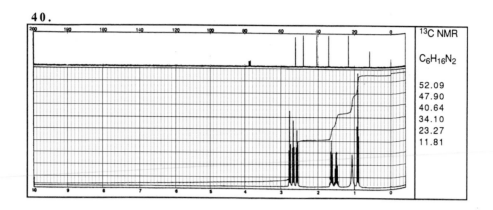

¹³C NMR

$C_6H_{16}N_2$

52.09
47.90
40.64
34.10
23.27
11.81

41.

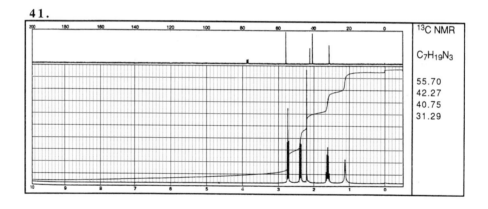

¹³C NMR

$C_7H_{19}N_3$

55.70
42.27
40.75
31.29

42.

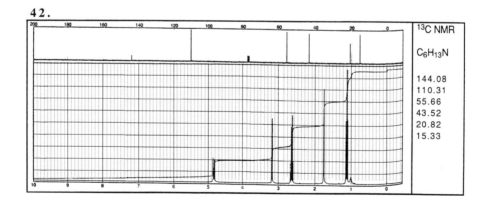

¹³C NMR

$C_6H_{13}N$

144.08
110.31
55.66
43.52
20.82
15.33

43.

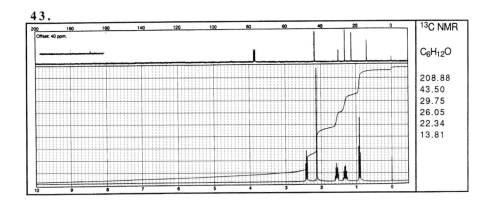

^{13}C NMR

$C_6H_{12}O$

208.88
43.50
29.75
26.05
22.34
13.81

44.

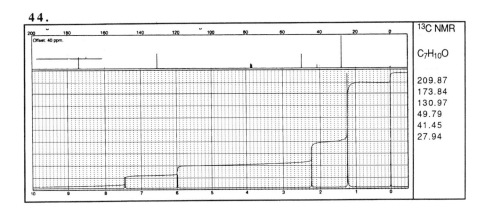

^{13}C NMR

$C_7H_{10}O$

209.87
173.84
130.97
49.79
41.45
27.94

45.

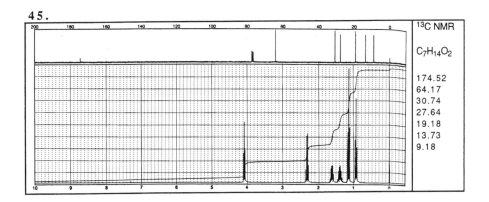

^{13}C NMR

$C_7H_{14}O_2$

174.52
64.17
30.74
27.64
19.18
13.73
9.18

46.

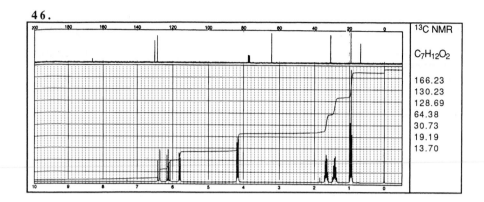

¹³C NMR

$C_7H_{12}O_2$

166.23
130.23
128.69
64.38
30.73
19.19
13.70

47.

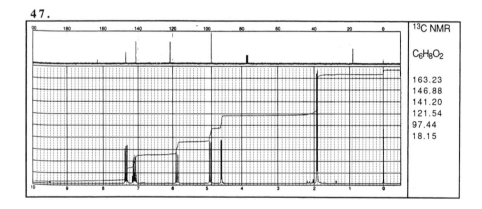

¹³C NMR

$C_6H_8O_2$

163.23
146.88
141.20
121.54
97.44
18.15

48.

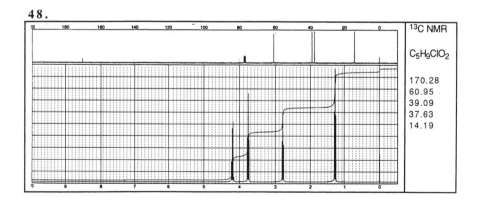

¹³C NMR

$C_5H_9ClO_2$

170.28
60.95
39.09
37.63
14.19

49.

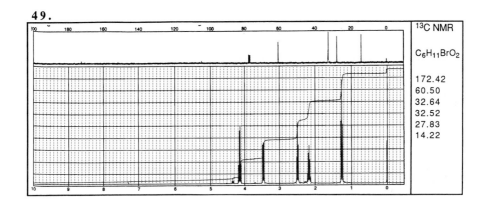

13C NMR

$C_6H_{11}BrO_2$

172.42
60.50
32.64
32.52
27.83
14.22

50.

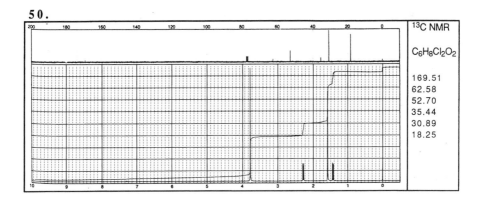

13C NMR

$C_6H_8Cl_2O_2$

169.51
62.58
52.70
35.44
30.89
18.25

51.

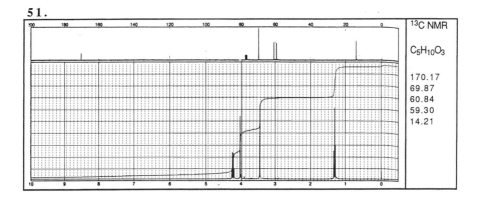

13C NMR

$C_5H_{10}O_3$

170.17
69.87
60.84
59.30
14.21

52

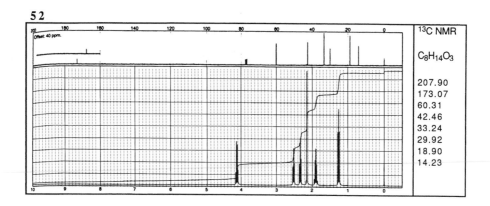

¹³C NMR

$C_8H_{14}O_3$

207.90
173.07
60.31
42.46
33.24
29.92
18.90
14.23

53.

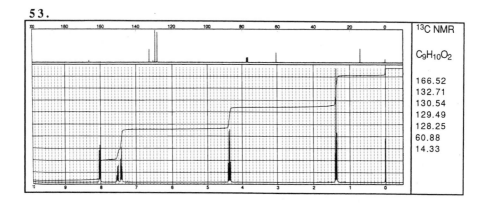

¹³C NMR

$C_9H_{10}O_2$

166.52
132.71
130.54
129.49
128.25
60.88
14.33

54.

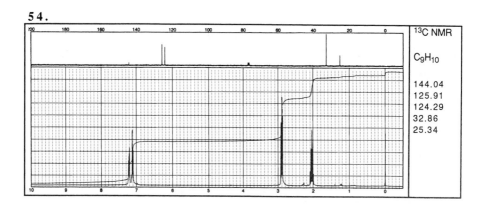

¹³C NMR

C_9H_{10}

144.04
125.91
124.29
32.86
25.34

55.

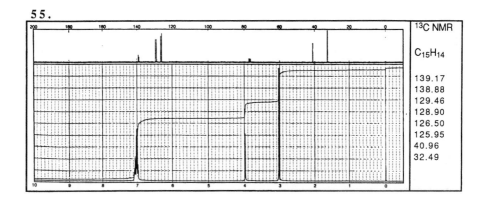

¹³C NMR

C₁₅H₁₄

139.17
138.88
129.46
128.90
126.50
125.95
40.96
32.49

56.

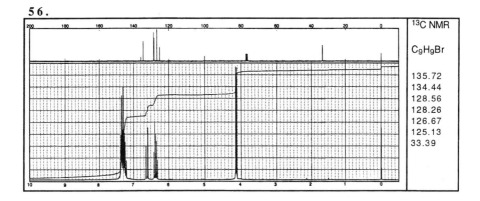

¹³C NMR

C₉H₉Br

135.72
134.44
128.56
128.26
126.67
125.13
33.39

57.

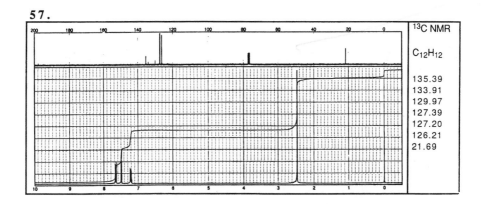

¹³C NMR

C₁₂H₁₂

135.39
133.91
129.97
127.39
127.20
126.21
21.69

58.

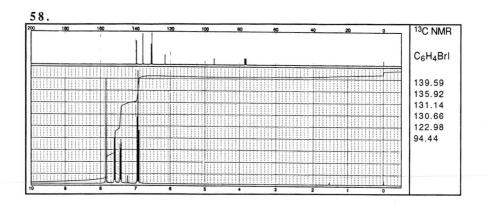

¹³C NMR

C₆H₄BrI

139.59
135.92
131.14
130.66
122.98
94.44

59.

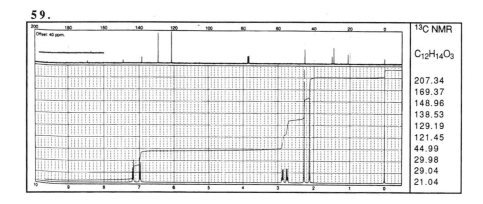

¹³C NMR

C₁₂H₁₄O₃

207.34
169.37
148.96
138.53
129.19
121.45
44.99
29.98
29.04
21.04

60.

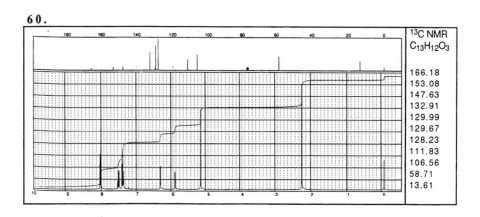

¹³C NMR
C₁₃H₁₂O₃

166.18
153.08
147.63
132.91
129.99
129.67
128.23
111.83
106.56
58.71
13.61

– 6 –

Spin Systems: Analysis of the ^{1}H-NMR Spectra

6.1 SECOND-ORDER SPECTRA

In the previous chapters we discussed only first-order spectra, from which the coupling constants and chemical shifts can be directly obtained. All general rules concerning the splitting, intensity distribution and integration of the peaks can be applied to first-order NMR spectra. To obtain a first-order NMR spectrum, the chemical shift difference ($\Delta\delta$) between the coupled protons in hertz has to be much larger than the coupling constant (see page 32)

$$\frac{\delta_{H_a} - \delta_{H_b}}{J_{ab}} \geq 10 \tag{25}$$

If this ratio is greater than 10, the spectra are considered to be *first-order* spectra. If this condition is not met, the term *second-order spectrum* is used.

The differences between first-order and second-order NMR spectra are as follows:

First-order spectra	Second-order spectra
The multiplicity of the signal splitting can be determined by the number of neighboring protons (the $n + 1$ rule)	The multiplicity of the signal splitting cannot be determined by the number of neighboring protons. The $(n + 1)$ is not valid. There is increased multiplicity
The relative intensity distribution can be determined by Pascal's triangle only	There are no definitive rules to determine the peak intensities. Intensity distribution is altered
Spectra are simple	Spectra are complex
Spectra can be analyzed directly	The analysis of spectra is complicated. It can be performed only by a computer

As one can see from the above table, second-order NMR spectra are more complex and their interpretation is much more complicated. Before going in detail, let us analyze the ^{1}H-NMR spectrum of acrylonitrile obtained on spectrometers operating at three different frequencies (Figure 79) [62].

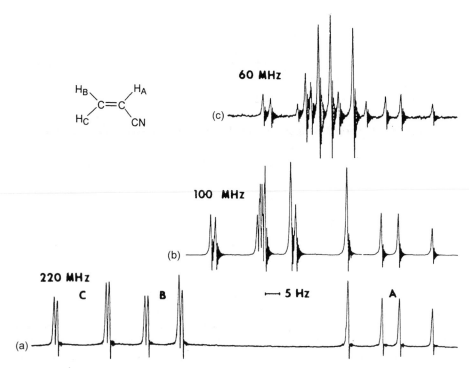

Figure 79 ^{1}H-NMR spectra of acrylonitril recorded on (a) 220 , (b) 100 , and (c) 60 MHz NMR instruments. (Reprinted with permission of American Chemical Society from *Anal. Chem.*, 1971, **42**, 29A.)

Acrylonitrile has three different olefinic protons. All of these protons have different chemical shifts and different coupling constants due to their configuration (*trans, cis* and *geminal*). They will resonate as a doublet of doublets. The 220 MHz NMR spectrum (Figure 79a) shows the resonances of three protons, which are sufficiently resolved, so that all coupling constants and proton chemical shifts can be extracted. In the 100 MHz NMR spectrum of the same compound, the signal groups of protons H$_B$ and H$_C$ approach each other and some signals overlap, whereas H$_A$ proton signal intensities are changed. However, quite a different situation is found in the 60 MHz NMR spectrum. It has a completely different appearance (instead of the expected doublet of doublets), where neither the number of lines nor their intensities are in accordance with the rules of the first-order NMR spectra that we have seen so far. This kind of NMR signal belongs to the class of second-order NMR spectra and their analysis can be done with the help of suitable computer programs. Comparison of these NMR spectra recorded at three different magnetic fields leads to the conclusion that the number of signals and their intensities of a given compound depend on the strength of the applied magnetic field. One point that always has to be kept in mind is that changes in magnetic field strength do not affect coupling constants or chemical shifts. The chemical shift difference in hertz between two protons changes upon changing the strength of the magnetic field. The chemical shift difference in ppm is not affected by the magnetic field. Since the coupling constant is not affected by the magnetic field, consequently, any increase in magnetic field strength must increase the ratio $\Delta\delta/J$ and therefore result in first-order splitting for spectra that are second order at lower field strengths.

To interpret NMR spectra more easily, first-order spectra have to be recorded by increasing the chemical shift difference between the coupled protons by using a spectrometer with a higher magnetic field strength. Therefore, NMR machines operating at higher magnetic fields are always required. In the next chapter we will classify second-order NMR spectra and study them in detail. For now, we will limit ourselves to the consideration of the more frequently encountered spin systems.

6.2 TWO-SPIN SYSTEMS

In this section, we shall begin our discussion of simple spin systems. Since the simplest spin system is formed between two different protons, we will examine these systems first.

6.2.1 A_2, AB and AX spin systems

The symbols designating the spin systems are employed wherein the letter A is used to represent a nucleus appearing at the lowest field, while the letters B, C, X represent nuclei appearing at progressively higher fields. When two coupled protons have the same chemical shifts, i.e. the protons are equal, they form the simplest two-spin system, an A_2 system, which gives rise to a single absorption line. In the case of an A_2 system, the protons can be bonded to the same carbon atom, as well as to two carbon atoms. Some examples of A_2 systems are illustrated in Table 6.1.

If the chemical shifts of the two coupled protons become slightly different, then we identify these protons by the letters A and B. B indicates that the chemical shift of proton H_B does not differ much from the chemical shift of proton H_A. In cases where the chemical shift difference $\Delta\delta$ between the coupled protons is much greater than the coupling constant J,

Table 6.1

Selected examples for A_2 spin systems

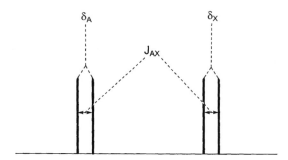

Figure 80 Doublet lines of an AX system and representation of the chemical shifts (δ_A and δ_X) and the coupling constant $^3J_{AX}$.

they are represented by letters well separated in the alphabet, A and X. Simple coupled systems, consisting of two protons, are called either *AB systems* or *AX systems* depending on the chemical shift difference.

Since the chemical shift difference between the coupled protons in AX systems are much greater than the coupling constant J, these systems can be analyzed by the *first-order* spectrum rules. As expected, the coupled protons H_A and H_X will resonate as two distinct doublets. As we have seen previously, the analysis of an AX system, doublet lines, is quite simple. The distance between individual peaks in frequency units gives the coupling constant J_{AX}. The center of the doublet lines gives the exact chemical shift of the resonating proton (Figure 80).

As the difference between the chemical shifts of the coupled protons approaches the size $\Delta\delta/J < 10$, the AX system approaches an AB system. The AB system also consists of a pair of doublets as in the AX system. However, the intensities of the four lines are no longer equal. The intensities of the inner pair of lines of the four-line spectrum increase at the expense of the outer pair. This is called the *roof effect*. This effect has considerable diagnostic value in the recognition of the coupled protons (Figure 81).

Figure 82 shows how the signal intensities of the four lines of an AB system change upon changing the chemical shift difference by a coupling constant of $J_{ab} = 10$ Hz. When the ratio $\Delta\delta/J$ is equal to zero, it means that the chemical shift difference $\Delta\delta$ between the coupled protons becomes zero, the middle peaks coalesce to give a single peak, and

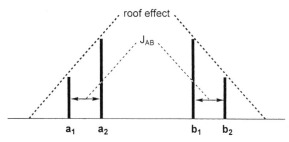

Figure 81 Doublet lines of an AB system and representation of the coupling constant $^3J_{AB}$ and the *roof effect*.

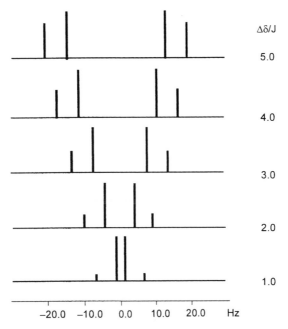

Figure 82   Calculated AB spectra. J_{AB} was set at 10 Hz and resonance frequencies were varied.

the end peaks vanish. This situation is encountered with two equal protons. The AB system changes into an A_2 system.

The separation between two lines in each pair ($a_1 - a_2$ and $b_1 - b_2$) is exactly equal and is also exactly equal to the magnitude of J_{AB} (Figure 81). The different intensities of lines do not affect the coupling constant. The chemical shift of the individual protons of an AX system is the midpoint of the doublet lines. However, in AB systems, the chemical shift position of each proton is no longer the midpoint between its two peaks, but it is at the 'center of gravity'. When the chemical shift difference between the coupled protons decreases, the intensity of the inner lines increases, and the center of gravity of each doublet moves towards its inner component. Although there are a number of different methods for analyzing an AB system (to determine the chemical shifts) we will provide only one equation. The lines of the spectrum relabeled with a_1, a_2, b_1 and b_2 are shown in Figure 81. The chemical shift difference between protons A and B is given by the following equation [63]:

$$a_1 - b_1 = a_2 - b_2 = 2C \tag{31}$$

$$\Delta\nu = \sqrt{4C^2 - J^2} \tag{32}$$

$$\Delta\nu = \sqrt{(2C - J)(2C + J)} \tag{33}$$

$$\Delta\nu = \sqrt{(a_2 - b_1)(a_1 - b_2)} \tag{34}$$

where $\Delta\nu$ is the chemical shift difference between two proton resonances in hertz.

The difference between the chemical shift of protons H_A and H_B can be determined by these equations. The actual values of ν_A and ν_B can then be obtained by adding and subtracting $\Delta\nu/2$ from the center M of the AB quartet.

The chemical shift for the proton H_A:

$$\nu_A = M + \frac{1}{2}\Delta\nu \tag{35}$$

The chemical shift for the proton H_B:

$$\nu_B = M - \frac{1}{2}\Delta\nu \tag{36}$$

AB systems are the most frequently encountered spin systems and they can easily be recognized by the *roof effect*. For example, the ^{1}H-NMR spectrum of the following urazol derivative **93** consists of three different AB systems (Figure 83) [25]:

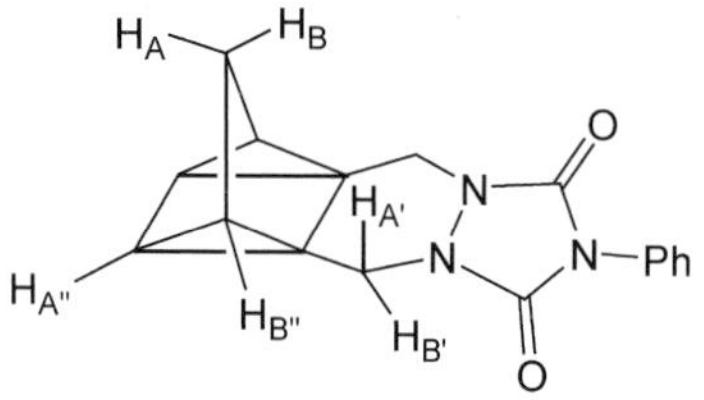

93

The neighboring methylene protons to nitrogen are nonequal, due to the rigid geometry, and give rise to an AB system. The signals observed at 3.8–4.1 ppm in Figure 83 belong to the protons $H_{A'}$ and $H_{B'}$. The AB system appearing at 2.0–2.2 ppm arises from the bridge proton (H_A and H_B) resonances. The interesting feature of this AB system, compared to the first one, is the line broadening. The quartet lines of this system should be further split into triplets by way of coupling with the adjacent bridgehead protons $H_{B''}$. Since this coupling value is so small, the expected line splitting appears as line broadening. The third AB system is formed between the cyclopropane protons $H_{A''}$ and $H_{B''}$ and appears at 1.6–1.9 ppm. The lines on the right-hand side (B-part of the AB system) are broader than the other lines, which can be explained by the fact that protons $H_{A''}$ and $H_{B''}$ have different coupling to the adjacent protons.

In fact, an AB system consists of four lines. However, if one or both of the protons have further couplings to the adjacent protons with different size couplings, then the AB system will not exhibit symmetry. One part of the AB system will be different.

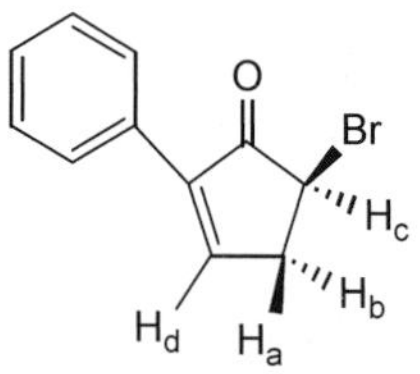

94

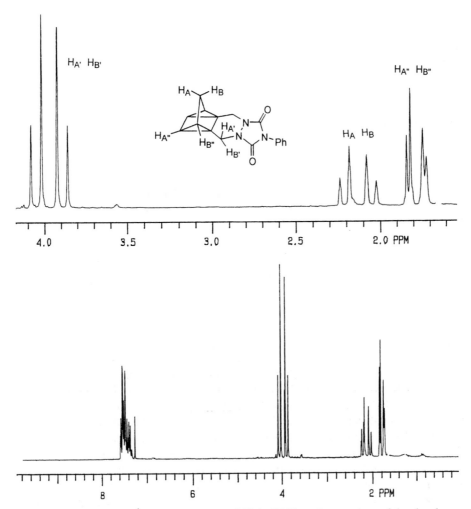

Figure 83 200 MHz ^{1}H-NMR spectrum of **93** in CDCl$_3$ and expansions of the signals.

For example, let us analyze the signal splitting of the methylene protons of compound **94** (Figure 84) [64]. The methylenic protons H$_a$ and H$_b$ are not equal because of the bromine atoms. Proton H$_a$ has the *cis* configuration and proton H$_b$ has the *trans* configuration compared with the location of bromine atom. Since these protons are located in different magnetic environments, it is expected that they will have different chemical shifts and give rise to an AB system.

The methylenic protons have two neighboring protons (H$_c$ and H$_d$) on the adjacent carbon atoms. These protons (H$_c$ and H$_d$) will also interact with the protons of the AB system and will cause further splitting in the lines of the AB system. One would expect both parts of the AB system to be split into a doublet of doublets. Inspection of the ^{1}H-NMR spectrum of **94** shows that the A-part of the AB system, resonating at $\delta = 3.41$ ppm, consists of eight lines. However, the B-part, resonating at $\delta = 3.00$ ppm,

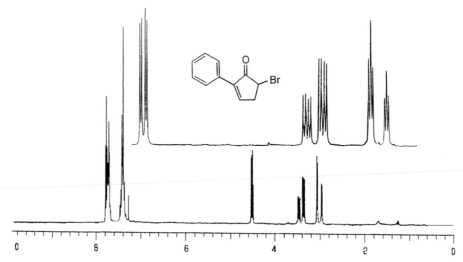

Figure 84 200 MHz ^{1}H-NMR spectrum of 2-phenyl-5-bromocylopent-2-enone (**94**) in CDCl$_3$ and expansions of the signals.

appears as a doublet of triplets. This different splitting pattern requires explanation. At the first instance, we cannot distinguish between these protons. For a correct assignment we have to measure the coupling constants. Since one of these protons, H$_a$ or H$_b$, couples to the adjacent protons (H$_c$ or H$_d$) with the same size coupling constant, the B-part of the AB system is further split into triplets. On the other hand, the coupling constants between the other proton (H$_a$ or H$_b$) and adjacent protons (H$_c$ and H$_d$) have quite different values, so that the A-part of the AB system is split into a doublet of doublets. The different coupling of the methylenic protons is also shown in the signals of proton H$_c$, which resonates as a doublet of doublets due to the splitting by the adjacent methylenic protons.

If the chemical shift difference between the interacting protons of an AB system is very small ($\Delta\delta \approx 0.1$ ppm) and these protons have further coupling with the adjacent protons, then at low field (60–100 MHz) recorded ^{1}H-NMR spectra cannot be analyzed. For the simplification of the spectra, the sample has to be recorded on a high-field NMR spectrometer. As an example, let us analyze the ^{1}H-NMR spectra of the bisepoxide **95** recorded on 60 and 400 MHz NMR spectrometers [65]. We will restrict our discussion to the resonance signal of the olefinic protons.

95

Since olefinic protons H_a and H_b have different chemical environments, it is expected that these protons should appear as an AB system. Furthermore, these protons will couple with epoxide proton H_c and the adjacent methylenic protons H_d and H_e. The 60 MHz ^{1}H-NMR spectrum of this compound shows that the olefinic protons appear at $\delta = 5.5-6.0$ ppm as a broad multiplet with overlapped lines, which cannot be analyzed (Figure 85a). Similar multiplets can also be observed for the resonances of the other part of the molecule.

The 400 MHz ^{1}H-NMR spectrum of this compound is shown in Figure 85b. Since the magnetic field strength is increased, the chemical shift difference in hertz between the two interacting protons is also increased so that the system approaches the border of the first-order-spectra. Now the olefinic proton resonances can be interpreted. The expanded signals of the olefinic proton resonances are presented in Figure 86.

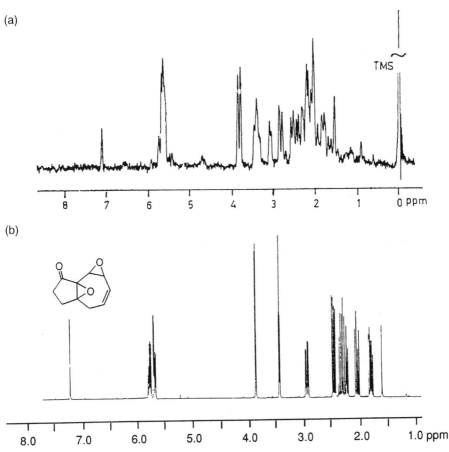

Figure 85 ^{1}H-NMR spectra of the bisepoxide (**95**) in CDCl$_3$ recorded on (a) 60 MHz and (b) 400 MHz NMR instruments.

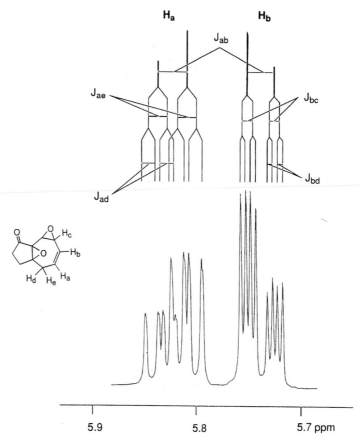

Figure 86   Expansions of the olefinic proton resonances of **95**.

As expected, the olefinic protons resonate as an AB system, which can be easily recognized by the *roof effect*. The olefinic proton H_a will couple beside the other olefinic
proton H_b, with the adjacent methylenic protons H_d and H_e. Therefore, the signal width
of proton H_a will be larger than that of H_b. Generally, the signal width of a proton
is equal to the sum of all possible couplings arising from the corresponding
proton. Based on this assumption, we can assign the low-field part of this AB system
to proton H_a and the high-field part to proton H_b. As expected, the A-part (low-
field resonance) is further split into a doublet of doublets by the adjacent vicinal protons
H_d and H_e. Since the dihedral angles between protons H_a and H_d and between H_a and
H_e are different, the coupling constants J_{ad} and J_{ae} have different values. Consequently, the A-part of AB system is split into a doublet of doublets. If these
coupling constants had the same magnitudes $J_{ad} = J_{ae}$, triplet splitting would occur.
Careful inspection reveals that each of these eight lines has further splitting
($J < 1.0$ Hz), which is not well resolved. This coupling probably arises from long-
range coupling ($^4J_{ac}$) with the epoxide proton H_c. The B-part of the AB system is

split into a doublet of doublets. In this case, proton H_b couples vicinally to proton H_c and allylically to one of the protons H_d or H_e. In spite of the presence of two allylic protons, proton H_b couples only to one of these protons. The measured allylic coupling constant is $^4J = 2.3$ Hz. This value is quite high for allylic coupling and indicates that the C–H bond is perpendicular to the plane formed by the C=C double bond. Since proton H_e is located in the plane of the C=C double bond, long-range coupling arises from the coupling $^4J_{bd}$.

With the above-mentioned example we have attempted to demonstrate that the protons of an AB system can have further coupling to adjacent protons. Both parts of an AB system may exhibit completely different signal splitting patterns. Now, with the next example, we will show a case in which only one part of an AB system exhibits coupling.

85

The bridge protons H_a and H_b in tribromo compound **85** are not equivalent and they resonate as an AB system (see Figure 65). The resonance signal of H_a proton (B-part of AB system) is further split into triplets by the bridgehead protons H_c and the triplets formed are further split into triplets by two H_d protons, according to the 'M' or 'W' arrangement of the coupled protons, so that proton H_a appears as a doublet of triplets of triplets. However, careful analysis of the lines shows that some of the signals overlap. On the other hand, the bridge proton H_b does not show any coupling to the protons H_c and H_d. The lack of coupling between protons H_b and H_c arises from the dihedral angle, which is nearly 80–85°. This AB system can be recognized by the *roof effect* present.

6.2.2 AB systems and molecular geometry

To generate an AB system, the interacting protons must have different chemical shifts and the ratio $\Delta\delta/J$ must be smaller than 10. Whether the interacting protons have different chemical shifts or not depends only on molecular geometry. For example, cyclopentadiene and dihydronaphthalene derivatives **96** and **97** have a plane of symmetry and the methylenic protons are equal. Therefore, they resonate as a singlet (we ignore the long-range couplings). The geminal coupling constant between these protons cannot be directly determined. In the case of diester **98** the situation is different. Because of the cyclopropane ring, the molecule does not have a plane of symmetry going through the plane of the benzene ring. Consequently, the methylenic protons are not equivalent.

96   **97**   **98**

As one can see from structure **98**, proton H_a is aligned in the same direction as the cyclopropane ring, with proton H_b in the opposite direction. Since these protons are in different environments, it is expected they should give rise to an AB system. Of course, they will resonate as an AX system, if the chemical shift difference is large.

99   **100**

In ketone **99** we encounter a similar case. The methylenic protons located between the benzene ring and the carbonyl group are equivalent due to the plane of symmetry. The expected resonance will be a singlet. When we reduce the other carbonyl group to alcohol **100**, the symmetry in the molecule is destroyed. The methylenic protons will now resonate as an AB system, since one of the protons will be located in the same direction as the alcohol group and the other one in the opposite direction.

101   **102**

The bridge protons in 1,4-dimethylnorbornane (**101**) are equal and their resonance signal appears as a singlet. However, the introduction of a double bond into molecule **101** will change the equivalency of these protons. Proton H_a in **102** that is aligned over the double bond is under the influence of the magnetic anisotropy generated by the double bond, whereas proton H_b is not. This situation leads to different chemical shifts. Finally, these protons resonate as an AB system.

The geometrical structure of a given compound determines whether the two interacting protons resonate as a singlet or as an AB system. Conversely, an observed

singlet or AB system provides us with very important information about the geometry
of the compound to be analyzed.

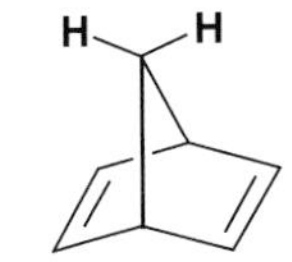

103

For example, the bridge protons of norbornadiene (**103**) resonate as a broad singlet,
which clearly shows the symmetry present in the molecule. There is coupling between the
bridge and bridgehead protons. Since the magnitude of this coupling is small, it appears
as line broadening.

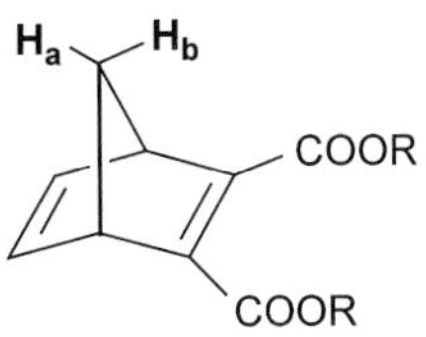

104a 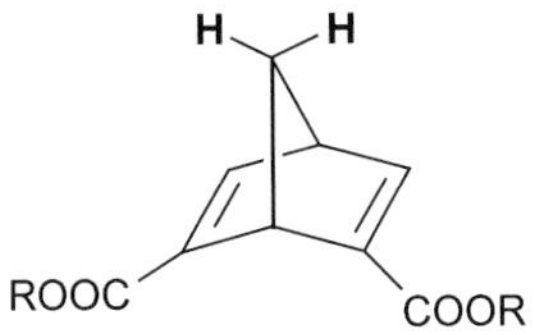   **104b** 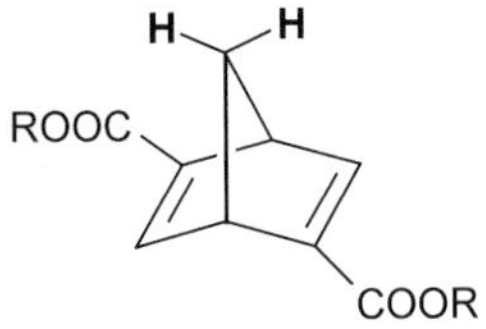   **104c**

Now let us replace two olefinic protons in the norbornadiene (**103**) with two ester
groups. There are three possible isomers. Now we shall discuss the appearance of the
bridge protons in relation to the location of the ester groups. In isomer **104a**, the
substituents are on one side of the molecule. The compound does not have a plane of
symmetry going through the bridgehead carbon atoms. The bridge protons will resonate
in this case as an AB system. In isomers **104b** and **104c**, the bridge protons are equal
and therefore they will resonate as singlets if we ignore the coupling between the bridge
and bridgehead protons. Just by looking at the resonance signals of bridge protons we
can easily characterize isomer **104a**. To distinguish between the other isomers we can
look at the bridgehead proton resonances. In **104c** they are equal, whereas in **104b** they
are not.

Influence of optically active centers on neighboring methylenic protons

Since cyclic systems have rigid structures, it is easy to determine whether the geminal
protons resonate as an AB system or not. In an acyclic system, the situation is different.
Let us now investigate the coupling pattern of a methylenic group, shown below.

$$R-C-CH_2-X$$

We assume that the methylenic protons do not have any adjacent protons for coupling.
At the first instance one would expect that these protons should resonate as a singlet due
to the free rotation about the C–C bond. This is actually not the case. The resonance
signal of these methylenic protons depends on the structure of the R group. For example,

the methylenic protons in **105** resonate as a singlet whereas the protons in **106** resonate as an AB quartet.

105   **106**

If the molecule contains an asymmetric center, which is shown by an asterisk in **106**, the methylenic protons are not equivalent and they resonate as an AB quartet. This asymmetric center may be connected directly to the methylene carbon or it may be in a part of the molecule. At this stage it is not easy to understand the nonequivalency of these methylenic protons in spite of the free rotation about the C–C bond. To understand this phenomenon we have to illustrate the molecule with a Newman projection and study the different conformers. We start with a system where the methylenic protons are directly attached to a nonchiral carbon atom. We refer to the substituents by the letters A and B.

Newman Projection

Three different conformations (I–III) of this compound are represented as Newman projections below. A rapid rotation about the carbon–carbon single bond takes place and the protons rapidly change position. The time spent in any one rotamer is short, because the energy barrier for rotation is small. Now we will analyze the chemical environments of protons H$_1$ and H$_2$ during the interconversion of the various conformations of the molecule.

I   II   III

In the first rotamer I, two protons H$_1$ and H$_2$ are located between the substituents A and B. Both protons have the same chemical environment so they are expected to be equivalent. In the second rotamer II, these protons have completely different locations. Proton H$_1$ is located between the substituents A and B, whereas proton H$_2$ is between A and A. In a frozen conformation, these protons would resonate as an AB system.

In the last conformation III, the protons are again in different chemical environments: proton H_1 is located between the substituents A and A, and proton H_2 is between A and B.

$H_1 \longrightarrow$ between A and B   between A and A $\longleftarrow H_1$

$H_2 \longrightarrow$ between A and A   between A and B $\longleftarrow H_2$

Conformation II   Conformation III

However, a careful inspection of rotamers II and III shows that the environments of H_1 and H_2 are averaged. This will result in an A_2 spectrum (singlet) for all of the population distributions of rapidly interconverting rotamers (fast rotation). Since these protons are equivalent, they are called *enantiotopic* protons.

When the methylenic protons are connected to an asymmetric center (a chiral carbon atom), the situation is different. In this case, we refer to the substituents by the letters A, B and C. Three different conformations (I–III) of this compound are represented as Newman projections below. Now we may analyze again the chemical environments of the protons H_1 and H_2 during the interconversion of the various conformations of the molecule.

I   II   III

In the first rotamer I, proton H_1 is between A and C, and H_2 is between B and C. Both protons are in different environments. In the second rotamer II, proton H_1 is between B and C, and H_2 is between A and B. In the first example (nonchiral center), we saw that the protons change positions exactly. However, the environments of H_1 in (I) and H_2 in (II) are not *exactly* the same and hence the chemical shifts of H_1 and H_2 will not be expected to be exactly averaged. This will result in an AB spectrum for these methylenic protons.

$H_1 \longrightarrow$ between A and C   between B and C $\longleftarrow H_1$

$H_2 \longrightarrow$ between B and C   between A and B $\longleftarrow H_2$

Conformation I   Conformation II

When we compare the locations of protons H_1 and H_2 in rotamers II and III, we also see that these protons do not change positions exactly. While H_1 is between B and C, H_2 is between A and B (rotamer II). However, when H_1 is between A and B, then H_2 is between A and C (rotamer III).

H_1 ⟶ between B and C   between A and B ⟵ H_1

H_2 ⟶ between A and B   between A and C ⟵ H_2

Conformation II   Conformation III

Unequal chemical environments and equal conformer populations will cause nonequivalence between H_1 and H_2. As a result, these protons will give rise to an AB system in spite of the fact that there is a fast rotation about the carbon–carbon single bond. These protons are called *diastereotopic* protons.

As we have mentioned before, a methylene group can be either directly attached to an asymmetric center or have some bonds removed in order to be diastereotopic. Interestingly, the nonequivalence does not decrease automatically with the distance from the asymmetric center. Systematic investigations have been carried out in order to determine the factors associated with the nonequivalence of methylene protons. It was found that the chemical shift difference between H_1 and H_2 depends on the size, proximity and type of substituents.

OCH_2CH_3 OCH_2CH_3

107 enantiotop   **108** diastereotop

In ethoxynaphthalene (**107**), the methylenic protons are enantiotopic, since the molecule does not have any asymmetric center. However, in [10]annulene derivative **108**, the symmetry is destroyed because of the methylene bridge, and therefore, the methylenic protons are diastereotopic.

One of the best examples demonstrating the nonequivalence of the methylene protons is acetaldehyde diethylacetal. Furthermore, this example shows that the nonequivalence of the methylene protons is not limited to molecules with optically active carbon atoms. The ^{1}H-NMR spectrum of acetaldehyde diethylacetal is shown in Figure 87.

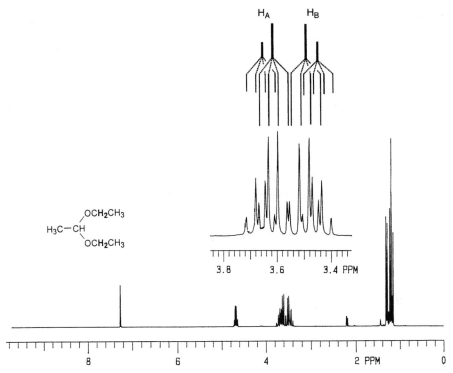

Figure 87   200 MHz ^{1}H-NMR spectrum of diethyl acetal (**109**) in CDCl$_3$ and expansions of the signals.

109

The quartet at 4.6 ppm corresponds to the CH proton adjacent to the CH$_3$ group. The multiplet observed between 3.4 and 3.8 ppm corresponds to the methylenic protons adjacent to the CH$_3$ group. In the case of enantiotopic protons, they will resonate as a quartet due to the coupling with CH$_3$ protons. Careful inspection of this multiplet clearly reveals that there is an AB system whose A- and B-parts are further split into quartets by the coupling with the adjacent methyl protons. The methylenic protons CH$_2$ are diastereotopic. They are bonded (separated by two bonds) to a carbon atom bearing three different substituents (H, CH$_3$ and OCH$_2$CH$_3$). As we have seen with the rotamers, three different substituents cause different environments for methylenic protons.

Diastereotopic protons are encountered very frequently in NMR spectra. For example, the ^{1}H-NMR spectrum (Figure 88) of dibromide **110** shows the existence of diastereotopic protons [66].

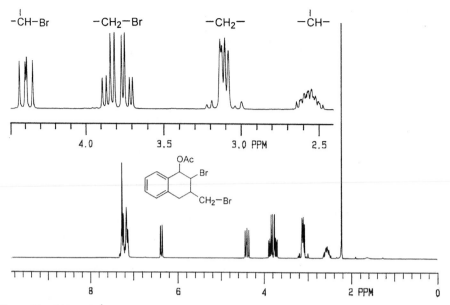

Figure 88   200 MHz ¹H-NMR spectrum of dibromo acetate (**110**) in CDCl₃ and expansions of the signals.

110

The methylenic protons adjacent to the bromine atom resonate as an AB system. This methylene group is directly attached to a chiral carbon atom that causes nonequivalence of the methylene protons. The lines of the AB system are further split into doublets, which arises from the coupling with CH proton.

The nonequivalence of the protons discussed above is not limited to methylene groups attached to optically active carbon atoms. A dimethyl carbon ($-C(CH_3)_2X$) group attached to a chiral carbon will have diastereotopic methyl groups. Methyl groups resonate as two distinct singlets (for some examples see Table 6.2).

diastereotopic methyl groups

Table 6.2

Selected examples for enantiotop and diastereotop protons

| enantiotop | diastereotop | diastereotop |

| enantiotop | diastereotop | diastereotop |

Another common situation involves an isopropyl group attached to an asymmetric center. For example, the methyl groups in a system such as $Me_2CH-CABC$ appear as two pairs of doublets, with each doublet showing vicinal coupling due to the methine protons.

6.2.3 Three-spin systems: A_3, A_2B, (AB_2), A_2X, (AX_2), ABC and ABX systems

The simplest three-spin system is the A_3 system, consisting of three equivalent protons. An isolated methyl group or three equal protons bonded to three carbon atoms form this system and they resonate as a singlet. It is not possible to observe the coupling constant between the protons. Some selected examples for the A_3 system are given below.

If one of the protons becomes largely different in chemical shift relative to the other two protons, to where $\Delta\delta$ is larger than the coupling constant J ($\Delta\delta/J > 10$), we have an AX_2 or A_2X system. We can interpret these spectra quite easily by using the rules of the *first-order* spectra. The A-part will give rise to a symmetrical triplet

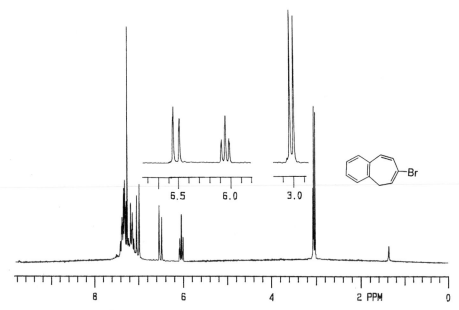

Figure 89 200 MHz ¹H-NMR spectrum of bromobenzocycloheptatriene (**111**) in CDCl₃ and expansions of the signals.

with spacings equal to J_{ax} and the X-part to a symmetrical doublet with spacings also equal to J_{ax}. For example, bromobenzocycloheptatriene **111** has an AX₂ system arising from the coupling of the methylenic protons (H$_x$) with the olefinic proton (H$_a$) (Figure 89) [67].

111

Seven-membered ring olefinic protons (H$_b$ and H$_c$) resonate as an AB system, whereas the other olefinic proton H$_a$ and methylenic protons H$_x$ form an AX₂ system. As one can see from the spectrum (Figure 89), there is a large chemical shift difference ($\Delta\delta = 3.0$ ppm) between A and X protons. Proton H$_a$ resonates at 6.0 ppm as a triplet and H$_x$ protons at 3.0 ppm as a doublet.

If one of the three equivalent protons becomes slightly different in chemical shift relative to the other protons ($\Delta\delta/J < 10$), the system formed is called an AB₂ system (or A₂B) [68]. Some examples of AB₂ systems are given below.

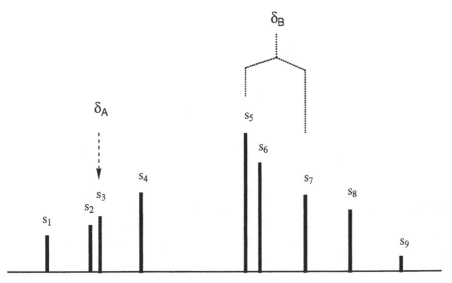

112   113   114

From quantum mechanical calculations it has been shown that this type of spin system can have a maximum of nine lines (Figure 90). One of these lines is a weak combination line and is often not observable. From an AB_2 system we can expect two different coupling constants, J_{ab} and J_{bb}. Since H_b protons are chemically equivalent, the coupling between H_b protons will therefore not be observable in the spectrum. Consequently, the shape of AB_2 spectra and relative intensities and line separation are not affected by this coupling J_{bb}. The shape of AB_2 spectra is affected only by the ratio $J_{ab}/\nu_a - \nu_b$.

The spectra of such systems do not possess any symmetry and cannot be readily analyzed in the way described for AB, AX and AX_2 systems. However, there are some equations and rules that permit analysis of AB_2 spectra. The first thing to be done is to number the lines (1–9), starting from the left. The chemical shift of proton H_a is given by the position of line 3 (see Figure 90). The chemical shift of H_b protons is the midpoint

Figure 90 A calculated AB_2 spectrum. Numbering of the lines and representation of the chemical shifts (δ_A and δ_B).

between lines 5 and 7. The coupling constant can be determined by eq. 39.

$$\delta_A = s_3 \tag{37}$$

$$\delta_B = \frac{1}{2}(s_5 + s_7) \tag{38}$$

$$J_{ab} = \frac{1}{3}(s_1 - s_4 + s_6 - s_8) \tag{39}$$

It is possible that in some cases not all expected lines can be resolved. Therefore, analysis by computer is recommended. It is generally suggested to simulate the spectrum in order to support the data extracted from direct analysis.

If the chemical shift difference between the interacting protons increases or the coupling decreases, the lines s_2 and s_3, s_5 and s_6, and s_7 and s_8 approach each other and the shape of the AB_2 spectrum changes progressively into an AX_2 spectrum. Figure 91 shows a number of calculated AB_2 spectra with different values of J_{ab} with $\delta_a = 2.33$ ppm and $\delta_b = 2.17$ ppm.

Figure 92 shows a number of calculated AB_2 spectra with different values of δ_a with $J_{ab} = 0.5$ Hz and $\delta_b = 1.20$ ppm. In both cases the transition from an AB_2 system to an AX_2 system can be clearly observed.

Examples of the types discussed here are those of dichloro phenol and homo-benzoquinone derivative [69] (Figure 93). The aromatic proton resonances of dichloro phenol are given in expanded form. The eight allowed transitions can be observed in the spectrum. Line 9 in the AB_2 spectrum is in this case not detectable.

ABX system

As the next three-spin system, let us analyze the ABX system. As the notation indicates, these systems have two protons with similar chemical shifts and a third proton with a chemical shift very different from the others (Table 6.3).

The ABX system has six different parameters: δ_a, δ_b, δ_x, J_{ab}, J_{ax}, and J_{bx}. We might expect a 12-line spectrum for an ABX system. First, we couple H_A and H_B protons independently of proton H_X, and we derive the typical AB system. The original four AB lines will be split into doublets by proton H_X. The eight-line portion of the spectrum is then made up of two AB-subspectra with equal intensities. Proton H_X should appear as a doublet of doublets. In fact, the ABX system frequently consists of 12 lines. In theory, ABX spectra can have 14 lines. Two of these are usually too weak to be observed and they appear in the X-part of the spectrum. Consequently, the X-part of an ABX system consists of four lines; however, it cannot be interpreted as a *first-order* spectrum (Figure 94). In fact, the X-part allows us at best to only determine the sum of the coupling constants ($J_{ax} + J_{bx}$).

The shape of ABX systems alters with the coupling constants and chemical shifts. It is possible that some of the signals overlap. In that case, the ABX system cannot be directly analyzed. Figures 95 and 96 show calculated ABX spectra. In the former (Figure 95), the chemical shifts of all three protons and the coupling constants J_{ax} and J_{bx} were kept constant and the coupling constant J_{ab} was varied. In the latter (Figure 96), ABX spectra are presented where only one parameter, the chemical shift of H_x, was changed.

We encounter an ABC system if all three protons have slightly different chemical shifts. The ABC spin system is considerably more complicated than those discussed

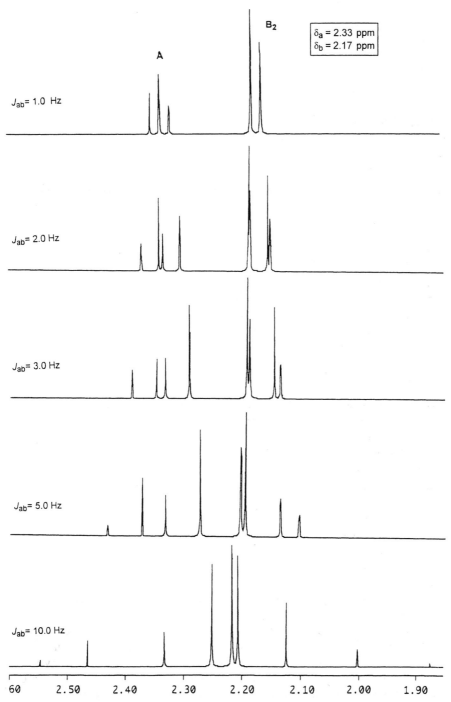

Figure 91 Calculated AB_2 spectra. The chemical shifts (δ_a and δ_b) were set at 2.33 and 2.17 ppm where the coupling constant J_{ab} was varied.

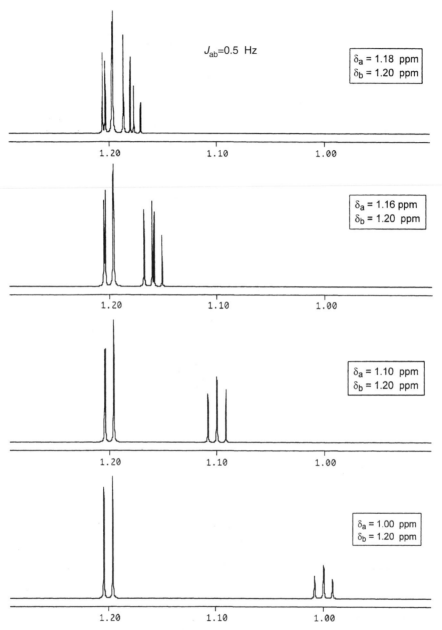

Figure 92 Calculated AB$_2$ spectra. The chemical shift δ_B and the coupling constant J_{AB} were set at 1.20 ppm and 0.5 Hz, where the chemical shift δ_A was varied.

above and no useful information can be obtained from ABC spectra by applying *first-order* analysis. In fact, computer analysis and simulation have to be applied to solve ABC spin systems. In Figure 96, we have moved the chemical shift of the X-part towards the chemical shifts of protons A and B. In other words, we have generated

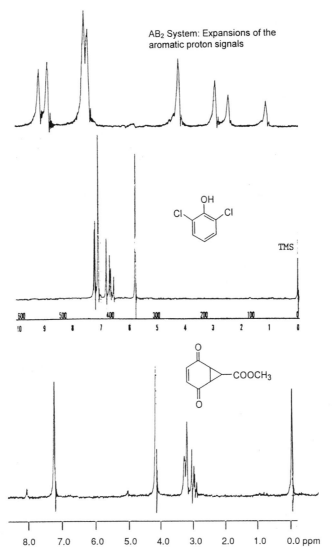

Figure 93 60 MHz ^{1}H-NMR spectra of 2,6-dichlorophenol and homobenzoquinone derivative in CDCl$_3$ and expansions of the aromatic signals.

an ABC system starting from an ABX system. It is easy to recognize that this ABC system has a complex structure that does not permit direct analysis.

6.2.4 Four-spin systems: A$_2$B$_2$, A$_2$X$_2$, AA'BB' and AA'XX' systems [70]

When we consider the interaction of four protons, there are many different combinations. Of course, we will not treat all possible spin systems, only those that are frequently encountered. For example, the most complex system is the ABCD system, which cannot

Table 6.3

Selected examples for the ABX system

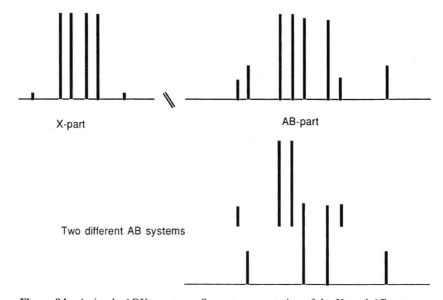

be easily analyzed. We will concern ourselves only with the spectra of four protons consisting of two pairs of protons (A_2B_2, A_2X_2, $AA'BB'$ and $AA'XX'$ systems).

The spectra of four-spin systems consisting of two pairs of *chemically* and *magnetically* equivalent protons are called A_2B_2 or A_2X_2, depending on the chemical shift difference of the interacting protons. We have to briefly discuss the concept of *magnetic equivalence*. If the protons in one pair couple equally to any proton (probe nucleus) in the same spin system, they are magnetically equivalent.

X-part

AB-part

Two different AB systems

Figure 94   A simple ABX spectrum. Separate presentation of the X- and AB-parts.

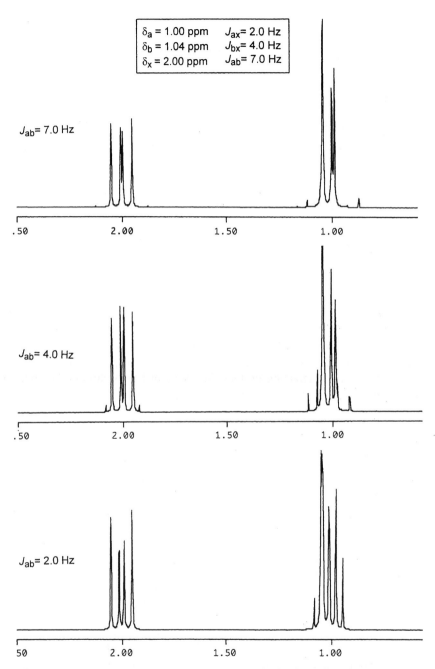

Figure 95   Calculated ABX spectra. The chemical shifts δ_a, δ_b, δ_x and the coupling constants J_{ax}, J_{bx} were set constant, where the coupling constant J_{ab} was varied.

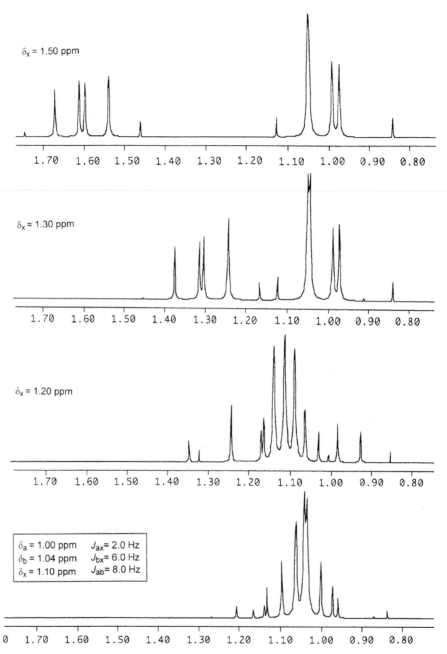

Figure 96 Calculated ABX spectra. The chemical shifts δ_A, δ_B, and the coupling constants J_{ab}, J_{ax}, J_{bx} were set constant, where the chemical shift δ_x was varied.

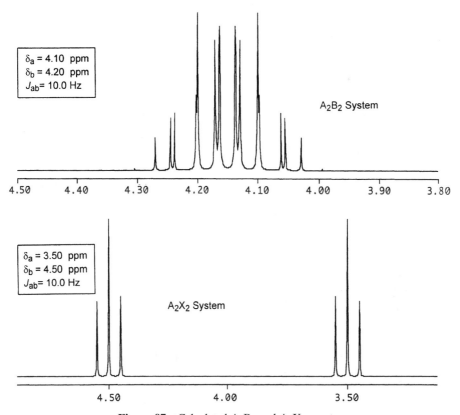

Figure 97   Calculated A_2B_2 and A_2X_2 spectra.

A_2B_2 or A_2X_2 systems are governed by three different parameters ν_A, ν_B (or ν_X) and J_{AB} (or J_{AX}). If the chemical shift difference of these pairs is large, then we are dealing with an A_2X_2 system. The open-chain, conformationally mobile compounds of this type consist of two sets of protons coupled to each other.

A_2X_2 System

A_2X_2 System

The groups H_3CCO and OCH_3 contain no chiral element and no protons that couple to the two sets of protons shown. Thus, each half on the A_2X_2 spectrum consists of a triplet (Figure 97), which can be analyzed by the *first-order* rules.

If the chemical shift difference of these pairs is of the same order of magnitude as their coupling constants, then the system becomes an A_2B_2 system, which would show a complex, *second-order* spectrum.

$$Br-CH_2-CH_2-Cl$$

A_2B_2 System

$$H_3C-\overset{\overset{\displaystyle O}{\|}}{C}-O-CH_2-CH_2-OCH_3$$

A_2B_2 System

The A_2B_2 spectrum has up to eight lines in each half of the spectrum. One of the typical features of an A_2B_2 system that makes them instantly recognizable is the total symmetry of the spectra. The signals of A protons are the mirror image of those of B protons. Figure 97 shows two calculated A_2X_2 and A_2B_2 spectra. The appearance of the spectra in each individual case depends on the ratio $\Delta\nu/J_{AB(AX)}$. The geminal coupling constants J_{AA} and J_{BB} have no effect on the spectra. The A_2B_2 and A_2X_2 spectra in Figure 97 have been calculated with the same coupling constant of $J_{AB(AX)} = 10$ Hz. The chemical shift difference of $\Delta\nu = 200$ Hz for A_2X_2 system has been changed to $\Delta\nu = 20$ Hz in the case of the A_2B_2 system.

The most frequently encountered four-spin systems are AA′BB′ and AA′XX′ systems. First let us discuss the difference between these systems (AA′BB′ and AA′XX′) and A_2B_2 and A_2X_2 systems. We have shown above that A_2B_2 and A_2X_2 systems are encountered in open-chain, conformationally mobile compounds. Because of the free rotation about the carbon–carbon single bond, *there is only one coupling between A and B (or X) protons*. However, in the case of AA′BB′ and AA′XX′ systems, there are two different couplings between A and B (or X) protons. The notation shows the existence of two different coupling constants J_{AB} and $J_{AB'}$ (J_{AX} and $J_{AX'}$). To explain this phenomenon let us analyze the coupling modes in *o*-dichlorobenzene.

121
two different couplings
between H$_A$ and H$_B$ protons

Thus, in *o*-dichlorobenzene the protons *ortho* to chlorine atoms (H$_A$ and H$_{A'}$) are chemically equivalent to each other because of the plane of symmetry that goes through the compound. The remaining protons (H$_B$ and H$_{B'}$) are also chemical shift equivalent to each other. As one can easily recognize from the structure, there are two different couplings between H$_A$ and H$_B$ protons: *ortho* coupling $^3J_{AB}$ and *meta* coupling $^4J_{AB}$.

In freely rotating molecules of the type $X-CH_2-CH_2Y$ there is only one coupling between the chemically nonequal protons. Two protons are considered *magnetically equivalent*, if they are chemically equivalent and if they couple equally to any other proton (probe proton) in the same spin system. However, the protons are considered *magnetically nonequivalent* if they are chemically equivalent but they couple *nonequally* to any other proton (probe proton) in the same spin system. In *ortho*-dichlorobenzene let us analyze the coupling behavior of H_A protons. Let us consider the couplings of H_B (probe nucleus) to two H_A protons. In one case we have an *ortho* coupling, and in the other case a *para* coupling. This behavior shows that the H_A protons are not completely identical. Because of the different couplings of these protons to a probe proton, they are considered *chemically equivalent* but *magnetically nonequivalent*. To distinguish between these protons, one is assigned the letter A and the other A'. We can check in the same way the magnetic behavior of H_B protons. In this case we have to take one of the H_A protons as a probe nucleus and look at the coupling mode with H_B protons. We notice that one of the H_B protons will have an *ortho* coupling to the probe proton and the other a *meta* coupling. To show these *magnetically nonequivalent* protons, the designation $AA'BB'$ or $AA'XX'$ is used. When the chemical shift difference between the interacting protons is large, the four-spin system is called an $AA'XX'$ system. As the chemical shift difference between the interacting protons approaches the size of the coupling constant, the system is called an $AA'BB'$ system. $AA'BB'$ and $AA'XX'$ systems appear generally in conformationally restricted systems. We may classify these into three different groups.

(1) In $AA'BB'$ ($AA'XX'$) systems all four protons are bonded to four carbon atoms. The common examples are *para* disubstituted (different substituents) and *ortho* disubstituted (identical substituents) benzene derivatives, furan, pyrrole, etc. Examples are given in Table 6.4.

(2) The four protons of $AA'BB'$ ($AA'XX'$) systems can be bonded to two adjacent equal carbon atoms. Of course, these carbon atoms have to be a part of a cyclic system. Furthermore, the system must have a plane of symmetry that goes through the carbon–carbon bonds connecting the carbon atoms bearing the protons. A protons are located in one face of the compound and B protons in the opposite face. Some examples are given below. In these systems, it can be easily recognized that there are two coupling constants between A and B protons: the geminal coupling $^2J_{AB}$ and the *trans* coupling $^3J_{AB'}$.

(3) The third class of $AA'BB'$ ($AA'XX'$) systems are those where the four protons are connected to the two nonequal carbon atoms. In this case, the systems also have a plane of symmetry that goes through the plane bisecting the whole compound. A protons are bonded to one carbon atom and B protons to the other carbon

Table 6.4

Selected examples for AA'BB' and AA'XX' systems

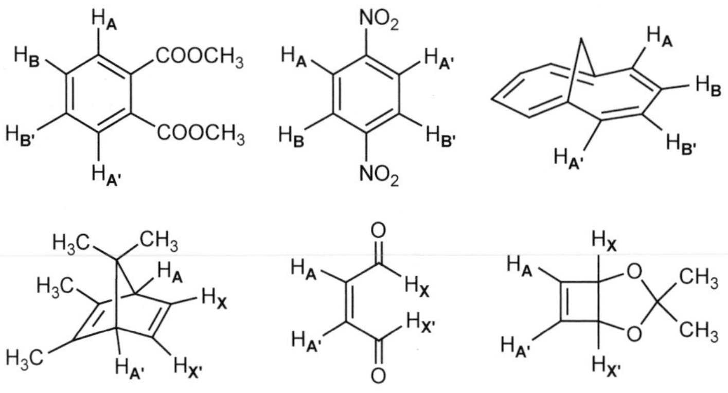

atom. Some examples are illustrated below. These examples also show two different couplings between A and B protons: the *cis* coupling $^3J_{AB}$ and the *trans* coupling $^3J_{AB'}$.

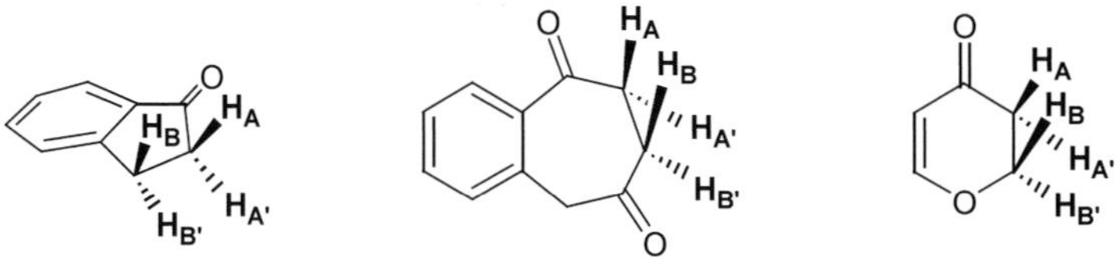

After showing in which systems AA'BB' (AA'XX') systems can be encountered, we now turn our attention to the appearance and the parameters of these systems.

AA'BB' and AA'XX' systems are governed by six different parameters, which are necessary for the analysis of these systems.

Chemical shifts
δ_A The chemical shift of proton H_A
δ_B The chemical shift of proton H_B

Coupling constants
$J_{AA'}$
$J_{BB'}$
$J_{AB} = J_{A'B'}$
$J_{AB'} = J_{A'B}$

AA'BB' (AA'XX') systems have a maximum of 24 (20) lines and the total symmetrical distribution of these lines makes them instantly recognizable. Thus, for example, in an AA'BB' system, a maximum of 12 lines caused by A protons is matched by a mirror image reflection of 12 lines due to B protons. When the number of lines is less than 24, the symmetry is always preserved. The 200 MHz ^{1}H-NMR spectrum of dichlorobenzene is given in Figure 98.

This AA'BB' system consists of a total of 12 lines. Six of them belong to H_A protons and the other six to H_B protons. It is not possible to directly determine from the spectrum as to which parts (low field or high field) of the signals belong to H_A and which to H_B protons. In some cases we can distinguish between H_A and H_B protons by analyzing the chemical shifts. However, if the chemical shift difference is small, special NMR experiments (analysis of the satellite spectra) have to be carried out to make a clear-cut distinction between H_A and H_B protons.

Even though there are some special equations for analyzing AA'BB' and AA'XX' systems, in most cases, straightforward direct analysis is not possible. Therefore, analysis of these systems has to be done with computer programs. In addition, the results obtained

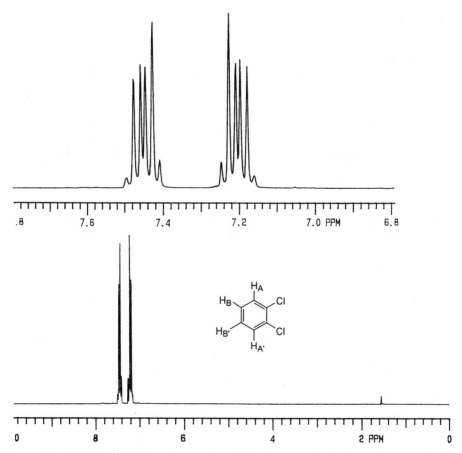

Figure 98   200 MHz ^{1}H-NMR spectrum of o-dichlorobenzene in CDCl$_3$ and expansion of the aromatic signals.

by direct analysis of a spin system always have to be checked by comparing a calculated spectrum with the experimental spectrum. This process is called spectrum simulation. If the calculated spectrum exactly matches the experimental spectrum, then the analysis is correct.

A series of calculated AA′BB′ and AA′XX′ systems is given in Figures 99 and 100. These examples clearly demonstrate that the appearance of the spectra depends on the parameters (chemical shifts and coupling constants) and the 24 lines cannot be observed every time.

Let us analyze first how the strength of the magnetic fields influences the shape of the AA′BB′ system. Figure 99 shows three different spectra, which are calculated at different magnetic fields. In all of the cases constant parameters are used. The chemical shift difference between protons H_A and H_B was taken as $\Delta\delta = 0.3$ ppm. At 60 MHz the calculated NMR spectrum shows typical patterns of an AA′BB′ system. All 24 lines can be observed separately. By going from 60 to 100 MHz (and not changing any parameters), the chemical shift difference between the coupled protons H_A and H_B is changed in hertz (*not in ppm*) from $\Delta\nu = 18$ Hz (at 60 MHz) to $\Delta\nu = 30$ Hz (at 100 MHz) which of course affects the ratio of $\Delta\nu/J$. At 400 MHz the calculated spectrum is an AA′XX′ system. In this case only 12 lines are observed.

An example of the change in appearance of an AA′BB′ spectrum with the variation of $\Delta\nu/J$ is given in Figure 100, where the chemical shifts of the two protons are kept constant, while the coupling constants J_{ab}, $J_{bb'}$, $J_{ab'}$, $J_{aa'}$ are changed. These spectra also show that the shapes of the spectra are drastically changed depending on the magnitudes of the coupling constants. However, the AA′ and BB′ (or XX′) parts are in all cases symmetrical multiplets.

We have already discussed the ^{1}H-NMR spectrum of the dichlorobenzene in Figure 98. In this case the protons are bonded to four carbon atoms. Now let us discuss the ^{1}H-NMR spectrum of a compound where the protons are bonded only to two equal carbon atoms. The ^{1}H-NMR spectrum of the bicyclic endoperoxide **122** is given in Figure 101.

122

H_A protons here are located in the same direction as the peroxide linkage. They have a *syn* configuration (with respect to the peroxide bridge). H_B protons have the *anti* configuration. There are two different couplings between H_A and H_B protons: *cis* coupling $J_{AA'}$ and *trans* coupling $J_{AB'}$. Therefore, this system should resonate as an AA′BB′ system. However, inspection of the expanded resonance signals of the methylene protons in Figure 101 (bottom spectrum, right-hand side) does not show any symmetry between the two parts of the signal groups.

We have previously mentioned that an AA′BB′ system can be recognized from the existing symmetry of the signals. The question arises of whether one part of an AA′BB′ or AA′XX′ system can be disturbed. In the discussion of AB systems we saw that one part of

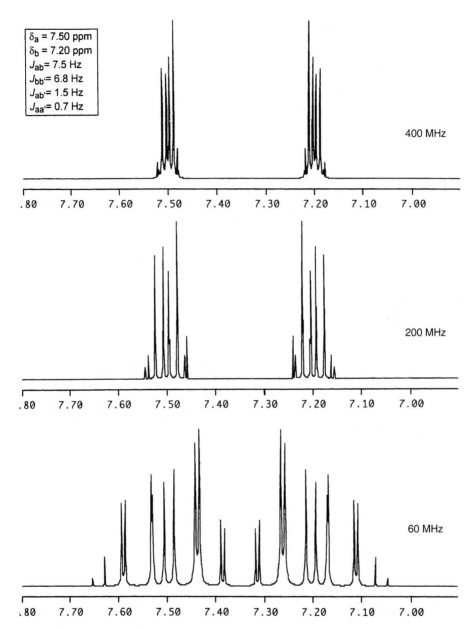

Figure 99 Calculated AA'BB' spectra with constant parameters (δ_a, δ_b, J_{ab}, J_{bb}, $J_{ab'}$, $J_{aa'}$) at different magnetic fields (60, 200 and 400 MHz).

an AB system can have further coupling to the other protons so that the appearance of a symmetrical AB system can no longer be expected. Here we encounter a similar case. The A-part, as well as the B-part, of an AA'BB' system can also have further coupling to other protons. Consequently, the symmetrical distribution of the signal pattern will be

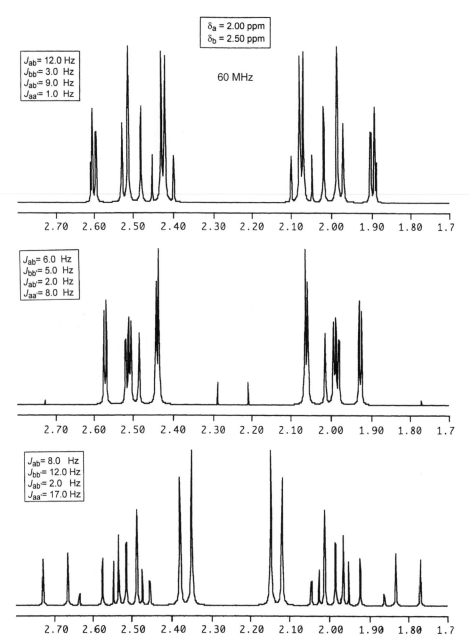

Figure 100   At 60 MHz calculated AA′BB′ spectra where the chemical shifts (δ_a, δ_b) set constant and where all coupling constants ($J_{ab}, J_{bb}, J_{ab'}, J_{aa'}$) varied.

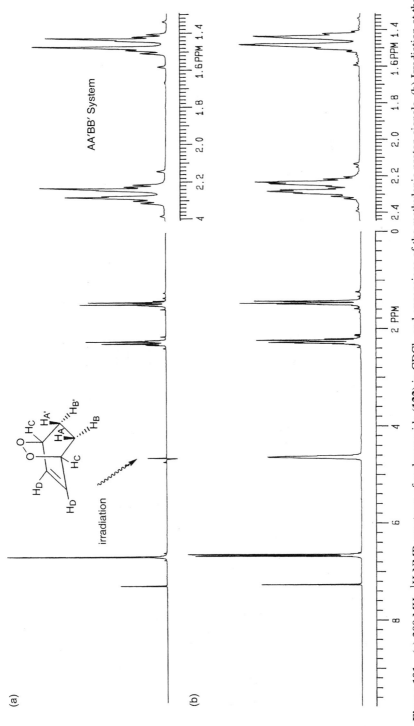

Figure 101 (a) 200 MHz ^{1}H-NMR spectrum of endoperoxide (**122**) in CDCl$_3$ and expansions of the methylenic proton signals. (b) Irradiation at the resonance frequency of bridgehead protons and expansions of the methylenic proton signals.

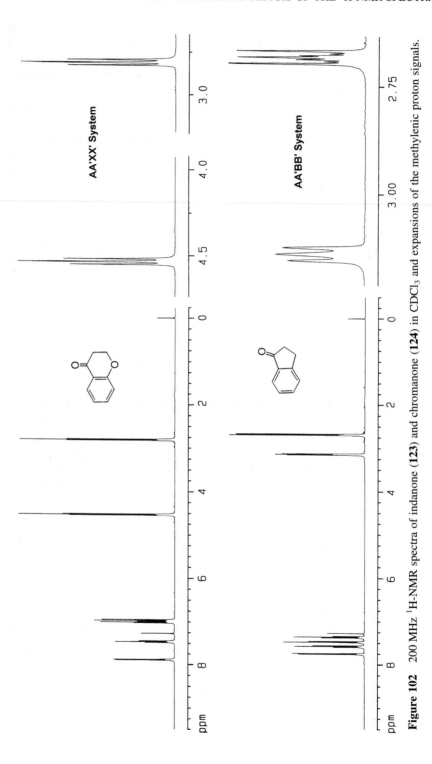

Figure 102   200 MHz ^{1}H-NMR spectra of indanone (**123**) and chromanone (**124**) in CDCl$_3$ and expansions of the methylenic proton signals.

lost. In the peroxide **122**, H$_A$ and H$_B$ protons have additional coupling to the bridgehead protons H$_C$. However, the magnitudes of these couplings are different $J_{AC} \neq J_{BC}$, due to the different dihedral angle between the coupled protons. Finally, lines of the A-part as well as of the B-part of this AA'BB' system will be further split by the proton H$_C$ by couplings of different size. Since the splitting in the signals will be different, of course the symmetry of the AA'BB' system will also be lost. If these couplings $J_{AC} = J_{BC}$ had been equal to each other, then both parts of the spectrum would have been split equally. Then the symmetry of the signals would have been preserved. To observe a symmetrical AA'BB' system of the methylenic protons in **122**, the couplings between bridgehead proton H$_C$ and methylenic protons H$_A$ and H$_B$ have to be removed. These couplings can be removed by a double resonance experiment, upon irradiation at the resonance frequency of the bridgehead protons (see the upper spectrum in Figure 101). Double resonance experiments are the subject of Chapter 7. By careful examination of the AA'BB' signals, after removing the coupling, the expected symmetry can be observed. This observation is completely in agreement with the chemical structure of this peroxide (**122**).

123   124

As the last example we will discuss the ^{1}H-NMR spectra of indanone (**123**) and chromanone (**124**). These compounds also have ethylene linkages that should resonate as an AA'BB' and an AA'XX' system, respectively. The 400 MHz ^{1}H-NMR spectra are given in Figure 102. The expanded spectrum of the methylene protons of indanone does not show the symmetrical distribution of the lines. The low-field part appears as a broad triplet, however, the high-field signals show additional fine splitting. How can we explain this phenomenon? It is clearly seen that the low-field signals are broadened due to the existence of some long-range couplings. These couplings are so small that they cause line broadening instead of splitting, and they arise from the interaction between the benzylic protons and *ortho* aromatic proton. As a consequence of this coupling, the signal of the AA' part of this AA'BB' system is broadened, which leads to the disappearance of the expected symmetry.

To observe an undisturbed, symmetrical AA'BB' or AA'XX' system, the protons of the system should not have any additional coupling to a third proton. To remove the coupling between the benzylic protons and aromatic *ortho* proton in **123**, the number of bonds between these protons has to be increased. In chromanone (**124**) the long-range couplings between the methylenic protons next to the oxygen atom and the aromatic protons are not observed. The ^{1}H-NMR spectrum shown in Figure 102 (lower spectrum) consists of a symmetrical AA'XX' system. The symmetry is preserved due to the lack of long-range couplings. Since the one set of methylenic protons is attached directly to the oxygen atom, there is a large chemical shift difference ($\Delta\delta = 1.6$ ppm) between the coupled protons. Therefore, these protons resonate as an AA'XX' system.

– 7 –

NMR Shift Reagents and Double Resonance Experiments: Simplification of the NMR Spectra

---

The complete analysis of some NMR spectra is often obscured by the presence of overlapping patterns, along with complications arising from *non-first-order* multiplets. Such spectra have to be simplified before they can be analyzed. There are several methods of simplifying NMR spectra:

(1) recording the compound on higher field NMR instruments;
(2) the use of NMR shift reagents;
(3) double resonance experiments.

We have previously discussed how NMR spectra are simplified when recorded at higher magnetic fields due to the increase in the chemical shift difference in hertz between the coupled protons. The *second-order* spectra are changed into *first-order* spectra, and hence the interpretation is much easier. This process is frequently not an option for many investigators. Even when the instrumentation is available, many spectra cannot be sufficiently simplified for a complete analysis. In this section we will focus on NMR shift reagents and the double resonance technique, and show how the spectra are simplified by the application of these methods.

7.1 SHIFT REAGENTS

If the NMR spectrum of an organic molecule having nucleophilic groups (halogen, alcohol, carbonyl group, sulfur, nitrogen, etc.) capable of complexation is recorded in the presence of paramagnetic reagents, large changes in the chemical shifts of the protons are produced. For example, proton resonances appearing in a narrow range (2.0–3.0 ppm) can be spread over a broad area (2.0–6.0 ppm). The complex *second-order* spectra can be easily changed into *first-order* spectra, which can be interpreted. Let us first discuss the properties of these paramagnetic salts. Most commercially available shift reagents are paramagnetic chelates of europium (Eu) and ytterbium (Yb). These chelates are

the anions of the β-diketones 2,2,6,6-tetramethyl-3,5-heptanedione and 1,1,1,2,2,3,
3-heptafluoro-7,7-dimethyl-4,6-heptanedione, and camphor derivatives. These complexes are paramagnetic due to the magnetic moment of the unpaired electron. The
addition of these reagents reduces the diamagnetic shielding of all the protons, and then
all signals are shifted to lower field. Therefore, these compounds are called shift reagents
[71]. They are soluble in CCl_4 and $CDCl_3$ and thus can be added directly to the solution
being observed by NMR. The shifts depend on the proximity of the proton to the
complexing center. The chemical structures of some commonly used paramagnetic shift
reagents are given below.

Eu(dpm)₃

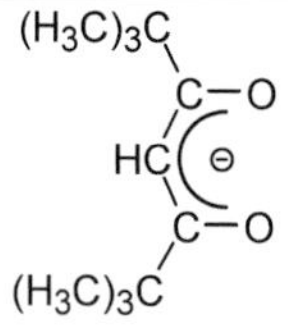

dpm: 2,2,6,6-tetramethyl-3,5,heptanedione

Eu(fod)₃

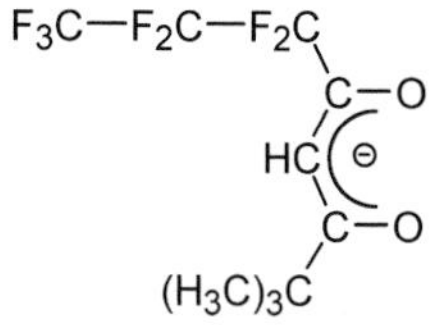

fod:6,6,7,7,8,8,8-heptafluoro-2,2-dimethyl-
octa-3,5-dione

Eu(facam)₃

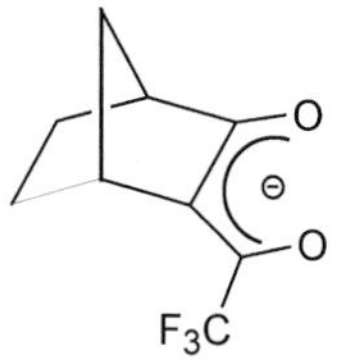

facam: Trifluoroacetyl-D-camphor

Eu(hfbc)₃

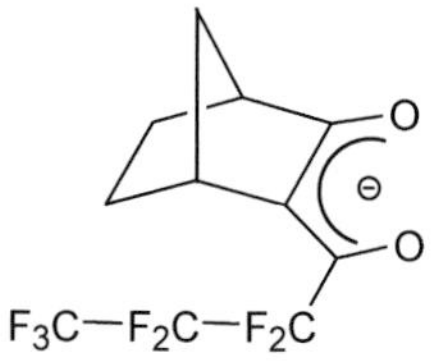

hfbc: 3-heptafluorobutyryl-D-camphor

By the mixing of, for example, an alcohol with a shift reagent, a reversible complex is
formed, as shown below. The equilibrium is very fast on the NMR time scale; we observe
a time-averaged spectrum.

$$Eu(fod)_3 + R\text{--}OH \rightleftharpoons \begin{array}{c} R \\ \diagdown \\ O\text{--}Eu(fod)_3 \\ \diagup \\ H \end{array}$$

As increased amounts of shift reagent are added, the above equilibrium is progressively
shifted to the right. Consequently, the chemical shifts move more to low field.

Before running an NMR spectrum with the shift reagents, a normal NMR spectrum
without shift reagent has to be recorded. As the next step, several spectra are recorded by
adding different amounts of shift reagents. The chemical shift differences of the
individual protons are analyzed in connection with the concentration of the added shift

reagents. It is useful to plot, in a single diagram, the chemical shifts for all the signals against the concentration of the added shift reagents. The proton closer to the complexing center will be the most affected, and its resonance will move more to low field. To understand this effect better, consider an application of paramagnetic salts to a molecule.

The ^{1}H-NMR spectrum of *n*-butanol is given in Figure 103. The methylenic protons bonded to carbon atoms C_2 and C_3 resonate at 1.3–1.7 ppm as a multiplet, whereas the methyl protons and C_1 methylenic protons directly attached to oxygen resonate separately at 0.95 and 3.65 ppm, respectively. The expanded signals for the C_2 and C_3 methylenic protons are given on the right side of the spectra. The NMR spectra recorded

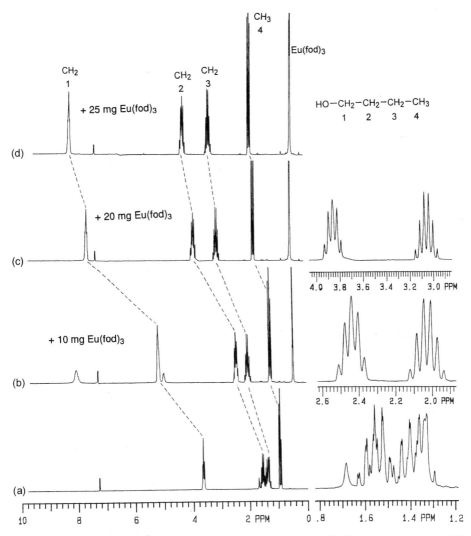

Figure 103   (a) 200 MHz ^{1}H-NMR spectrum of *n*-butanol in CDCl$_3$ and expansions of the methylenic proton signals. (b) Recorded spectrum after adding 10 mg Eu(fod)$_3$, (c) 20 mg Eu(fod)$_3$, (d) 25 mg Eu(fod)$_3$ and expansions of the methylenic proton signals.

after the addition of the shift reagent $Eu(fod)_3$ in amounts of 10, 20, and 25 mg are recorded. These spectra clearly indicate that all of the signals are shifted to lower field and at the same time all CH_2 peaks become separated. The methylenic protons most affected are the C_1 protons. The shifts increase with the proximity of the protons to the oxygen atom of alcohol, which is the complexing center to the shift reagent. Now we can understand why C_1 methylene protons are affected the most. These protons are under the strong influence of the paramagnetism generated by the shift reagent. Since the methyl protons are far from the complexing center, they are less affected. Methylene protons C_2 and C_3 are now especially well resolved, and they resonate as a quintet and sextet, as expected from the *first-order* spectra.

This example demonstrates that the overlapping signals can be separated and the spectrum can be changed to the *first-order* spectra upon treatment with the shift reagents. This method is not restricted to resolving the overlapped signals. In most cases, this method can also be successfully used to determine the exact configuration (*cis/trans*, *endo/exo*, *syn/anti*, etc.) in isomeric molecules.

The chemical shift difference caused by the shift reagents depends on the distance between the affected proton and the complexing center. The interaction between those centers is transmitted through space. The resulting shift can be described by the following equation:

$$\Delta\delta = K \frac{3 \cos^2 \Theta_a - 1}{r^3}$$

where K is the the magnetic moment of lanthanide, r is the distance between the paramagnetic ligand and the proton being observed, and Θ_a is the angle between the magnetic axes and the line L–H.

As one can see from the above equation, the chemical shift differences of the observed protons are inversely proportional to the distance r^3.

For example, the configuration (*endo* or *exo*) of alcohol **125** can be determined exactly.

125

If this compound has an *exo* configuration, as shown above, the most affected proton chemical shift will be that of the bridge proton H_a. This observation will help to assign the correct configuration.

The development of the high-resolution NMR spectrometer has reduced the application of the shift reagents. Nonetheless, one can always find examples where the shift reagents can be successfully used for different kinds of problem. For example, the ¹H-NMR spectrum of biphenylenequinone **126** consists of two separate singlets belonging to olefinic and aromatic protons (Figure 104).

However, there are two sets of different aromatic protons, which should resonate as an AA'BB' system. Aromatic protons appear accidentally as a singlet at 7.75 ppm. From a singlet we cannot determine the coupling constants between protons H$_A$ and H$_B$. Those coupling constants may be important in questions concerning the electronic configuration of the aromatic ring, such as whether or not there is any bond fixation. However, singlet resonances do not allow us to determine the coupling constants between the aromatic protons.

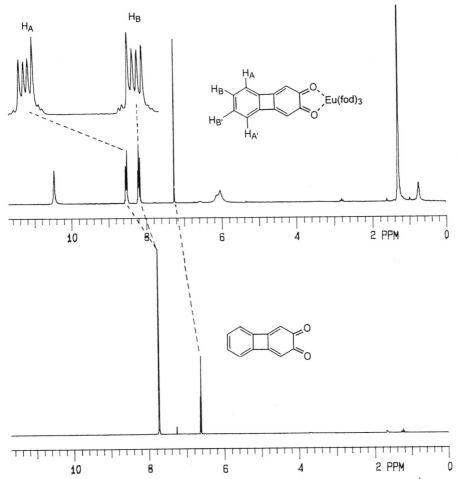

Figure 104   (a) 200 MHz ¹H-NMR spectrum of 2,3-biphenylenequinone in CDCl₃. (b) ¹H-NMR spectrum recorded after addition of Eu(fod)₃ and expansions of the aromatic proton resonances.

126   **127**

The ^{1}H-NMR spectrum of this quinone has been recorded in the presence of a shift reagent (Eu(fod)$_3$) [19a]. Since the distances between protons H$_A$ and H$_B$ and the complexing center are different, these proton signals are now separated through the effect of the shift reagent, the shift effect being greatest for the H$_A$ protons. There is a chemical shift difference between the aromatic protons. Consequently, their resonances appear now as an AA′BB′ system, which can be analyzed. It is likely that this AA′BB′ system cannot be observed, even by the application of a higher magnetic field. This example shows how successful the shift reagents can be when applied to special systems.

128   **129**

We have pointed out that we can differentiate between the possible isomers by using the shift reagents. For example, the isomeric dihydroazulenones **128** and **129** cannot be differentiated from each other by the simple ^{1}H- and ^{13}C-NMR spectra. We can, however, easily distinguish between these compounds by using the shift reagents.

Figure 105 shows the spectrum of **128** (or **129**). The ^{1}H-NMR spectra have been recorded after adding Eu(fod)$_3$. In the upper spectrum, the concentration of added shift reagent has been increased. It can be seen that all of the signals are shifted to lower field. The shift reagent will form a coordination complex with the carbonyl group. If this compound has structure **128**, then the proton most affected in the seven-membered ring will be the olefinic proton H$_2$. However, when the compound has structure **129**, then the methylene protons in the seven-membered ring will be the most affected. The measurements show that the olefinic proton signal is the signal shifted to the least field. This observation is only in agreement with structure **128**.

With these three examples we have attempted to provide a short overview of the shift reagents and their applicability. In cases where the structure or the correct configuration cannot be determined, one should always keep in mind whether or not the shift reagents can be applied.

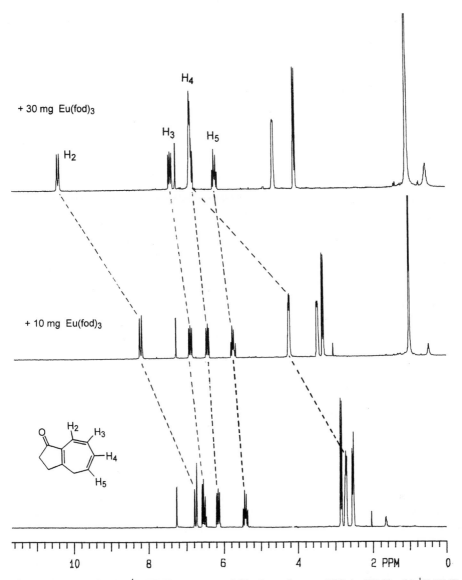

Figure 105 (a) 200 MHz ^{1}H-NMR spectrum of dihydroazulenone (**128**) in CDCl$_3$. (b) ^{1}H-NMR spectra recorded after addition of 10 and 30 mg of Eu(fod)$_3$.

7.2 DOUBLE RESONANCE EXPERIMENTS

The second technique for simplifying NMR spectra is the spin–spin decoupling, or double resonance, experiment. The underlying principle of double resonance is the effective removal of some couplings from NMR spectra. To explain this technique, let us go back to the simple spin–spin coupling process. An AX system is the simplest example of spin–spin coupling. We consider two coupled protons H_a and H_x with a large chemical shift difference. Both of these protons resonate as a doublet.

$$-\overset{|}{\underset{|}{C}}-\overset{|}{\underset{|}{C}}-$$
$$\quad H_a \quad H_x$$

When discussing the coupling mechanism (Chapter 4) we saw that a different alignment of the magnetic moment of the neighboring proton causes a splitting in the signal of the resonating proton. Under normal NMR recording conditions, the field sweep method is used, in which the protons are brought into resonance one by one. In the pulse method all of the protons in the sample are simultaneously excited by a radiofrequency pulse. In double resonance experiments, a particular proton (or set of protons) is irradiated with a strong continuous wave frequency at its resonance frequency ν_2 while the other proton is observed with the conventional ν_1 pulse. Since the samples are irradiated from two different sources, this process is called a double resonance experiment. To carry out a double resonance experiment (for an AX system), we first determine the exact resonance frequency of proton H_a (or H_x). Then we irradiate the system at exactly the resonance frequency ν_a from a second source. This process is called irradiation. The irradiation at ν_a induces transitions between the two spin states of proton H_a. If sufficient irradiating power is applied, proton H_a flips back and forth between two energy levels. A rapid exchange between the two energy levels will take place so that proton H_x can no longer distinguish the separate orientations (parallel and antiparallel) of H_a and 'sees' only a time average orientation of H_a. As a consequence, a single resonance line (a singlet) results rather than a doublet. In other words, when we irradiate the system at resonance frequency ν_a, proton H_x resonates as a singlet. In the same manner, if we irradiate at resonance frequency ν_x, proton H_a will also resonate as a singlet (Figure 106).

The decoupling of a simple AX system such as this is of no diagnostic value by itself; the double resonance experiment is of great use in instances where the spectra are relatively complex and it may be quite difficult to determine which nuclei are spin coupled. In such cases the removal of the coupling will simplify the spectrum and give us an idea about the coupled protons, which will help in determining the molecular structure.

The ^{1}H-NMR spectrum of methyl crotonaldehyde is given in Figure 107. The olefinic protons H_2 and H_3 resonate as a complex AB system where the A and B parts of this system have further couplings to the aldehyde and methyl protons. These protons have been irradiated separately at their resonance frequencies and the changes in the spectrum have been observed. Upon irradiation at the resonance frequency of the aldehyde proton there is no change (Figure 107, spectrum c) in the A-part of the AB system. However, decoupling causes a doublet of doublets of quartets (the B-part at 6.1 ppm) to collapse to a doublet of quartets. This observation provides evidence that the aldehyde proton has direct coupling to proton H_2, resonating at high field. Furthermore, we determine that the olefinic proton (H_2) next to the aldehyde group resonates at higher field. The second spectrum (Figure 107, spectrum b) shows the irradiation experiment at the resonance frequency ν_{CH_3} of the methyl protons. Decoupling the doublet (actually a doublet of doublets) at 2.0 ppm partially collapses the doublet of doublets of quartets (the B-part of

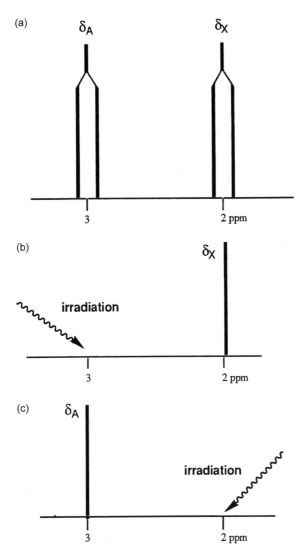

Figure 106 (a) Representation of a simple AX system. (b) Irradiation at the resonance frequency of δ_A. (c) Irradiation at the resonance frequency of δ_X.

the AB system) at 6.8 ppm to a doublet of doublets, and the doublet of quartets (A-part of AB system) to a doublet. In both parts of the AB system, the quartet couplings arising from methyl protons are removed and the spectrum is highly simplified. This experiment clearly shows that methyl protons have couplings to both olefinic protons H_2 and H_3. In the high-field part, small splittings are removed. We can assume from this observation that the methyl protons have *long-range* couplings to the B-part of the AB system.

   In the next example we will discuss the ^{1}H-NMR spectrum of compound **130**, which has a complex structure. We will demonstrate the application of the double resonance experiments for the determination of the molecular structure.

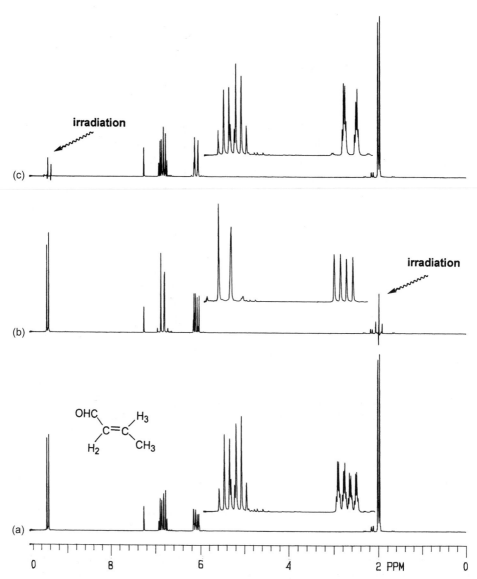

Figure 107   (a) 200 MHz ^{1}H-NMR spectrum of crotonaldehyde in CDCl$_{3}$. (b) The recorded NMR spectrum during irradiation at the resonance frequency of the methyl group and expansions of the resonance signals. (c) The recorded NMR spectrum during irradiation at the resonance frequency of the aldehyde proton resonance frequency and expansions of the resonance signals.

130

Let us first discuss the simple ^{1}H-NMR spectrum (Figure 108A) [72]. There are three separate AB systems. The doublet at δ 9.55 (not shown in the spectrum) belongs to the aldehyde proton and shows the presence of a single proton next to the aldehyde group. The AB pattern in the olefinic region (6.0–7.0 ppm) arises from the olefinic protons. The low-field part of this signal represents olefinic proton H_I (α,β-enone) and the high-field part H_K. Both parts of the AB system are further split into doublets by the adjacent protons H_A and H_L. The fine splitting observed in the high-field part (δ 6.29) arises from the allylic coupling between protons H_A and H_K. The extracted main coupling of this AB system is $J = 15.8$ Hz, indicating the *trans* configuration of the double bond. The methylenic protons H_B and H_C of the five-membered ring resonate at 1.77 and 2.46 ppm as an AB system. The low-field part (the A-part, H_C) is further split into a doublet due to the coupling with proton H_A. Since the dihedral angle between protons H_C and H_D is near 90°, there is only a small splitting in the signal lines of proton H_C. High-field signals (1.77 ppm) appear as a doublet of triplets. Doublet splitting is due to the geminal coupling of protons H_B and H_C. Triplet splitting arises from the coupling between the methylenic proton H_B and the vicinal protons H_A and H_D. In this particular case, the couplings J_{AB} and J_{BD} are accidentally equal. The third AB system appears at 2.97–3.10 ppm and it belongs to the four-membered ring methylene protons. Analysis of this system reveals a geminal coupling constant of $J_{EF} = 19.1$ Hz, which is in complete agreement with the suggested structure. This large coupling can be explained only by the presence of a carbonyl group (π-bond). We have already explained that the π-bonds in the α-position increase the magnitude of the geminal coupling constants. Both parts of the AB system are further split into triplets. These triplet splittings are caused by vicinal coupling with proton H_G and *long-range coupling* with proton H_D. The remaining protons, H_G, H_A, and H_D, resonate at 5.06, 4.81, and 3.91 ppm, respectively. To secure this structure and the above-mentioned assignments, some double resonance experiments have been carried out. More accurately, the structure has been determined after recording the double resonance spectra illustrated in Figure 108A and B. Irradiation at the resonance position of H_G at 5.06 ppm removes the triplet splitting in the AB system lines at 2.97–3.10 ppm and leaves an AB system with doublet splitting. This decoupling experiment strongly supports the vicinal position of proton H_G to the methylenic protons H_E and H_F. Furthermore, the multiplet at δ 3.91 (H_D) collapses to a quintet, which also supports the structure. Decoupling the multiplet at δ 4.81 (H_A) causes the doublet of doublets (the A-part of the olefinic resonances) at δ 6.80 (H_I) to collapse to a doublet, and the doublet of doublets of doublets at δ 6.29 (H_K) to a doublet of doublets. Furthermore, the signal lines of the AB system at δ 1.76–2.46 are simplified. This experiment secures the position of proton H_A, which is then located between the double bond and the methylenic protons H_B and H_C. Decoupling the aldehyde proton (the aldehyde signal is not shown in

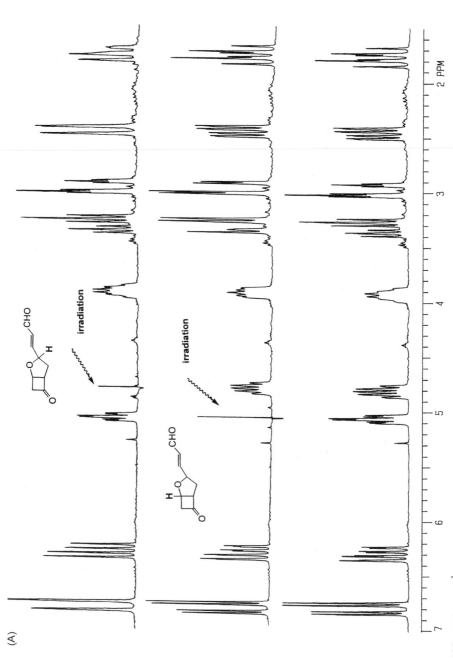

Figure 108 200 MHz ^{1}H-NMR spectrum of the aldehyde (**130**) and the double resonance experiments obtained after irradiation at resonance frequencies of different protons.

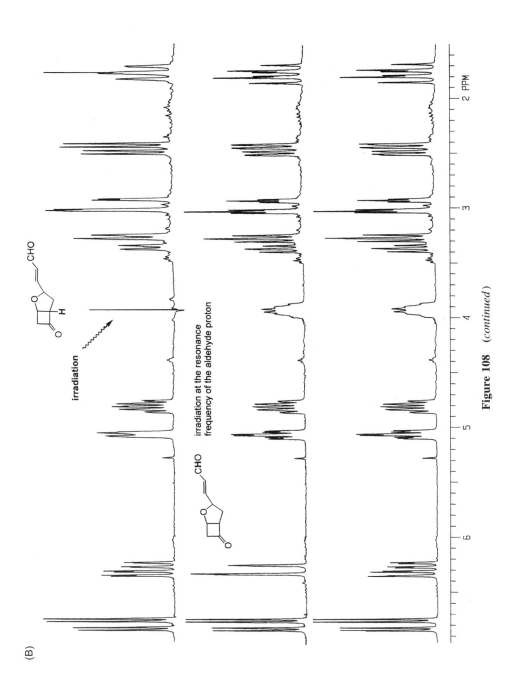

Figure 108   (*continued*)

the spectrum) at δ 9.55 (H_L) collapses the lines at δ 6.29 (H_K) to doublets, which contain small *long-range* coupling. Decoupling the proton at δ 3.91 (H_D) causes changes in many signals (Figure 108B, upper spectrum). Because of the simplification of many resonance signals caused by irradiation of proton H_D, we recognize that this proton is coupled to protons H_B, H_C, H_G and H_E, as well as to H_F.

Protons can be readily decoupled provided that their resonance frequencies are more than 100 Hz apart. On a 60 MHz NMR instrument the chemical shift difference of the protons to be decoupled has to be at least 1 ppm. However, on a 400 MHz instrument the protons with a chemical shift difference of 0.2–0.3 ppm can be successfully decoupled.

The technique of double resonance can be extended by the use of two audio oscillators to accomplish the irradiation of two different protons simultaneously. For example, a multiplet caused by coupling to two sets of nonequivalent protons can be completely collapsed into a singlet by the irradiation of the coupled proton frequencies, allowing a further simplification of the spectra. This process is called *triple resonance*.

In addition to spectrum simplification, double resonance experiments provide very important information about the connection of the protons, so that this technique can always be used in the analysis of highly complex spectra.

Nowadays, double resonance experiments are generally replaced by the two-dimensional measurement technique, COSY (see Chapter 16). However, if one is interested in measuring the size of coupling constants in a multiplet or complex lines, then double resonance experiments are recommended.

– 8 –

Dynamic NMR Spectroscopy

NMR spectroscopy is not only a spectroscopic method of determining the chemical structure of the unknown compounds; at the same time, it is also a powerful tool for observing the dynamic processes that may be occurring within or between molecules: bond rotation about bond axes, ring inversion, and tautomerism (intramolecular and intermolecular exchange of nuclei between functional groups). All of these dynamic processes result in changes of the chemical environment. These changes appear in NMR spectra as changes in chemical shifts and coupling constants. The most obvious way of altering the rate of these dynamic processes is to alter the temperature. With variable temperature probes as standard accessories to modern NMR spectrometers, such experiments are easily performed. With these experiments very important physical parameters can be determined, such as the rate of the dynamic processes, and the activation parameters (E_a, $\Delta H^{\ddagger}$, $\Delta S^{\ddagger}$, and $\Delta G^{\ddagger}$) of equilibrating systems. The appearance of the NMR spectra of an equilibrating system is a function of the rate of interconversion of the molecule.

To determine the existence of a dynamic process between the two molecules A and B, the NMR spectra of this system are recorded at different temperatures. Then the observed changes in the spectra are the subject of interpretation.

$$A \rightleftharpoons B$$

By lowering the temperature, the internal dynamic processes are slowed down, and by increasing the temperature they are accelerated. Let us assume that the activation energy of an interconverting system ($A \rightleftharpoons B$) is 25 kcal/mol. Components A and B can be separately observed by NMR spectroscopy at room temperature. On raising the temperature the activation barrier is overcome, and if the rate of the interconverting becomes sufficiently rapid, compounds A and B can no longer be distinguished by NMR spectroscopy. Only one signal is then observed in the spectrum. There are some dynamic processes in which the activation barriers are much lower. For example, the activation barriers for the rotation about the C–C single bond of the substituted ethanes are between 5 and 15 kcal/mol. This kind of 'fast process' can be observed directly on cooling the system. Most spectrometers allow the NMR spectra to be measured in a range of +200 to − 150 °C.

8.1 BASIC THEORIES [73]

Let us assume a system (A $\rightleftharpoons$ B) that is relatively fast on the NMR time scale. This process can be an electrocyclic reaction: a cycloheptatriene–norcaradiene system (56/57) or the ring inversion of cyclohexanes (131/132). Another example is keto–enol tautomerism (133/134), which involves intramolecular proton transfer from one atom to another.

56

57

131

equatorial

132 X

axial

133

134

In an NMR experiment, the systems shown above have two separate signals if interconversion between these systems is slow. In a fast reaction, in which we have a fast dynamic equilibrium, we describe two rate constants k and k' for the forward and reverse reactions. Of course, the concentrations of A and B will be different (although they may be equal accidentally). The concentrations of A and B are described with the mole fractions n_A and n_B.

$$A_{n_A} \underset{k''}{\overset{k}{\rightleftharpoons}} B_{n_B}, \qquad n_A + n_B = 1$$

where n_A and n_B are the mole fractions of A and B, respectively.

This equilibrium can be shifted to the left as well as to the right. The position of the equilibrium is determined by ΔG, the free energy of the process.

$$\frac{n_A}{n_B} = e^{-\Delta G/RT} \tag{40}$$

The rate constant of the interconversion is determined by the well-known *Eyring* equation.

$$k = \frac{RT}{Nh} e^{-\Delta G^{\#}/RT} \tag{41}$$

where $\Delta G^{\#}$ is the free activation energy, N is the Loschmidt number, and h is the Planck constant.

We have to consider two different cases:

Slow exchange: As we have discussed above, if the rate of the interconversion of A and B is slow on the NMR time scale, then we will observe separate signals for A and B. The measurement of the relative intensities of the signals will directly give the mole fractions n_A and n_B and, therefore, ΔG.

Fast exchange: If the rate of interconversion is fast, we will observe an average NMR spectrum in which the position of the signal will be determined by the mole fractions n_A and n_B. The chemical shift of the signal is given by the following equation:

$$\nu_{obs} = n_A \nu_A + n_B \nu_B \tag{42}$$

Since

$$n_A + n_B = 1$$

we have

$$\nu_{obs} = n_A \nu_A + (1 - n_A)\nu_B$$

With the aid of this equation, the mole fractions of the interconverting system can be determined easily provided that we know the exact chemical shifts of A and B, which can be determined by freezing the system.

Figure 109 shows the temperature-dependent NMR spectra of an interconverting system. We assume that the components A and B resonate as singlets. In the range of

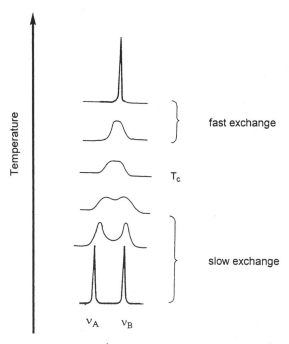

Figure 109 Variable temperature ^{1}H-NMR spectra of an equilibrating system A $\leftrightharpoons$ B.

slow exchange we can separately observe the individual compounds A and B resonating as singlets. However, by raising the temperature, the NMR spectrum will change. When the barrier of this equilibrium is overcome, and the rate of interconverting accelerates, the signals start to broaden in the intermediate temperature range, finally collapsing into a single line. The temperature at which the individual resonance lines merge into a broad resonance line is referred to as the *coalescence temperature*. For this *coalescence temperature* T_C the rate constant of this interconverting system is given by the following equation:

$$k_{T_C} = \frac{\pi}{\sqrt{2}}(\Delta \nu) = \frac{\pi}{\sqrt{2}}(\nu_A - \nu_B) \qquad (43)$$

Here $\Delta \nu$ is the difference in hertz between the two signals in the absence of exchange. This equation shows that the rate constant at the coalescence temperature depends only on the chemical shift difference $\Delta \nu$. Since this difference varies with the strength of

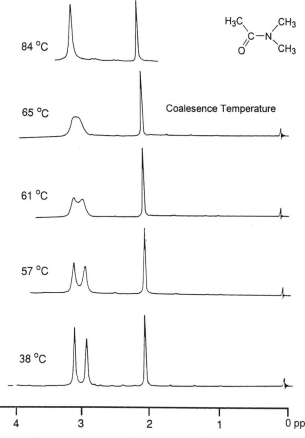

Figure 110 60 MHz ^{1}H-NMR spectra of dimethyl acetamide, recorded at different temperatures. (Reprinted with permission of John Wiley & Sons, Inc. from R.J. Abraham and P. Loftus, *Proton and Carbon-13 NMR Spectroscopy: An Integrated Approach*, 1978.)

the magnetic field, it is expected that the *coalescence temperature* for a given system will vary with the strength of the magnetic field. By replacing k in the *Eyring equation* (eq. 41) we obtain the following equation:

$$\frac{\pi}{\sqrt{2}}(\nu_A - \nu_B) = \frac{RT_C}{Nh}e^{-\Delta G^{\#}/RT_C} \tag{44}$$

and

$$\Delta G^{\#} = RT_C \ln\frac{RT_C\sqrt{2}}{\pi Nh(\nu_A - \nu_B)} \tag{45}$$

Measurement of T_C and the resonance frequencies in hertz can then provide the free activation energy $\Delta G^{\#}$:

$$\Delta G^{\#} = 19 \times 10^{-3}T_C(9.97 + \log T_C - \log(\nu_A - \nu_B)) \tag{46}$$

The temperature-dependent ^{1}H-NMR spectra of dimethylformamide **135** are given in Figure 110. The C–N bond between the carbonyl group and the nitrogen atom has a significant double bond character. Rotation of the dimethylamino group is restricted at room temperature. The protons of the two methyl groups are in different chemical environments and therefore resonate at different frequencies.

When the temperature is raised, the high energy barrier of rotation is overcome, and then two methyl groups exchange position (*cis* and *trans* to carbonyl group) so fast that they cannot be distinguished by NMR spectroscopy. At first, the signals broaden and finally, at temperatures above 85 °C, coalesce to a single line. It is interesting to observe that the acetyl methyl resonance remains sharp during the line broadening of the amide methyl resonances. At the coalescence temperature, the rate constant of this dynamic process can easily be determined by using Eq. (43).

8.2 EXERCISES 61–101

Determine the structure of the compounds whose NMR spectra and molecular formulae are given below.

All the spectra (61–101) given in this section are reprinted with permission of Aldrich Chemical Co., Inc. from C.J. Pouchert and J. Behnke, *The Aldrich Library of 13C and 1H FT-NMR Spectra*, 1992.

61.

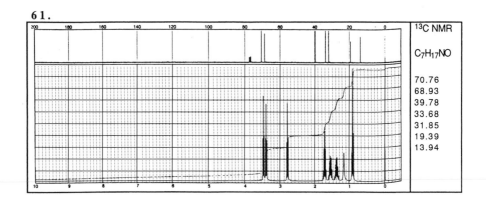

¹³C NMR

$C_7H_{17}NO$

70.76
68.93
39.78
33.68
31.85
19.39
13.94

62.

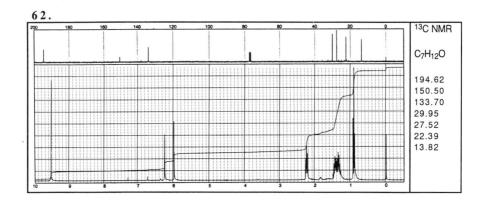

¹³C NMR

$C_7H_{12}O$

194.62
150.50
133.70
29.95
27.52
22.39
13.82

63.

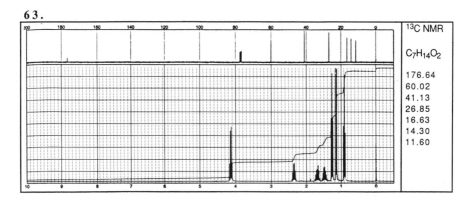

¹³C NMR

$C_7H_{14}O_2$

176.64
60.02
41.13
26.85
16.63
14.30
11.60

64.

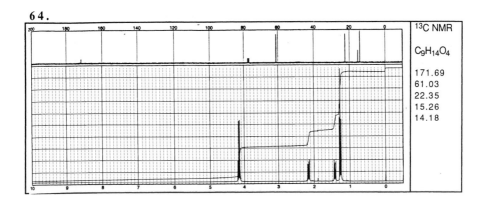

¹³C NMR

C$_9$H$_{14}$O$_4$

171.69
61.03
22.35
15.26
14.18

65.

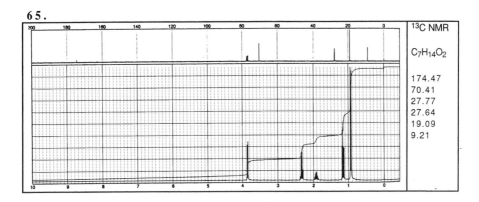

¹³C NMR

C$_7$H$_{14}$O$_2$

174.47
70.41
27.77
27.64
19.09
9.21

66.

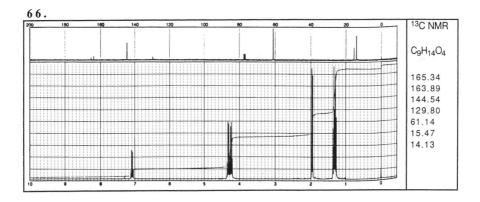

¹³C NMR

C$_9$H$_{14}$O$_4$

165.34
163.89
144.54
129.80
61.14
15.47
14.13

67.

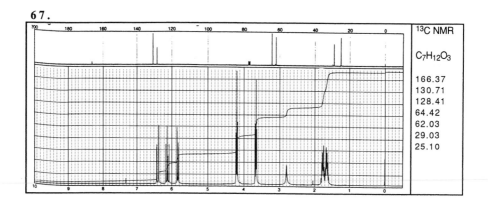

¹³C NMR

$C_7H_{12}O_3$

166.37
130.71
128.41
64.42
62.03
29.03
25.10

68.

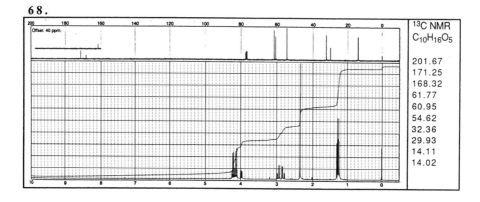

¹³C NMR
$C_{10}H_{16}O_5$

201.67
171.25
168.32
61.77
60.95
54.62
32.36
29.93
14.11
14.02

69.

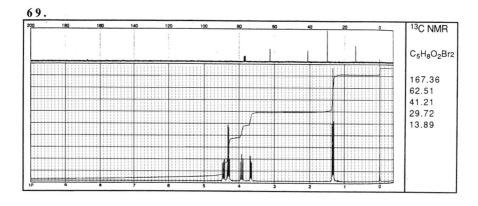

¹³C NMR

$C_5H_8O_2Br2$

167.36
62.51
41.21
29.72
13.89

70.

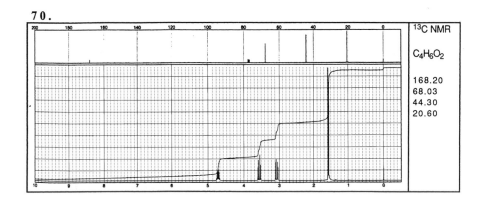

¹³C NMR

$C_4H_6O_2$

168.20
68.03
44.30
20.60

71.

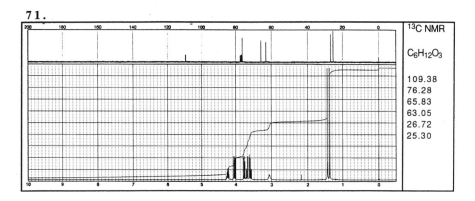

¹³C NMR

$C_6H_{12}O_3$

109.38
76.28
65.83
63.05
26.72
25.30

72.

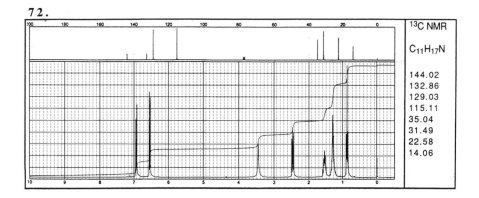

¹³C NMR

$C_{11}H_{17}N$

144.02
132.86
129.03
115.11
35.04
31.49
22.58
14.06

7 3.

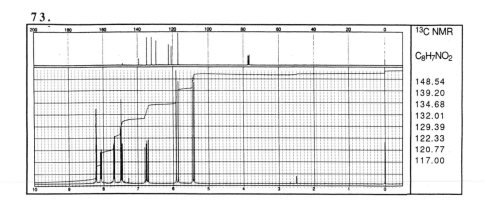

```
                                    ¹³C NMR

                                    C₈H₇NO₂

                                    148.54
                                    139.20
                                    134.68
                                    132.01
                                    129.39
                                    122.33
                                    120.77
                                    117.00
```

7 4.

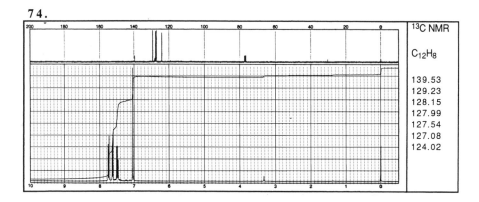

```
                                    ¹³C NMR

                                    C₁₂H₈

                                    139.53
                                    129.23
                                    128.15
                                    127.99
                                    127.54
                                    127.08
                                    124.02
```

7 5.

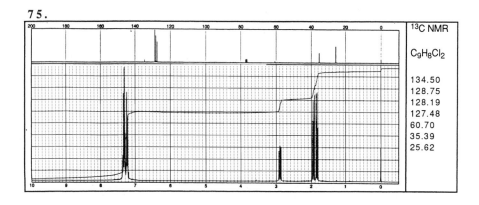

```
                                    ¹³C NMR

                                    C₉H₈Cl₂

                                    134.50
                                    128.75
                                    128.19
                                    127.48
                                    60.70
                                    35.39
                                    25.62
```

76.

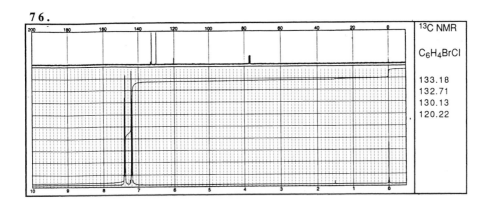

¹³C NMR

C₆H₄BrCl

133.18
132.71
130.13
120.22

77.

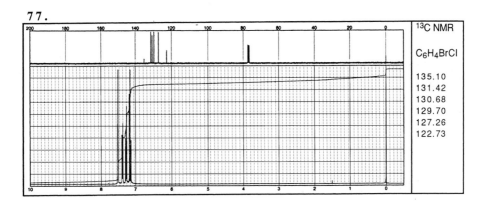

¹³C NMR

C₆H₄BrCl

135.10
131.42
130.68
129.70
127.26
122.73

78.

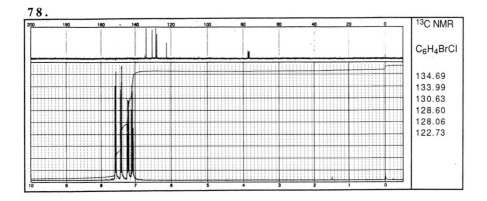

¹³C NMR

C₆H₄BrCl

134.69
133.99
130.63
128.60
128.06
122.73

79.

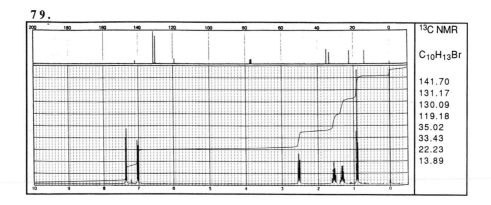

¹³C NMR

C₁₀H₁₃Br

141.70
131.17
130.09
119.18
35.02
33.43
22.23
13.89

80.

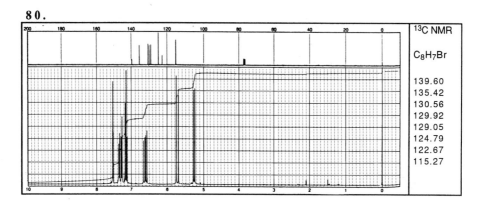

¹³C NMR

C₈H₇Br

139.60
135.42
130.56
129.92
129.05
124.79
122.67
115.27

81.

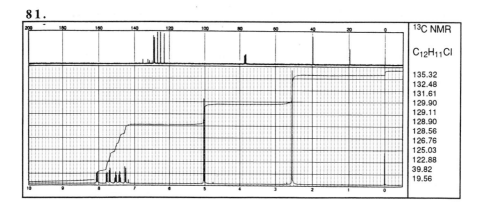

¹³C NMR

C₁₂H₁₁Cl

135.32
132.48
131.61
129.90
129.11
128.90
128.56
126.76
125.03
122.88
39.82
19.56

82.

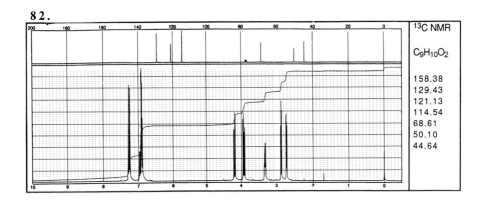

¹³C NMR

C₉H₁₀O₂

158.38
129.43
121.13
114.54
68.61
50.10
44.64

83.

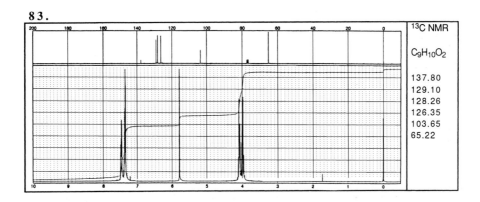

¹³C NMR

C₉H₁₀O₂

137.80
129.10
128.26
126.35
103.65
65.22

84.

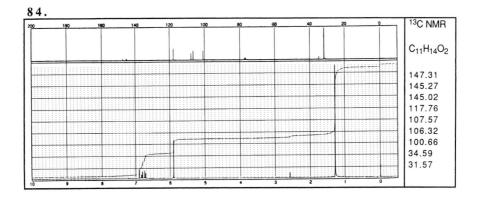

¹³C NMR

C₁₁H₁₄O₂

147.31
145.27
145.02
117.76
107.57
106.32
100.66
34.59
31.57

85.

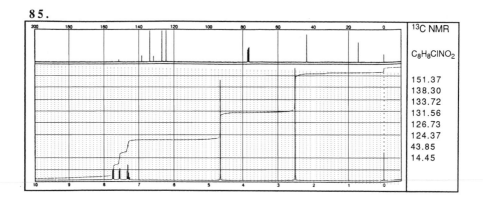

¹³C NMR

C₈H₈ClNO₂

151.37
138.30
133.72
131.56
126.73
124.37
43.85
14.45

86.

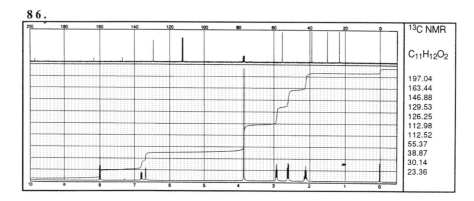

¹³C NMR

C₁₁H₁₂O₂

197.04
163.44
146.88
129.53
126.25
112.98
112.52
55.37
38.87
30.14
23.36

87.

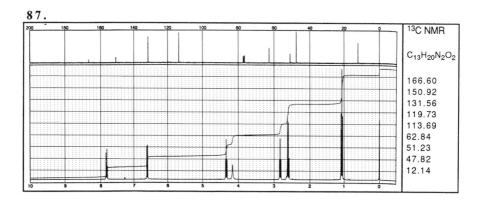

¹³C NMR

C₁₃H₂₀N₂O₂

166.60
150.92
131.56
119.73
113.69
62.84
51.23
47.82
12.14

88.

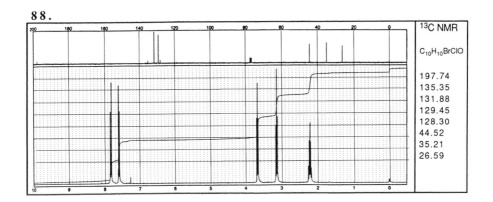

¹³C NMR

C₁₀H₁₀BrClO

197.74
135.35
131.88
129.45
128.30
44.52
35.21
26.59

89.

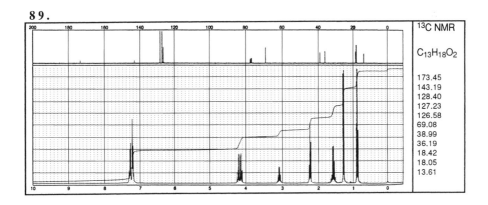

¹³C NMR

C₁₃H₁₈O₂

173.45
143.19
128.40
127.23
126.58
69.08
38.99
36.19
18.42
18.05
13.61

90.

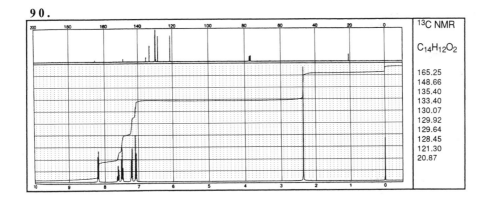

¹³C NMR

C₁₄H₁₂O₂

165.25
148.66
135.40
133.40
130.07
129.92
129.64
128.45
121.30
20.87

91.

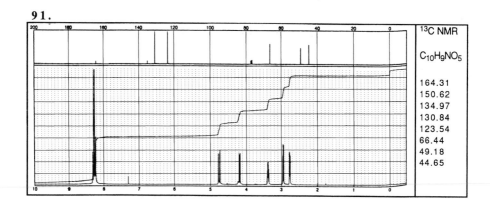

13C NMR

C10H9NO5

164.31
150.62
134.97
130.84
123.54
66.44
49.18
44.65

92.

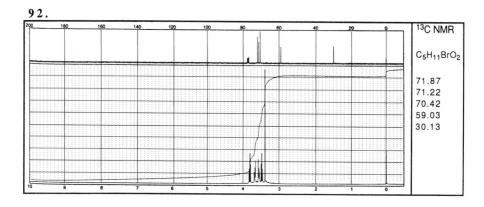

13C NMR

C5H11BrO2

71.87
71.22
70.42
59.03
30.13

93.

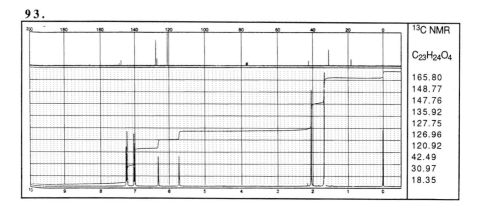

13C NMR

C23H24O4

165.80
148.77
147.76
135.92
127.75
126.96
120.92
42.49
30.97
18.35

94.

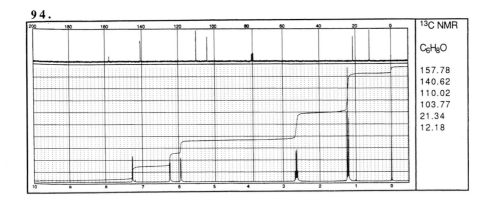

¹³C NMR

C₆H₈O

157.78
140.62
110.02
103.77
21.34
12.18

95.

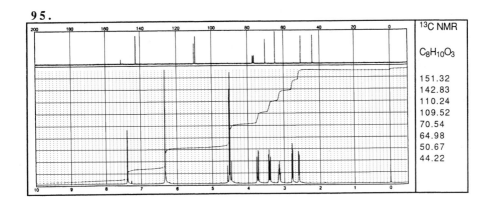

¹³C NMR

C₈H₁₀O₃

151.32
142.83
110.24
109.52
70.54
64.98
50.67
44.22

96.

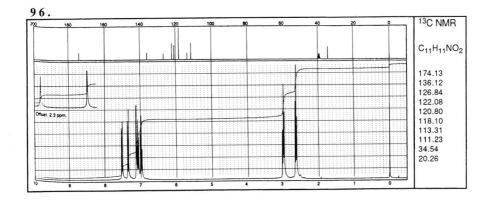

¹³C NMR

C₁₁H₁₁NO₂

174.13
136.12
126.84
122.08
120.80
118.10
113.31
111.23
34.54
20.26

97.

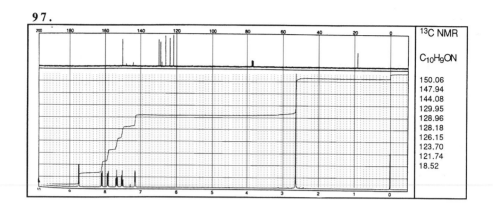

¹³C NMR

$C_{10}H_9ON$

150.06
147.94
144.08
129.95
128.96
128.18
126.15
123.70
121.74
18.52

98.

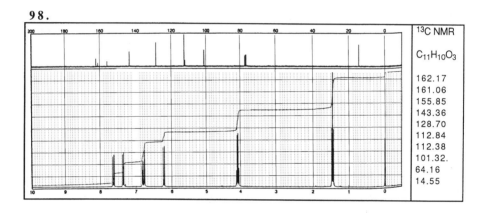

¹³C NMR

$C_{11}H_{10}O_3$

162.17
161.06
155.85
143.36
128.70
112.84
112.38
101.32.
64.16
14.55

99.

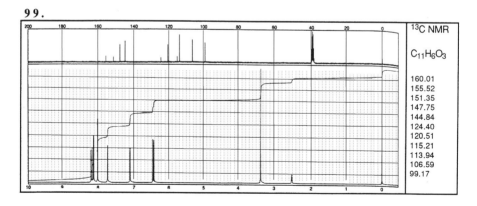

¹³C NMR

$C_{11}H_6O_3$

160.01
155.52
151.35
147.75
144.84
124.40
120.51
115.21
113.94
106.59
99.17

100.

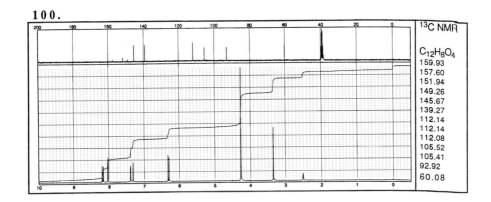

^{13}C NMR

$C_{12}H_8O_4$
159.93
157.60
151.94
149.26
145.67
139.27
112.14
112.14
112.08
105.52
105.41
92.92
60.08

101.

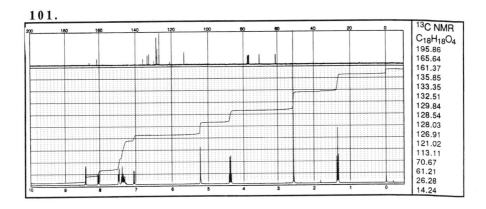

^{13}C NMR
$C_{18}H_{18}O_4$
195.86
165.64
161.37
135.85
133.35
132.51
129.84
128.54
128.03
126.91
121.02
113.11
70.67
61.21
26.28
14.24

Part II

^{13}C-NMR Spectroscopy

– 9 –

Introduction

9.1 DEVELOPMENT OF ^{13}C-NMR SPECTROSCOPY [74]

Although the first NMR signal was observed in 1945, the first structural analysis of a compound was not carried out until 1951. At first, proton resonances were studied. The first ^{13}C-signal was detected by Lauterbur in 1957. The greater problems associated with ^{13}C-NMR, due to the much lower sensitivity of this nucleus in natural abundance compared with the proton, delayed the development of ^{13}C-NMR. In the early 1960s the first ^{13}C-NMR spectra of some organic compounds were recorded. However, the real progress in ^{13}C-NMR spectroscopy started in 1965. A breakthrough in this field came after the introduction of commercial Fourier transform (FT) NMR spectrometers around 1970.

9.2 COMPARISON OF THE ^{1}H AND ^{13}C NUCLEUS

As mentioned above, the development of ^{13}C-NMR came approximately 20 years later than that of the ^{1}H-NMR spectroscopy. The reason can be found in the nature of the ^{13}C nuclei. Before looking at the differences between these nuclei, we have to determine first the common properties of these nuclei.

Any nucleus must have a spin quantum number differing from zero ($I \neq 0$) in order to be active in NMR spectroscopy. Both ^{13}C and ^{1}H nuclei have a common spin quantum number $I = 1/2$.

According to the quantum condition

$$m = (2I + 1) \tag{3}$$

^{13}C nuclei have two ($m = 2$) magnetic quantum numbers:

$$m_1 = +\frac{1}{2} \quad \text{and} \quad m_2 = -\frac{1}{2}$$

This means that there will be two possible orientations of the nuclear spin in an external magnetic field.

9.3   THE FACTORS INFLUENCING THE SENSITIVITY
OF THE ^{13}C NUCLEUS

9.3.1   Natural abundance of the ^{13}C and ^{1}H nucleus

Carbon has two important isotopes in nature:

$$^{12}C = 98.7\%, \qquad ^{13}C = 1.108\%$$

As ^{12}C nuclei have a spin quantum number $I = 0$, they are not accessible to NMR spectroscopy. Since the ^{12}C nucleus has a magnetic quantum number of $m = 1$, the nuclear spin of ^{12}C will only have one possible orientation in an external magnetic field, and there will be no energy separation. Consequently, only ^{13}C nuclei are active in NMR spectroscopy. The low natural abundance of ^{13}C nuclei is the reason why the development of ^{13}C-NMR was delayed for a while. The comparison of the natural abundance of ^{13}C and ^{1}H nuclei clearly indicates that the ^{1}H nucleus is approximately 100 times more sensitive than that of ^{13}C.

9.3.2   Gyromagnetic constants of ^{1}H and ^{13}C nuclei

We have seen in the ^{1}H-NMR section that the gyromagnetic constants are important parameters for NMR phenomena. As one can see from the resonance condition equation:

$$\nu = \frac{\gamma H_0}{2\pi} \tag{13}$$

the gyromagnetic constant determines the resonance frequency of any given nucleus. For example, two different nuclei having different gyromagnetic constants will have different resonance frequencies.

Let us compare the ^{13}C and ^{1}H nucleus:

$$\gamma_{^1H} = 2.674 \times 10^8 \ s^{-1} \ T^{-1}, \qquad \gamma_{^{13}C} = 0.672 \times 10^8 \ s^{-1} \ T^{-1}$$

From the above values we see that the ratio of these gyromagnetic constants is approximately 4:1. According to the resonance condition equation (eq. 13), ^{1}H nuclei need an energy level of 60.00 MHz in a magnetic field of 14,100 G (1.4 T).

$$\nu = \frac{\gamma H_0}{2\pi} = \frac{2.674 \times 10^8 \times 14,100}{2\pi} = 60.00 \ \text{MHz}$$

However, an energy level of 15.00 MHz is sufficient to bring ^{13}C nuclei into resonance in the same magnetic field. This shows us that ^{1}H and ^{13}C nuclei resonate at different regions in a given constant magnetic field. If we talk about a 400 MHz NMR instrument it means that for the resonance of the ^{1}H nuclei we need an energy level of 400.00 MHz. However, ^{13}C nuclei need a lower energy level (100.00 MHz) for the resonance in the same NMR instrument. In summary, gyromagnetic constants determine the resonance frequencies of all nuclei in a given uniform magnetic field.

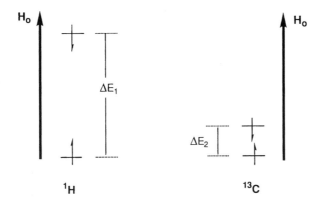

Figure 111   Energy levels for carbon and hydrogen nuclei in a magnetic field H_0.

We have already mentioned that, as in the case of a proton, there are two energy levels for the carbon that will be established in an external magnetic field. According to the Boltzmann distribution law, the population difference between these energy levels depends on two parameters:

$$\frac{N_\alpha}{N_\beta} = e^{-\Delta E/kT} \tag{14}$$

These parameters are temperature T and the energy difference ΔE between the energy levels. For more details see Section 2.3.4. ΔE depends on a uniform magnetic field on the gyromagnetic constant γ. Therefore, the energy difference between two energy levels of protons will be four times more than that of carbon in a given magnetic field. Different energy levels for carbon and protons are illustrated in Figure 111.

Let us attempt to answer the question of how the differences between the energy levels can affect the NMR spectra. The ratio of the populations between two energy levels is given by the Boltzmann equation: The greater the energy difference between the energy levels, the larger the population difference between the nuclei in the upper and lower energy states. The greater the population excess is at the lower energy level, the greater the probability for resonance. If the resonance probability is in some way increased, the signal intensity is also increased. We can come to the following conclusion: The gyromagnetic constant of a given nucleus is directly responsible for the sensitivity of the corresponding nucleus. The ratio between the gyromagnetic constants of protons and carbon is given by the following equation:

$$\gamma_C = 0.2514\gamma_H \tag{47}$$

Since the sensitivity of a nucleus at a constant magnetic field is proportional to the cube of γ, a ^{13}C nucleus gives rise to a $(0.2514)^3 = 0.0159$ less intense signal that a proton nucleus would yield on excitation. That means if the ^{13}C nuclei distribution was as much as that of the proton ^{1}H, the signal intensity of the carbon nuclei would be approximately 63 times less than the intensity of the protons.

9.3.3   Nuclear spin relaxation

In the ^{1}H-NMR section we showed that the spin–lattice relaxation T_1 of protons plays a very important role in restoring the original equilibrium condition among the energy levels. For protons in the usual nonviscous solution, the relaxation times, T_1, are short. However, for carbon nuclei, the relaxation times must be considered since they are longer than those of protons and vary widely. A long spin–lattice relaxation time indicates that the carbon nucleus has no efficient relaxation pathway and is thus easy to saturate. Care must be taken in the NMR experiment to avoid saturation.

All of the following factors

(a)   low natural abundance of ^{13}C nuclei;
(b)   lower gyromagnetic constant γ_C;
(c)   longer relaxation times.

together cause a lowering of the sensitivity by a factor of 5700 in a ^{13}C experiment relative to a ^{1}H experiment.

The greater problems associated with ^{13}C-NMR spectroscopy were not satisfactorily overcome until the introduction of commercial FT spectrometers around 1970 (see the ^{1}H-NMR section, Section 2.5). Since ^{13}C-NMR spectroscopy is a very important tool for organic chemists, these problems with sensitivity had to be solved. In the next section we will briefly discuss what has been done in the past in order to routinely obtain ^{13}C-NMR spectra.

9.4   THE FACTORS INCREASING THE SENSITIVITY IN ^{13}C-NMR SPECTROSCOPY

9.4.1   Increasing the amount of the sample

Since the signal intensity of a sample is directly proportional to the number of the resonating nuclei, the amount of the sample has to be increased. Obviously, if one has unlimited amounts of a sample, then it becomes profitable to use as large a sample tube and consequently sample volume as possible. For ^{13}C measurements, much larger sample tubes (up to 20–25 mm diameter) were used at the beginning. Nowadays, only a 5 mm sample tube is used. Therefore, an increase of the sample concentration in a 5 mm sample tube is limited. On the other hand, high concentration of the sample can cause line broadening (see ^{1}H-NMR part, Section 3.3).

9.4.2   Temperature

As we have discussed in detail (see Section 2.3), sensitivity can be slightly increased by lowering the temperature to where the relative population of the lower energy level will increase, which will subsequently be reflected in the increase in signal intensity. The problems encountered such as the solubility of the sample, increase in the viscosity, and the freezing of some dynamic processes hinder the application of low temperature NMR measurements to increase the peak intensity.

9.4.3 Magnetic field strength

^{13}C-NMR spectroscopy at higher magnetic fields is generally characterized by increased sensitivity. When the magnetic field strength is increased, the energy gap between the two energy levels will also increase (eq. 10), i.e. the number of protons populating the lower energy level will increase. Consequently, resonance probability will increase and a more intense signal will be obtained.

9.4.4 Spectra accumulation

We have shown the lower sensitivity of ^{13}C nuclei and discussed all of the possible parameters affecting the sensitivity and found that there are limitations to increasing the sensitivity. A large increase in sensitivity can be reached by way of spectra accumulation. The spectral region is divided into a number of channels so that during the measurement a corresponding number of data points can be stored. Repeated recording allows the accumulation of thousands of spectra by a digital computer. Since absorption signals are always positive, whereas signals arising from random noise vary in their intensity and especially in their sign (a noise signal can be an absorption signal as well as an emission), the signal-to-noise ratio will be improved upon the accumulation of many spectra. Most noise signals will cancel each other out. This process is known as the CAT (computer-averaged transients) method. The signal-to-noise (S/N) ratio of a spectrum improves with the square root of the number of scans or pulses:

$$\frac{S}{N} = \sqrt{n} \tag{48}$$

where S is the intensity of signal, N is the intensity of noise, and n is the number of scans or pulses.

According to this equation, recording 100 spectra of a sample will increase the signal-to-noise ratio 10 times (not 100 times):

$$\frac{S}{N} = \sqrt{100} = 10$$

However, improving the sensitivity by spectral accumulation of the CW spectra requires long measurement times (for the accumulation of 100 spectra we need approximately $5 \times 100 = 500$ min). Therefore, it is very important to reduce the measurement time of a single spectrum. The availability of pulsed FT instrumentation, which permits simultaneous irradiation of all ^{13}C nuclei, has dramatically reduced the recording time of spectra to an order of $10^3 - 10^4$. This method will be discussed later in Chapter 11.

– 10 –

Absorption and Resonance

10.1 CLASSICAL TREATMENT OF ABSORPTION AND RESONANCE

When we place a magnetic dipole (an atomic nucleus) in a static magnetic field, there will be an interaction between the magnetic moment of the dipole and the external magnetic field. The magnetic dipole in the homogeneous magnetic field H_0 will experience a torsional moment that will force the alignment of the magnetic moment with the direction of the field. The magnetic dipole of the nucleus will attempt to escape this effect and the angular momentum of the nucleus will cause a precessional motion. This means that the magnetic moment of the nucleus in a static magnetic field will not align exactly with the direction of the external magnetic field. There is a dynamic precessional motion (Figure 112). This process can be compared with the motion of a spinning top. An off-perpendicular motion of a spinning top will also cause a precessional motion under the influence of gravity.

The nuclei placed in a static magnetic field will align parallel and antiparallel with the external applied magnetic field. Antiparallel-aligned nuclei also experience a precessional motion just as in the parallel-aligned nuclei. The precessional motion of the parallel- and antiparallel-aligned nuclei is shown in Figure 112.

It is not sufficient to consider just the interaction of a single nucleus with the static magnetic field. We shall now attempt to extend our analysis to a large number of spins, a macroscopic sample. We have previously seen that the number of the nuclei with the parallel orientation with the external magnetic field is slightly in excess than those of the antiparallel-aligned nuclei. Since the nuclear moments do not rotate in a phase, they are randomly distributed over a conical envelope (Figure 113).

The nuclear moments of the nuclei rotate around the applied magnetic field. This is referred to as the *Larmor frequency*. Larmor frequency depends on the strength of the applied magnetic field and intrinsic properties of the nucleus reflected in its gyromagnetic ratio.

$$\omega_0 = \frac{\gamma H_0}{2\pi} \tag{49}$$

where ω_0 is the Larmor frequency, H_0 is the strength of the external magnetic field, and γ is the gyromagnetic constant.

The magnetic moment is a vector and it has two components. One of these components is in the z-direction, the second one is placed in the xy-plane. Since the magnetic moments

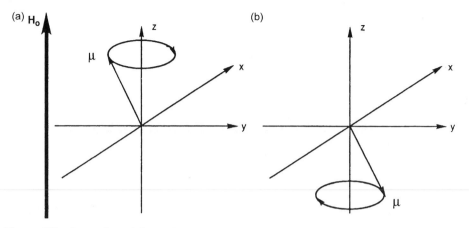

Figure 112 Precession of the nuclear moment μ of a (a) parallel and (b) antiparallel aligned nucleus in the external magnetic field H_0.

are statically distributed over a conical envelope, the xy-components of these nuclei will also be distributed statically on the xy-plane. Therefore, for a very large collection of spins at equilibrium there will be no net magnetization in the xy-plane since these components will cancel each other out:

$$\sum \mu_{x,y} = 0 \tag{50}$$

However, there will be a net magnetization in the z-direction (Figure 114).

Let us summarize: All nuclei in a macroscopic sample will generate a net magnetization in the z-direction due to the precessional motions of these nuclei about the applied magnetic field. However, there will be no magnetization in the xy-plane.

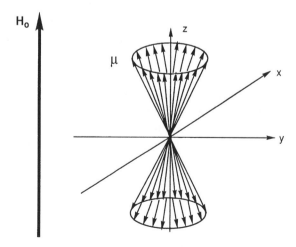

Figure 113 Parallel and antiparallel alignment of the nuclear moments μ of a macroscopic sample in the external magnetic field H_0.

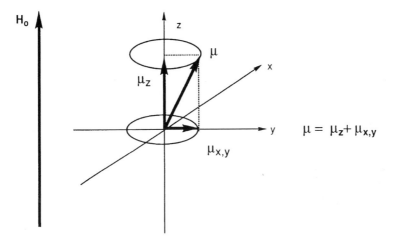

Figure 114 The components μ_z and μ_{xy} of the magnetic moment μ in the z-axis and xy-plane.

We can consider the antiparallel-aligned nuclei in the same manner. They will also have a net magnetization in the z-direction. In the ^{1}H-NMR chapter we have described the orientation of the nuclei in an external magnetic field with files (↑ parallel alignment, ↓ antiparallel alignment). In reality, we describe the orientation of the net magnetization with these files. Considering a macroscopic sample, different direction oriented nuclei will generate magnetization vectors at the opposite directions ($+z$ and $-z$ directions). Because there is a slight excess of the nuclei in the upper cone (parallel-aligned nuclei) than the nuclei in the lower cone (due to the Boltzmann distribution law), this entire collection of vectors will have a resultant magnetic moment M_0, called the *longitudinal magnetization M_0* (Figure 115).

After explaining the precessional motion of the nuclei and the longitudinal magnetization in an external magnetic field, let us move on to the classical resonance phenomena in the CW spectrometer. In order to observe an NMR signal, a sample containing nuclear spins is placed in a static magnetic field H_0. The magnetic moments

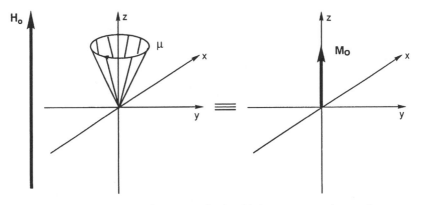

Figure 115 Longitudinal magnetization M_0 in a macroscopic sample.

will generate a net magnetization in the $+z$ direction. To change the orientation of the longitudinal magnetization, we must now provide the necessary torque in the form of a radiofrequency in a coil which is positioned along the x-axis, perpendicular to that of the magnet that lies along the x-axis. For the interaction of this radiofrequency field with the net magnetization vector M_0, this rf field has to have a component in the xy-direction with the Larmor frequency ω_0 in order to interact with M_0. How can a rotating radiofrequency field be generated in the xy-direction? We apply a linear polarized radiofrequency field in the x-direction. As we know from classical physics, a linear polarized (rotating) electromagnetic field can be resolved into two components rotating in opposite directions. One of these components is rotating in the same direction as the precessional orbit of the nuclei. The oppositely rotating magnetic field will be ineffective. When the condition is met, the frequency of the rotating magnetic field is equal to the Larmor frequency of the nuclei, an interaction between the rf field and magnetization M_0 will take place. Then, the net magnetization vector M_0 will be tipped toward the horizontal plane, producing a component in the xy-plane (Figure 116).

After the tipping of the magnetization vector M_0 towards the xy-plane, the magnetization vector M_0 will rotate in the xy-plane about the external magnetic field. The magnetic component thus generated in the xy-plane can be detected by the receiver coil as an NMR signal. We will explain this process later in more detail.

To understand this process better and simplify the complex motions, Trossey has introduced the concept of the *rotating frame*. The problem here is that our rf field is in motion, and while the net magnetization of the sample is static along the z-axis, if it becomes displaced from that axis, it will evidently have precessional motion about the static field. We choose a set of coordinates that rotate along Larmor frequency (Figure 117). In representing static and rotating frames it is customary to give the x- and y-axis different labels. In rotating frame we will label the axes with x' and y'. We will analyze the processes in the rotating frame for better comprehension.

In order to describe a rotating coordinate system, let us give the following example. We consider the motion of a car moving toward us from a fixed point. The car will approach us and then leave. We imagine a second car, which moves with the same speed next to the first car. We are sitting in the second car and observe the motion of the first car

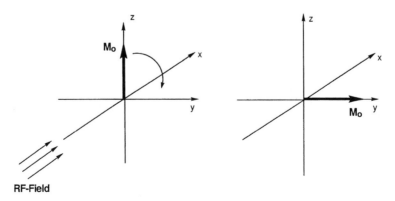

Figure 116 Tipping of the magnetization vector M_0 towards the y-axis after interaction with radiofrequency field.

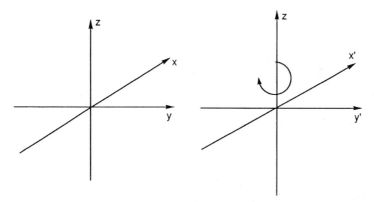

Figure 117 Fixed laboratory coordinate system and rotating coordinate system after Trossey.

from a different point. Since we shall move with the same speed and look at the first car, we shall realize that the second car does not move and is fixed (of course, if we look only at the car). In a rotating coordinate system, the magnetization vector M_y will be now stationary while M_y will rotate in the static coordinate system with the Larmor frequency. In summary, if we watch the motion of the magnetization vector from the y'-direction we notice that the net magnetization will be fixed in the y'-direction (Figure 118).

We call the tipping of the magnetization vector M_0 towards the y'-direction under the influence of the radiofrequency field *resonance*. We have previously explained that the magnetization vector M_0 does not have any component in the xy-plane. However, after the flip or tipping of the magnetization cone under the influence of the radiofrequency field, the net magnetization vector M_0 will have two components in the z and y'-direction, generating a magnetization in the y'-direction for the first time (Figures 119 and 120).

The induction of a magnetization in the y'-direction (which is an electrical current) will be detected as the NMR signal by a receiver. This current will be amplified and displayed as a function of frequency on a recorder (Figure 120). The NMR spectrometers are normally designed so that they can detect any induced magnetization along the

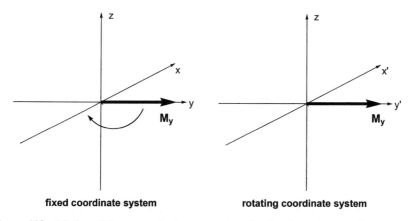

fixed coordinate system   **rotating coordinate system**

Figure 118 Motion of the magnetization vector in a fixed and rotating coordinate system.

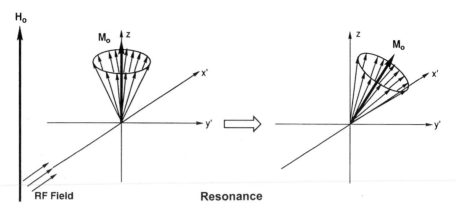

Figure 119   (a) Interaction of the magnetization vector M_0 with the radiofrequency field. (b) Tipping of the magnetization vector M_0 towards the y'-direction under the influence of the radiofrequency field (*resonance*).

y'-direction. The stronger the magnetization in the y'-direction, the stronger the intensity of the observed NMR signal in the receiver. As we will see later, the amount of the magnetization M_y in the y'-direction depends on the flip angle, which in turn can be influenced by the time period in which the rf-field generator is turned on.

In summary, resonance (observation of an NMR signal) can be achieved by the interaction of the net magnetization M_0 and rf-field which causes the tipping of the magnetization vector toward the y'-axis in order to generate a component of the magnetization in the y'-direction. The induced magnetization M_y in the y'-direction will be recorded as the NMR signal.

For recording a second NMR signal from an excited system, all spins have to lose their excess energy and go back towards their equilibrium condition that was established at the beginning. In the ^{1}H-NMR part of this book we briefly discussed the relaxation mechanism. In the next section we will discuss the two different relaxations and their mechanisms.

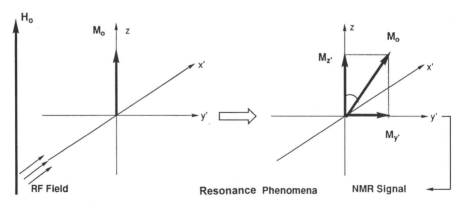

Figure 120   Generation of a magnetization in the y'-direction after interaction of the longitudinal magnetization M_0 with the radiofrequency field.

10.2   RELAXATION PROCESSES [75]

The relaxation process is one of the most important topics in NMR spectroscopy. As it has been shown in the preceding section, the NMR signal is obtained by measuring the decay of the magnetization in the y'-axis. In other words, we measure the relaxation of the nuclei. Therefore, relaxation times play an important role during NMR measurements. If we apply a radiofrequency field to the sample, the delay time that must be introduced between the pulses and the tipping angle have to be programmed in combination with the relaxation times. Relaxation times control the width of the peaks. Furthermore, relaxation times can provide us with some information about hydrogen bondings, relative motions of the molecule, steric effects and distances between the groups.

It was mentioned in the preceding section that many spectra are collected to obtain a good NMR spectrum. In order to record a second spectrum of a spin system, the excited spins must first return to the ground states. This process is called relaxation. In the resonance section we have shown how one can tip the magnetization cone M_0 toward the xy-plane (Figure 119). We now, need to discuss how M_0 returns back to the z-axis.

There are two relaxation processes:

(1)   spin–lattice relaxation T_1 (longitudinal relaxation);
(2)   spin–spin relaxation T_2 (transverse relaxation).

If the magnetization vector M_0 is tipped toward the xy-plane (Figure 120), the tipped magnetization will generate two components of the magnetization in the y'- and z-axes. However, the amount of the original magnetization in the z-axis will be reduced due to the tipping. Reduction of the magnetization in the z-axis will depend on the tipping angle (see Figure 120). The magnetization generated in the y'-axis is called *transverse magnetization*. Return of the z-component to its equilibrium (to the original position) value M_0 is referred to as *longitudinal relaxation* while the return of M_{xy} ($M_{y'}$) to zero is called *transverse relaxation*. Both processes can be characterized by the time T_1 and T_2, respectively. Let us first discuss the decay of the transverse magnetization T_2 to zero. The return of the magnetization to its original equilibrium value begins immediately after the pulse. But, the fact that the transverse magnetization returns to zero does not mean that the tipped magnetization cone has reached its original position. There are two processes that can induce transverse relaxation. We have already explained that the nuclei of the same type have the same Larmor frequency and they rotate with the same frequency along the external magnetic field. But, in fact not all of the nuclei rotate with the same frequency. Due to the inhomogeneities in the magnetic field, some spins will rotate slightly faster and some slower than the Larmor frequency. These slightly different frequencies will not affect the total amount of the net magnetization M_0. If the magnetization cone is tipped toward the xy-plane, the magnetization vector $M_{y'}$ will be static when all the spins rotate with the same resonance frequency (Figure 121a). In reality, these spins will start to fan out or dephase in the rotating frame, as well as in the static frame (Figure 121b and c) since they do not rotate with the same Larmor frequency.

In order to understand this process better, let us give an example. Imagine a convoy consisting of many cars. These cars have different speeds (10, 20, 30, 40 and 50 km/h). Consider the movement of these cars from a static point. When the cars begin to move, we will notice that all of them will be removed from the starting point at different speeds. This can be compared with the spreading out of the magnetic moments in a static frame. In the second case, we enter the car that has a speed of 30 km/h. Consider the movement of the cars after starting to move. If we ignore everything in the environment and observe only the cars, we notice that the cars spread out. This is exactly the case in the rotating frame. In the rotating frame, magnetic moments that are faster than the Larmor frequency rotate in one direction and the slower ones in the opposite direction. After a while all of the spins will fan out or dephase, producing a net xy-magnetization of zero ($M_{y'} = 0$).

The second process that is responsible for the spin–spin relaxation is the dipole–dipole interaction between nuclei of the same type. This mechanism is based on an energy transfer within the spin system (from one spin to another spin). Because nuclei of the same type have the same Larmor frequency, they actually generate magnetic fields of the correct frequency for relaxation. Thus, one nucleus will relax (will return to its ground state) and transfer the energy to another nucleus, which results in destroying the phase coherence (dephasing) and leads to transverse relaxation. Any process that causes loss of transverse magnetization contributes to T_2. All of the processes occurring during the transverse relaxation will not change the total energy of the sample. Since the total energy of the spin system does not change, spin–spin relaxation is classified as an entropy

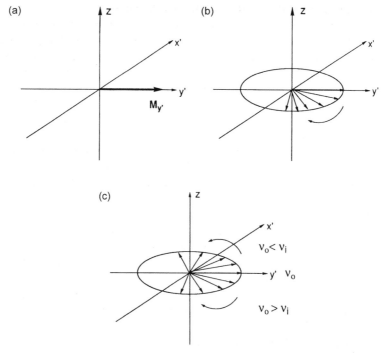

Figure 121   (a) Transverse magnetization; (b) motion of the magnetic moments in the fixed coordinate system; (c) motion of the magnetic moments in the rotating coordinate system.

process. When the transverse magnetization has reached a value of zero, it does not mean that the spin system has returned to its ground state.

We call any magnetization in the z-direction the longitudinal magnetization. The spin–lattice or longitudinal relaxation process designated by the time T_1 involves the transfer of energy from the excited nuclei to the environment, the so-called lattice. The spin–lattice process is an enthalpy process. The spin–lattice relaxation time is always greater than the spin–spin relaxation time. The reverse case is impossible. If the longitudinal relaxation has been completed, there cannot be any spin in the xy-plane.

$$T_1 > T_2$$

The process of relaxation takes place over a period ranging from milliseconds to minutes, depending on the nucleus and its environment. For protons in the usual nonviscous solutions, the relaxation times T_1 and T_2 are short, and equilibrium is re-established within seconds. As we will see later, the relaxation times of ^{13}C nuclei are longer than those of protons and vary widely. The establishment of equilibrium may require several minutes. In longitudinal relaxation, the nucleus has to release the absorbed energy in order to relax. The energy absorbed during the resonance process is in the form of electromagnetic radiation. The probability of releasing any energy in the form of electromagnetic radiation depends on the frequency of the energy. Since the frequency of the absorbed energy in the NMR processes is within the radiofrequencies, it cannot be released as electromagnetic radiation. Therefore, the excess energy of the spins will be transferred to the lattice. Any kind of molecules in the solvent (gas, liquid, solid), including the solvent molecule, are called the lattice. Thermal motion (translation, rotation, etc.) of these molecules will generate magnetic fields rotating with various frequencies, since all of these motions involve the movements of charged particles and electrons. Some of these frequencies will match the Larmor frequency of the nuclei, in order that energy transfer can occur. The magnetic energy received by the lattice is then transformed into thermal energy.

For example, the vibrational motions of the bond electrons generate magnetic fields. IR frequencies are in the range of 10^{13}–10^{15} Hz, which are not close to the Larmor frequencies (10^7–10^9) of most nuclei. Rotational and vibrational motions are usually characterized by their *correlation times* (τ_C). Correlation time is a measure of how rapidly the molecule undergoes reorientation in solution. Molecules do not have a defined correlation time, which depends on the temperature and size of the molecule, as well as the viscosity of the solution. Only those molecular motions whose 'frequencies' lie in the region of the Larmor frequencies lead to rapid relaxation of the molecule. If the viscosity of the solution is sufficiently low, small- and medium-sized molecules tumble very fast. The frequencies of their motion often exceed the Larmor frequencies and are in the range of 10^{-11}–10^{-13} s. For the very large molecules, correlation time may increase by about 10^{-8} s.

For interpreting the NMR spectra it is not necessary to know the detailed relaxation mechanism, but in order to understand the nuclear Overhauser effect in Chapter 11, it is recommended to continue reading this chapter.

10.2.1   Spin–lattice relaxation

There are a number of mechanisms which can contribute to spin–lattice relaxation in a molecule.

(1)   dipolar relaxation;
(2)   spin-rotation relaxation;
(3)   paramagnetic relaxation;
(4)   quadrupolar relaxation.

Spin–lattice relaxation occurs by way of the contribution of all the above-mentioned relaxations.

Dipolar relaxation

When an excited nucleus is directly bonded to a second nucleus possessing a magnetic spin, then there is the possibility of an efficient relaxation. In a specific case where a ^{13}C nucleus is directly bonded to a proton, this system can be considered to be a small dipole located at the center of ^{13}C and ^{1}H atoms. Consequently, the ^{13}C nucleus will experience a small field due to the dipolar interaction with the proton. This field will depend on the magnitude of the two dipoles (μ_C and μ_H) and the orientation angle ϕ of their line of interaction relative to the magnetic field of the spectrometer as shown in Figure 122. The magnetic field H_{DD} created at the ^{13}C nucleus is given by the following equation:

$$H_{DD} = \mu_H(3\cos^2\phi - 1)/4\pi r^3 \qquad (51)$$

Since the molecules tumble in solution, the variations in ϕ will cause fluctuations in H_{DD}. Relaxation can be induced by any oscillating electric or magnetic field which has a component at, or close to, the Larmor frequency of the nuclei.

For most of the carbon atoms bearing directly bonded protons, the dipolar mechanism plays the most important role and is often the only relaxation mechanism. As can be seen from eq. 51, the effect of the magnetic field is inversely proportional to r^3, and falls off rapidly with increasing distance. This is the reason why this effect is often only observed for carbon atoms carrying protons because of the short distance (C–H distance: 1.10 Å).

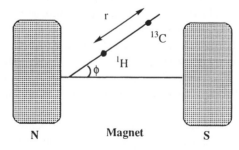

Figure 122   Alignment of a C–H bond in a magnetic field.

Quaternary carbon atoms do not have directly bonded protons and the dipolar relaxation for these carbon atoms is less effective (C–C distance: 1.54 Å). Increasing temperatures decrease the effect of the dipolar mechanism on the relaxation.

Spin-rotation relaxation

A careful look at any ^{13}C-NMR spectrum having CH_3 and CH_2 signals will show that the peaks of methyl carbons are similar in intensities or smaller to those of methylene carbon atoms. The similar or smaller intensities of the methyl carbon signals indicate a slower relaxation rate. Terminal methyl groups have more freedom of motion and therefore they rotate more rapidly and have a shorter correlation time. Shorter correlation times result in higher frequency rotating magnetic fields and less efficient relaxation. This spin-rotation process prevents the relaxation. We will talk about the spin-rotation process in the NOE section.

Paramagnetic relaxation

Another example of the dipole–dipole relaxation is paramagnetic relaxation. Unpaired electrons in the molecule generate local magnetic fields which are much stronger than that of a proton since the gyromagnetic constant of an electron is approximately 1000 times greater than that of a proton. Those magnetic fields can provide very efficient relaxation by a dipolar mechanism. The presence of a molecule having unpaired electrons, such as paramagnetic metals or oxygen which are nearly always dissolved in solvents, can drastically reduce the relaxation time. As a consequence, line broadening will appear and the fine splitting in the signals will disappear. Under these circumstances, it is very difficult to interpret the NMR spectra. Therefore, paramagnetic oxygen must be removed from the solvent whenever accurate T_1 values are to be measured. In some cases, the presence of such paramagnetic compounds in the solution can be helpful. For example, adding a paramagnetic compound such as $Cr(acac)_3$ can reduce the relaxation time of quaternary carbon atoms with very long relaxation times so that all carbon signals have equal intensities.

Quadrupolar relaxation

Nuclei with $I > 1/2$ possess a quadrupole moment Q which arises because the distribution of charge in the nucleus is not spherical but ellipsoidal. The charge distribution within the nucleus is either flattened or slightly elongated. Such nuclei give rise to an electric field gradient at the nucleus, providing a highly efficient relaxation mechanism. Let us consider a coupled system consisting of two different nuclei, A and X. If nucleus X relaxes faster than nucleus A (because of the quadrupole moment of nucleus X), then no signal splitting is observed between A and X. This mechanism is responsible for ensuring that there are no observable ^{1}H and ^{13}C couplings to chlorine and bromine despite the fact that they possess nuclear spins.

– 11 –

Pulse NMR Spectroscopy

11.1 INTRODUCTION TO THE PULSE NMR SPECTROSCOPY

As we have seen before, we have to collect many spectra in order to observe signals from the carbon atoms of the organic molecules. An average time to sweep the proton absorption range (*ca.* 10 ppm) is 5 min. This would require more than 8 h in order to collect 100 NMR spectra. In the case of ^{13}C-NMR spectra it will take more time since we scan a range of 250 ppm and we have to collect more spectra than in proton NMR. What can we do in order to shorten the length of time to collect 100 NMR spectra? As mentioned in the ^{1}H-NMR section, CW spectrometers use either a frequency-sweep technique or field-sweep technique. By the application of such techniques, each different kind of proton (or carbon) must be brought into resonance one by one and any change in the absorption will be recorded. The question is how can we shorten the time that is necessary for one scan. For example, we can increase the rate of scans from 5 to 1 min, but this would result in a decrease in resolution. Fast scanning is not a solution. The pulse technique was developed largely in response to the need for much higher sensitivity. This higher sensitivity is achieved by simultaneously exciting all of the nuclei. With this technique, scanning time is decreased from 5 min to a few seconds. Nowadays, one can collect up to 1000 scans in 10–15 min. This mode is called *Fourier transform* spectroscopy. Let us compare the CW (continuous wave) and FT (Fourier transform) techniques.

11.2 CW AND FT SPECTROSCOPY

In the conventional CW NMR experiment, a radiofrequency field is continuously applied to a sample in a magnetic field. An NMR experiment is obtained by sweeping the rf field. Most of the basic features of the FT spectrometer are the same as those of the CW spectrometers. The significant difference is in the means of sweeping the frequency or rf field. In FT spectrometers, all nuclei are simultaneously excited by using a high intensity rf pulse. In order to understand how this pulse excites nuclei with different Larmor frequencies, we need to briefly discuss the characteristics of a radiofrequency pulse. The characteristics of an rf pulse depend on the duration time of the pulse. If the duration time of a radiofrequency pulse is long enough, the generated pulse will be monochromatic. This means that it contains only one frequency. If we generate the pulse within a very

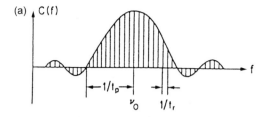

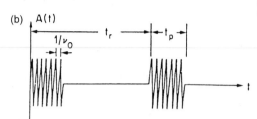

t_p = pulse width (duration time)
t_r = pulse repetition time

Figure 123 (a) The corresponding frequency components. (b) Sequence of radiofrequency pulses of frequency ν_0 with width t_p and repetition time t_r. (Reprinted with permission of John Wiley & Sons, Inc. from H. Günther, *NMR: An Introduction*, 1980.)

short time, a pulse of this type contains a wave of various frequencies that are necessary to excite all of the nuclei. This is illustrated in Figure 123.

Mathematically, it can be shown that the frequencies contained within this pulse are described by the following equation:

$$\nu_0 \pm \frac{1}{t_p} \tag{52}$$

where ν_0 is the carrier frequency and t_p is the pulse width (duration time).

We assume that the carrier frequency is 20 MHz and the duration time is 10 µs. From eq. 52 we can calculate that

$$\frac{1}{10^{-5}} = 100 \text{ KHz}$$

i.e. the radiofrequency pulse will contain an envelope of frequencies of 20 MHz ± 100 KHz, which is sufficient enough to excite the nuclei of different Larmor frequencies.

Let us calculate the duration time of a pulse that will be necessary to excite all of the carbon nuclei in a 200 MHz NMR spectrometer. Since the gyromagnetic constant of carbon is different to that of a proton, carbon nuclei will resonate at about 50 MHz. Most of the carbon nuclei will resonate in a range of 200 ppm ($200 \times 50 = 10{,}000$ Hz). Therefore, a pulse width t_p must be chosen to produce half the range required, which is 5000 Hz. The time that is necessary to generate this pulse can be calculated as

$$5000 = \frac{1}{t_p}, \qquad t_p = \frac{1}{5000} = 0.2 \text{ ms.}$$

From this calculation we arrive at the following conclusion: The duration time of the generated pulse must be at least 0.2 ms or less in order to cover the resonance range of the ^{13}C nuclei.

In the CW technique, the applied pulse is monochromatic and only one type of nucleus is brought into resonance. However, in the case of the FT technique all nuclei are brought into resonance at the same time, so that the average NMR recording time of 5 min for CW spectrometers is drastically decreased down to a few milliseconds.

Let us give the following example in order to better demonstrate the difference between the CW and FT techniques. Imagine for a moment that you are obliged to catch all of the fish in a sea. You can do this in two different ways. In the first case by using fishing tackle you can catch the fish one by one. Since you have to sweep the entire surface of the sea it will be very time consuming. This case is similar to the field-sweep or frequency-sweep technique in the CW spectrometers. In the second case you need a net that can cover the entire surface of the sea exactly. By throwing the net to cover the entire surface you will be able to catch all of the fish in a single action. This case is comparable to the FT technique in which all of the protons are brought into resonance at the same time.

After explaining the monochromatic radiofrequency and a pulse consisting of a bundle of frequencies, let us now focus on the interaction of these frequencies with the sample.

11.3 INTERACTION OF A MONOCHROMATIC RADIOFREQUENCY WITH THE SAMPLE

We now return to the resonance phenomenon and discuss it in more detail. Let us assume that there is only one type of proton in our sample, for example, $CHCl_3$. If we place $CHCl_3$ nuclei in a static magnetic field, the magnetic moment vectors of these nuclei will be randomly oriented in the precessional cone about H_0 and generate net magnetization M_0 in the z-direction because of the slight excess of nuclei in the upper case. We must now provide a radiofrequency field (B_1) along the x-axis for the resonance (Figure 124).

The angle α (the tip angle or the flip angle) through which the magnetization is tipped from the z-axis is easily calculated:

$$\alpha = \gamma B_1 t_p \qquad (53)$$

As one can see from eq. 53, the flip angle depends on the duration time t_p and the power of the B_1 field. We will show later that when the transmitter is on long enough, the magnetization will rotate onto the negative z'-axis. The flip of the bulk magnetization M_0 towards the y'-axis generates a component of M_0 in the y'-axis which will be detected as an NMR signal by the receiver. The signal intensity depends on the amount of the magnetization generated in the y'-axis. The more the magnetization is tipped towards the y'-axis, the stronger the signal intensity will be. Therefore, the transmitter has to be on a longer time to receive a stronger signal. The signal will reach a maximum value for a tip angle of 90° ($\pi/2$ pulse), and a null at 180° (π pulse). Now, we have to look at the magnetization generated in the y'-axis after the transmitter is turned off. As we have seen previously, the nuclei begin to lose energy to their environment and relax back to equilibrium. As the relaxation occurs, the component $M_{y'}$ of the magnetization along

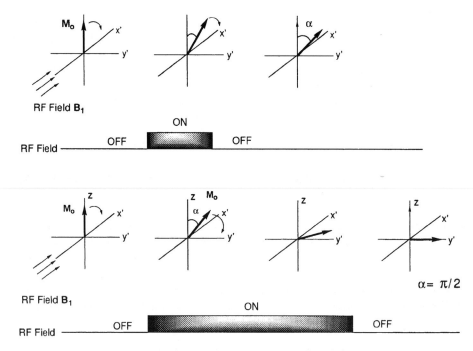

Figure 124 The effect of the radiofrequency field B_1 on the bulk magnetization at different duration times t_p.

the y'-axis decreases exponentially to zero. Decreasing the transverse magnetization to zero ($M_{y'} = 0$) does not mean that all of the nuclei have relaxed back to equilibrium. During that time some (or all) of the nuclei transfer the energy to the lattice and relax back. When the transverse magnetization decreases, longitudinal magnetization will increase, but not in the same amount (Figure 125).

The transverse magnetization $M_{y'}$ will be static in the rotating frame. Exponential decrease of the magnetization induces a current in the receiver coil that also decays exponentially (Figure 126).

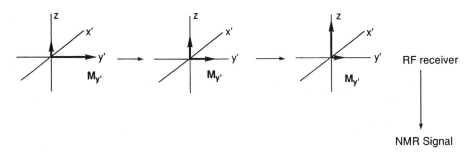

Figure 125 Decrease of the transverse magnetization and increase of the longitudinal magnetization.

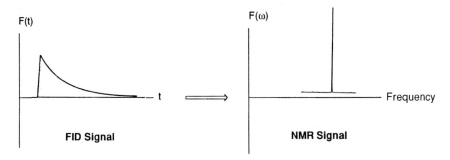

Figure 126   A typical FID (free induction decay) signal for radiofrequency field at resonance and its conversion to an NMR signal.

This process is recorded as a signal. As shown in Figure 126, for this exponential decay of the magnetization, the Larmor frequency of the nuclei must be equal to the rotating frequency of the rotating frame. The result is a so-called *free induction decay* (FID).

A FID signal shows the decay of the magnetization in the receiver coil when the transmitter is turned off. It gives rise to an exponential decay in the observed signal:

$$M_{y'(t)} = M_{y'(0)} \exp^{-(t/T_2)} \tag{54}$$

where T_2 is the spin–spin relaxation time, $M_{y'(t)}$ is the magnetization vector at the time t, and $M_{y'(0)}$ is the magnetization vector at the time $t = 0$.

We have already explained the exponential decrease of magnetization along the y'-axis in the rotating frame for cases where the Larmor frequency of nuclei is equal to the frequency of the rotating frame. Now we have to address the decay of the magnetization along the y-axis in the laboratory frame (static frame). Magnetization vector M_y will rotate at the Larmor frequency and it will decay in a spiral fashion back to equilibrium. The magnetization experienced by the receiver will actually be an exponential, sinusoidal decaying signal. The corresponding FID signal is given in Figure 127.

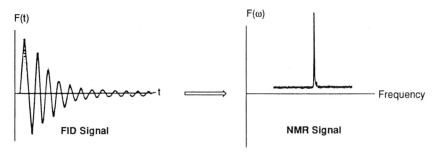

Figure 127   A typical FID (free induction decay) signal for radiofrequency field off resonance and its conversion to an NMR signal.

11.4 FOURIER TRANSFORMATION

So far we have considered a sample that has only one type of proton or carbon. In a realistic experiment, there will be more than one line. The question can be raised of how we can treat a sample containing more than one line. Consider just two lines, with some chemical shift difference $\Delta\nu$ between them. Let us suppose that one type of proton (or carbon) has a Larmor frequency of ω_0 and the other is precessing slightly faster at a frequency of ω_f. After the pulse, magnetization vectors of both the nuclei (let us assume that the pulse is 90°) will be tipped at the same time down onto the y'-axis. Of course, the applied pulse will contain the Larmor frequencies of both nuclei. If we choose the rotating frame frequency as the frequency of one of the two nuclei, then the magnetization arising from those nuclei will remain along the y'-axis. The magnetization of the other nuclei, although it arrives initially along the y'-axis, will precess at a different speed. In this case we observe two different FID signals. The first magnetization will decay exponentially as has been seen previously (Figure 126). The second one will decay sinusoidally and exponentially (Figure 127). The resulting FID signal will make a composite of the two, relaxing magnetization. If the sample contains more than two signals at different Larmor frequencies, the observed FID signals will be much more complex.

Let us return to the resonances of two different nuclei. We have seen that the magnetization vector of the nuclei remains along the y'-axis where the other one rotates about the z-axis. Now we raise the question: what is the rotation speed of the second magnetization vector? The second magnetization vector precesses about the z-axis at a frequency equal to its frequency difference from the other nuclei ($\omega_f - \omega_0$), so that this frequency difference will be reflected in the FID signal. The difference between two maximas will be equal to $1/(\omega_f - \omega_0)$. Generally, a FID signal contains the frequency difference (the chemical shift difference) of any signal from the reference. If we use a carrier frequency of the pulse as the reference, any FID signal will contain the exact difference (i.e. chemical shift difference) of the resonating signal from the carrier frequency (Figure 128).

Consider a macroscopic sample. All kinds of nuclei have different resonance frequencies so that every nucleus also has a different FID signal. All different FID signals contain resonance frequencies or, in other words, chemical shifts.

After a sample is pulsed a sufficient number of times, the FID signals obtained will be accumulated. Any observed signal in the NMR spectrum is derived from a FID signal. The view of the accumulated FID signals is complex because of the overlapping of many FID signals (Figure 129). The accumulated FID signal must be converted into a real spectrum. FID signals are time-dependent functions stored in a computer. However, NMR spectra are drawn by the plotting of signal intensity versus frequency, so that the data stored in computer memory must be converted from time dependency to frequency dependency. This conversion is accomplished by means of a mathematical process known as *Fourier transformation* (FT). The relation between the dependent function $f(t)$ and frequency dependent function $F(\omega_0)$ is given by the following equation:

$$F(\omega_0) = \int f(t)\, e^{-iwt}\, dt \tag{55}$$

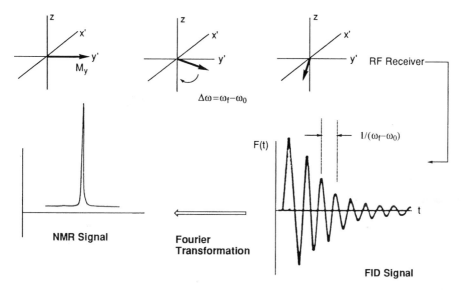

Figure 128   The exact frequency difference (chemical shift difference) between the resonance frequency of a signal from the carrier frequency.

The same computer in which the FID signal is stored generally also performs the Fourier transformation. Figure 129 shows a FID signal of a sample and the resulting NMR spectrum after its conversion.

So far we have explained the basic working principle of the FT-NMR instruments. The FID signals are then Fourier transformed by way of a computer into a conventional NMR spectrum. It is possible that the amount of a sample to be measured is not enough to record the NMR spectrum and interpret it, or the sensitivity of the nucleus to be measured can be low. These nuclei may require several hundred or thousand pulses to accumulate in the computer in order to improve the S/N ratio. The advantage of the FT technique is that rapid repetitive pulsing with signal accumulation is possible. In the classical CW-NMR spectrometer, a complete scan across the spectrum would require approximately 5 min. However, that time is reduced to a few seconds or fractions of a second by the FT-NMR technique. Therefore, in cases where we need many scans from the sample, we have to apply the FT technique. Nowadays, all of the instruments above 200 MHz on the market are based on helium-cooled superconducting magnets and operate in the pulse FT mode.

To improve the S/N ratio, the operator must select the number of pulses or scans. For example, it is possible to collect 15–20 pulses or more in 1 min from the ^{13}C nuclei. With the modern NMR instruments it is possible to record ^{13}C-NMR of a sample (20–50 mg) of modest molecular weight in 10–15 min.

In Figure 130, the 50 MHz ^{13}C-NMR spectra of dimethyl 1,6-cycloheptatriene carboxylate recorded with 1, 10, 100, and 1000 pulses are given [76]. As we can easily see from the ^{13}C-NMR spectra, there is a satisfactory S/N improvement between the spectra recorded with 1 pulse and 100 pulses. On the other hand, the improvement between the spectra recorded with 100 and 1000 pulses (a 900-pulse difference) is not as

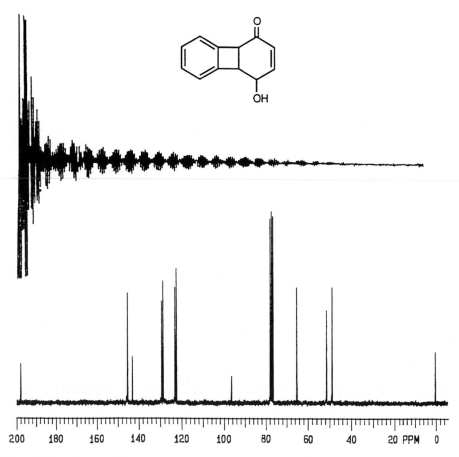

Figure 129   FID signal of the hydroxyketone and its 50 MHZ ^{13}C-NMR spectrum after Fourier transformation.

good as in the first case. It is very important to remember that the S/N ratio increases by the square root of the number of pulses:

$$\frac{S}{N} = \sqrt{n} \tag{48}$$

where S is the signal intensity, N is the noise intensity, and n is the number of the pulses.

We have already pointed out, for example, that when using FT instruments many FID signals have to be accumulated in order to increase the S/N ratio. After applying the first pulse, the transmitter is turned off and the decay of the magnetization M_y is recorded. The recorded FID signal is then stored in the computer. Before applying the second pulse there must be a delay time (pulse delay) τ between pulses which allows the magnetization to return back to its original equilibrium. If we apply a second pulse before the magnetization is fully recovered, the amount of the generated magnetization in the y'-axis will be decreased. If we repeat the experiments many times with a short delay time,

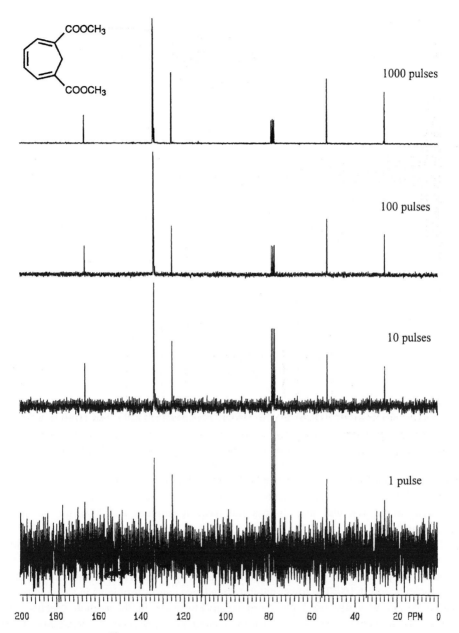

Figure 130 50 MHz ^{13}C-NMR spectra of dimethyl 1,6-cycloheptatriene carboxylate recorded with 1, 10, 100, and 1000 pulses.

we will not be able to increase the S/N ratio. Furthermore, it is possible that we will lose some signals in the NMR spectrum due to the saturation of some nuclei. Since the relaxation time of carbon nuclei is longer than that of protons, the delay time between pulses in the ^{13}C-NMR should be longer than in the ^{1}H-NMR.

Pulse width (duration time) t_p is another parameter which can increase the signal intensity in the NMR spectra. We will now go back to the following equation:

$$\alpha = \gamma B_1 t_p \tag{53}$$

We have seen previously that the tip angle α depends on the pulse width t_p. Since we can only detect a component of the magnetization along the y'-axis by the receiver, the maximum signal will be obtained when the tip angle is 90°. In Figure 131, the ^{1}H-NMR spectra of methyl homobenzobarrelene carboxylate recorded with one pulse are illustrated [77].

During the recording of these spectra, pulse widths are changed from 3 to 18 ms. As one can easily see, there is a remarkable increase in the signal intensities by a prolonged duration time of the pulse. Further increase of the duration time indicates the decrease of the signal intensities. This means that the 18 ms duration time generates the maximum intensity, in other words, the magnetization vector M_0 is tipped through 90° onto the y'-axis. For a single-pulse experiment, using a 90° pulse the maximum signal-to-noise ratio will be obtained. However, as shown earlier, the magnetization vectors take more time (actually $5T_1$) to return to its equilibrium value. Therefore, in multiple-pulse experiments, there should be a time delay of $5T_1$ between each pulse to allow the spins to recover. 90° pulses can be applied by ^{1}H-NMR spectra since the relaxation time of the protons is short. Unfortunately, T_1 values can be quite long for ^{13}C nuclei. The establishment of equilibrium may require 1–2 min, so that there must be a long time delay between each pulse. Rather than to wait for such a long time between pulses, it is more convenient to use smaller pulse angles (30–50°). This topic will be discussed later in more detail.

11.5 ROUTINE PULSED ^{13}C-NMR MEASUREMENT TECHNIQUES

The techniques applied to the measurement of ^{13}C nuclei are different to those for protons. The ^{13}C-NMR measurement techniques can be considered under two main groups. The first group is routinely applied and useful to record simple ^{13}C-NMR spectra. The second group of measurements provides us additional information besides the chemical shifts. We will first discuss the basic techniques and later the advanced techniques.

The ^{13}C nucleus has a spin quantum number of $I = 1/2$, just as it is for a proton. As we have seen before in ^{1}H-NMR spectroscopy, ^{13}C nuclei will also couple with the neighboring protons and give rise to complex peak splitting. The natural abundance of ^{13}C nuclei is only 1.1%. The lower natural abundance causes problems in the sensitivity. Although this is a disadvantage, there is an advantage in the carbon spectra. The probability of finding two ^{13}C nuclei in a fixed position is 10^{-4}. Therefore, C–C couplings will be observed as weak satellites (0.5% of the normal ^{13}C signal intensity) in ^{13}C-NMR spectra. This means the probability of finding two coupled ^{13}C nuclei is negligible. Consequently, ^{13}C–^{13}C couplings are normally not observed, which would lead to the complex ^{13}C-NMR spectra. On the other hand, we observe ^{13}C–^{1}H couplings between ^{13}C nuclei and protons. Since both nuclei have the spin quantum number of $I = 1/2$, they will couple with each other and resonate as doublets (Figure 132). This doublet splitting can be observed in the ^{13}C-NMR spectra. However, these couplings

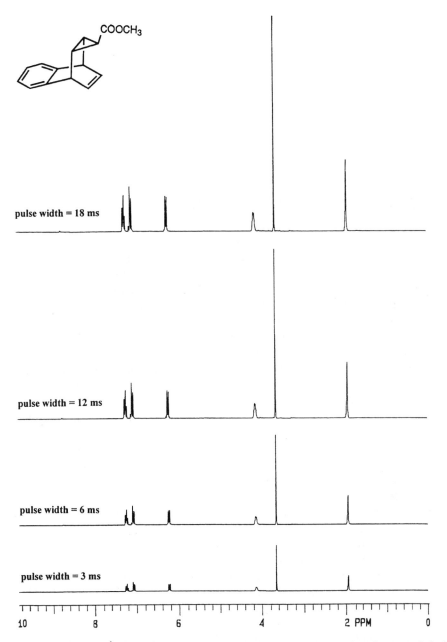

Figure 131   200 MHz ^{1}H-NMR spectra of methyl homobenzobarrelene carboxylate recorded with one pulse and different pulse widths (from 3 to 18 ms).

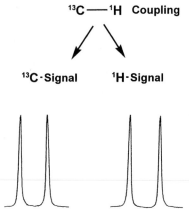

Figure 132   Appearance of ^{13}C–^{1}H couplings in the ^{1}H-NMR as well as in the ^{13}C-NMR spectra.

cannot be observed well in the ^{1}H-NMR spectra. Since organic compounds contain 1.1% ^{13}C isotope, only 1.1% of the protons in the compound will couple with carbon nuclei and give rise to the formation of the satellites around the main peaks in the ^{1}H-NMR spectroscopy. These signals are called satellite spectra (see Section 14.1).

Single bond carbon–proton coupling constants J_{CH} range from 100 to 200 Hz. Carbon-proton couplings over two and three bonds are also observed and are generally less than 20 Hz. This means that the ^{13}C-NMR spectra of organic compounds show complex overlapping multiplets, leading to considerable difficulties in the interpretation of the signal. In order to simplify the spectra and increase the peak intensity, those couplings between ^{13}C and ^{1}H nuclei should be removed. This can be achieved by using the broadband decoupling technique.

11.6   BROADBAND DECOUPLING

In the ^{1}H-NMR section we discussed homonuclear decoupling and have shown that the irradiation of one proton in a spin-coupled system removes its coupling to neighboring protons. To remove all of the possible ^{13}C–^{1}H couplings in the ^{13}C-NMR spectra, all proton resonances have to be simultaneously irradiated. For this reason, the decoupling field B_2, which covers the range of all proton Larmor frequencies, will be applied to the sample. In this case, all of the possible couplings between the ^{13}C and ^{1}H nuclei are removed and ^{13}C signals are converted into singlets. The complex ^{13}C-NMR spectra are then simplified by application of this technique and the spectrum interpretation will be easier. An additional advantage of this experiment is an intensity enhancement for the ^{13}C resonances. The collapse of multiplets into singlets improves the signal-to-noise ratio in a straightforward manner. Consider a ^{13}C nucleus that is coupled with only one proton and gives rise to a doublet. In other words, the resonance signal of ^{13}C nucleus is split into two lines, so that the intensity of the signal is reduced to half. Broadband decoupling collapses these doublet lines into singlets and the signal intensity increases twofold. Signal enhancement by CH_2

and CH_3 carbons is much more effective because of the collapse of a triplet and quartet into singlets.

Although decoupling simplifies the spectra and permits it to be obtained in substantially less time, valuable information on proton–carbon coupling values, and the multiplicity of the signals, are lost. The splitting pattern of the signals shows the substitution (primary, secondary, tertiary, and quaternary). This kind of information cannot be obtained.

In Figure 133, the coupled and broadband decoupled ^{13}C-NMR spectra of two different compounds are illustrated [78]. We shall now call the broadband decoupled ^{13}C-NMR spectra 'normal ^{13}C spectra'. Ethyl acetate contains four different carbon atoms. Figure 133b shows the ^{13}C broadband proton decoupled spectrum having four singlets. However, the coupled ^{13}C-NMR spectrum of the same compound shows a complex ^{13}C-NMR spectrum in spite of the simple structure of ethyl acetate. Furthermore, the signal-to-noise ratio of the second spectrum (Figure 133b) recorded with 500 scans is much better than that of the first spectrum (Figure 133a) recorded with 4000 scans.

The low natural abundance and sensitivity of the ^{13}C nucleus have forced the NMR scientist to develop techniques to improve the signal-to-noise ratio. Therefore, a wide variety of heteronuclear decoupling experiments are routinely used. The past three decades have witnessed the development of nuclear magnetic resonance into one of the most powerful analytical tools. Nowadays, by applying special measurement techniques such as attached proton test (APT) and distortionless enhancement by polarization transfer (DEPT), the peak assignments can be performed in a very short time, whereby peaks can be classified as representing CH_3, CH_2, CH, and C. This discussion is reserved for Section 15.2.1 'Special NMR Techniques'.

Careful inspection of the coupled and decoupled ^{13}C-NMR spectra in Figure 133 clearly indicates that the signal-to-noise improvement cannot solely arise by the collapse of the multiplets into singlets. The enhancement of the carbon signals is usually considerably greater than would be expected from the collapse of the multiplets. There are two important factors which are responsible for the intensity enhancement of the ^{13}C resonances:

(i)   collapse of the multiplets into singlets;
(ii) nuclear Overhauser effect (NOE).

The first effect has already been discussed in detail. Now, we will discuss the second effect, the NOE effect. The NOE is not only applied for the enhancement of signal intensities of ^{13}C resonances, but also has an important application in ^{1}H-NMR spectroscopy to determine the configuration of organic compounds.

11.7   NUCLEAR OVERHAUSER EFFECT

Before discussing NOE in detail, let us briefly describe it and later explain its basic mechanism. NOE immediately reminds us of the enhancement of the peak intensities. NOE is based on the fact that if one nucleus is selectively irradiated, the signal intensity of the neighboring nucleus will change (in most cases it will be increased). A proton that

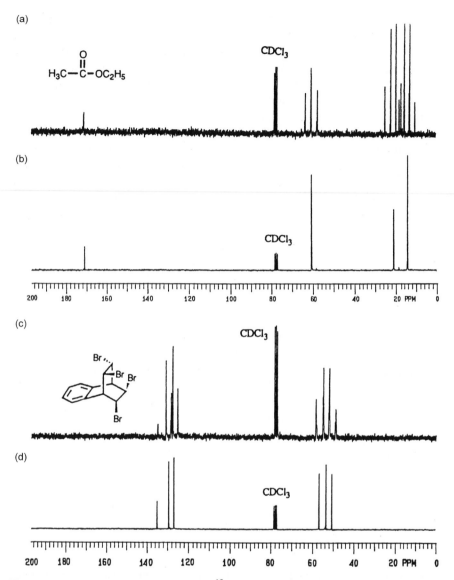

Figure 133 (a) 50 MHz proton-coupled ^{13}C-NMR spectrum of ethyl acetate recorded with 4000 pulses; (b) 50 MHz proton-decoupled ^{13}C-NMR spectrum of ethyl acetate recorded with 500 pulses; (c) 50 MHz proton-coupled ^{13}C-NMR spectrum of tetrabromo compound recorded with 14,000 pulses; (d) 50 MHz proton-decoupled ^{13}C-NMR spectrum of tetrabromo compound recorded with 2500 pulses.

is close in space to the irradiated proton is affected by NOE whether or not it is coupled to the irradiated protons. NOE operates through the bonds as well as through the space. If one of the two coupled (or close in proximity) protons is irradiated continuously this will cause a change in the population of the energy levels of the other proton. Any change in the population of the energy levels will be reflected as a change of the signal intensity. NOE is always encountered in ^{1}H-NMR, as well as in ^{13}C-NMR spectroscopy. In the ^{1}H-NMR, the NOE application determines the correct stereochemistry of the compounds; in ^{13}C-NMR spectroscopy, NOE plays an important role in the enhancement of signal intensity. Let us explain first the NOE application in ^{1}H-NMR spectroscopy and proceed to its mechanism of operation.

The ^{1}H-NMR spectrum of N,N-dimethylformamide is given in Figure 134. The characteristic feature of this NMR spectrum is two distinct proton resonances for the methyl groups at room temperature. Normally, a fast rotation about the C–N bond would result in one methyl resonance. However, the C–N bond between the carbonyl group and the nitrogen atom has a significant double bond character, represented by the contribution of the second resonance structures, so the free rotation about C–N bond is hindered.

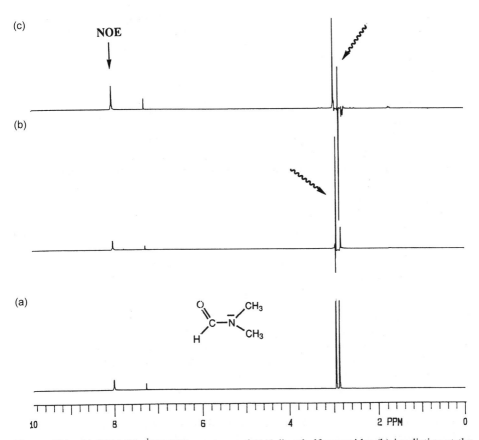

Figure 134 (a) 200 MHz ^{1}H-NMR spectrum of N,N-dimethylformamide; (b) irradiation at the resonance frequency of the low-field methyl group; (c) irradiation at the resonance frequency of the high field resonance frequency and NOE observation.

If the temperature is raised these two distinct N–CH$_3$ singlets coalesce into a single line. Now the question of how we can assign these distinct singlets can be raised. This question can be answered by NOE measurements. Experimentally, we proceed by irradiating low-field CH$_3$ resonance while we observe the aldehyde proton resonance. This is schematically illustrated in Figure 134b. There is no change in intensity of the aldehyde proton resonance. However, irradiation of the high-field CH$_3$ resonance (Figure 134a) produces a definite increase in the intensity of the aldehyde proton resonance. This peak enhancement is called 'nuclear Overhauser enhancement'. This increase in the aldehyde proton resonance clearly indicates for us that the high-field CH$_3$ resonance belongs to the methyl group, which is the *cis* to the aldehyde group or in close proximity to the aldehyde functional group. This experiment above shows that when any proton is continuously irradiated, this will change the population in the energy levels of the second nucleus, which is separated by a small distance from the irradiated one. Any change in the population results in a change of signal intensity.

If NOE is carried out on nuclei of a single type, the process is called the *homonuclear Overhauser effect*. The second and most common example of the NOE occurs in ^{13}C-NMR spectroscopy during the application of broadband decoupling. If NOE is observed between the different nuclei, it is called the *heteronuclear Overhauser effect*. If one is interested in the physical process behind NOE, it is suggested to continue reading. However, it is not so important to know the mechanism of NOE in order to interpret NMR spectra. A limited amount of information will be sufficient in order to understand the changes in the intensity of the signals.

The low natural abundance of ^{13}C nuclei was the starting point in looking for new measurement techniques in order to improve the signal-to-noise ratio. NOE is one of these measurement techniques. We have seen earlier that the intensity of a signal is directly proportional to the population of the energy levels of the nuclei in a given magnetic field. According to the Boltzmann distribution law, there is a slight excess of nuclei aligned with the external magnetic field H_0. The intensity of the NMR signals obtained is directly proportional to the number of the excess nuclei. The energy levels of protons and carbon are established after a thermal equilibrium. If we perturb this thermal equilibrium, for example, by increasing the number of the nuclei populating the lower energy level, the probability for resonance, i.e. the intensity of the signal, increases. The physical process behind NOE is based on the perturbation of the energy levels [79].

We will now consider a heteronuclear coupled system, such as the ^{13}C–^{1}H system. These nuclei have two possible orientations in a magnetic field (parallel and antiparallel). The state β, being of a lower energy level, will then contain an excess of nuclei, while α shows a higher energy level. According to this description, a nucleus resonates by transition from the energy level β to α. We consider ^{13}C and ^{1}H nuclei as an AX system. This system has four energy levels, which are shown in the Solomon diagram (Figure 135).

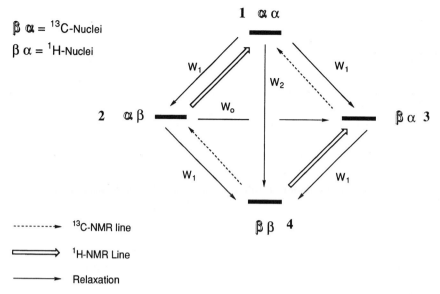

Figure 135 Energy level schemes (Solomon scheme) for a two-spin system (carbon and proton) and possible relaxation transitions.

These energy levels depend on the orientation of the ^{13}C and ^{1}H nuclei in the magnetic field. Those energy levels are given below:

(1) $\alpha\alpha$: both nuclei antiparallel;
(2) $\beta\alpha$: proton parallel, carbon antiparallel;
(3) $\alpha\beta$: proton antiparallel, carbon parallel;
(4) $\beta\beta$: both nuclei parallel.

Bold letters indicate the ^{13}C nucleus, the other ^{1}H. Transitions between those four different energy levels are the resonances. The broken lines show the ^{13}C resonances, and the solid lines the ^{1}H resonances. W_1 indicates relaxation transitions that are responsible for the maintenance of the Boltzmann distribution. Additionally, there are also the transitions W_2 and W_0 for the cases in which the spin of both nuclei flips simultaneously.

Now, if the proton resonance, i.e. the transitions (2) $\rightarrow$ (1) and (4) $\rightarrow$ (3) are irradiated with a strong decoupling field (as during broadband decoupling of the protons) this will induce rapid transitions between the spin states concerned, causing their populations to be equal. After this process (irradiation of the ^{1}H nuclei) the number of the nuclei populating the energy levels 1 and 3 increases while the number of the nuclei in the energy levels 4 and 2 decreases. Finally, the initially established equilibrium between the spin states is disturbed. The total number of nuclei in each of the four energy levels does not obey the Boltzmann distribution. The system must now attempt to regain a Boltzmann distribution. The simple way to restore the Boltzmann equilibrium is by relaxation of the excited protons. This relaxation (W_1) is irrelevant here, because the population difference across those transitions is fixed by the saturation of the corresponding resonances. There are other more effective mechanisms for the relaxation by which the $\alpha\alpha$ spin state relaxes directly back to the $\beta\beta$ state, which is labeled as W_2 on the Solomon scheme (Figure 135). This transition corresponds to a change in the total spin of two units ($\Delta m = 2$, double quantum transition) and would normally be forbidden by the selection rules. According to the rules of quantum mechanics, only those transitions with $\Delta m = \pm 1$ (m = total quantum number for the two nuclei) are allowed (single quantum transition). There are, however, many possible methods in which W_2 occurs. Thus, W_2 transition will decrease the population of the top and increase the population of the bottom. The important result of this repopulation of levels is that there is now a greater excess of ^{13}C nuclei in level 4 compared with level 2 and in level 3 compared with level 1. Thus, both transitions due to the ^{13}C nuclei will be more intense because of the greater excess of ^{13}C nuclei in levels 4 and 3. Furthermore, there is a population increase in the energy level 3 and a population decrease in the energy level 2. This process is called '*dynamic nuclear polarization*'. There is a second relaxation transition (W_0) between the energy levels 2 and 3 which is not as effective as W_2. This transition also does not obey the selection rule ($\Delta m = 0$, zero quantum transition).

There are, however, many possible methods by which W_0 and W_2 occur that do not involve the direct emission of radiation without violating the selection rules. Actually, we cannot predict which one of these transitions, W_0 or W_2, is dominating. However, we are already in a position to experimentally test which pathways are followed. If NOE occurs in practice, this indicates that W_2 and/or W_0 processes (cross-relaxation) are involved, and from the sign of the NOE we can discover which is dominant.

To understand the NOE better, let us explain the transitions between the energy levels with some numbers. We assume that levels 2 and 3 are degenerated.

As shown in Figure 136, we assume that the energy level 4 has 40 nuclei, the energy levels 2 and 3 have 30 nuclei and the energy level 1 has 20 nuclei. The difference between the energy levels is always 10. Thermal equilibrium is already set up at the beginning. Now we irradiate (saturate) the proton resonances (transition from 4 to 3 and from 2 to 1) and reach saturation (Figure 136). Now a new equilibrium is established where the energy levels 4 and 3 contain 35 nuclei and the energy levels 2 and 1 contain only 25 nuclei. The difference between the numbers of the nuclei at energy levels 4 and 1 was 20 at the beginning. But, after irradiation of the proton transition, it is decreased to 10. On the other hand, the energy levels 2 and 3, which had the same number of the nuclei at the beginning, have now a population difference of 10.

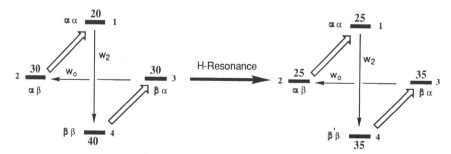

Figure 136 Population distribution in a two-spin system (carbon and proton) immediately after saturation of the proton resonances (broadband decoupling).

The initially established equilibrium is completely disturbed. To regain the Boltzmann distribution any relaxation of the protons from the energy levels 3 and 1 back to their origin is not realistic, since the transition is fixed between those energy levels by the saturation. We will now look at the differences between the nuclei in these energy levels carefully. We can see that the population difference between the energy levels 4 and 2, as well as between 3 and 1 is always 10, as is the case before irradiation, so that we cannot expect any transition between these mentioned energy levels. In general, there is no change in the population difference of the ^{13}C resonances. However, we have pointed out at the beginning that we are interested in affecting these energy levels which give rise to ^{13}C signals in order to obtain an NOE. Therefore, the forbidden transition can only reestablish the equilibrium. Let us first analyze the zero transition, W_0. The population difference between the energy levels 2 and 3 is now 10, whereas at equilibrium it was zero. If W_0 is the dominant relaxation pathway, we observe the transition from the energy level 3 to 2 to restore a population difference of 0 (Figure 137). After this transition, the energy levels 2 and 3 again have the same population as they had at the starting point. However, the population of the top (energy level 1) is increased and the population of the bottom (energy level 4) is decreased. Now, we have to analyze the probability of the ^{13}C transition. Transition from energy level 4 to 2 and from 3 to 1 produces the ^{13}C signal. However, now the population difference between these energy levels is decreased from 10 to 5. If we compare this population difference with that of the beginning, one can conclude that the probability for the ^{13}C resonance is decreased, when the W_0 transition plays the dominating role.

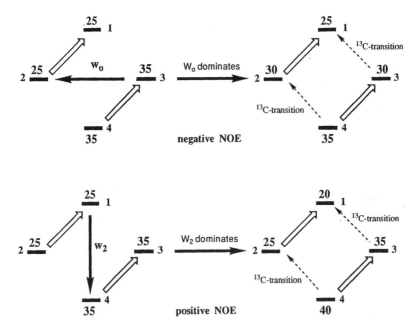

Figure 137 Population distribution where zero quantum transition is effective (W_0 transition, negative NOE) and population distribution where double quantum transition is effective (W_2 transition, positive NOE).

The second alternative to restore the equilibrium is the double quantum transition W_2. The difference between the energy levels 1 and 4 is now 10. At the beginning this difference was only 20. The transition from the energy level 1 to the energy level 4 can take place in order to restore the equilibrium. After relaxation through W_2, we now have the original populations at the energy levels 1 and 4 (20 and 40). However, the populations at the energy levels 2 and 3 are changed. If we now look at the differences in population between the energy levels 4 and 2 and the energy levels 3 and 1 (which are ^{13}C transition states) we see that the population difference is increased up to 15 compared to the difference of 10 in the absence of irradiation of the protons (broadband de-coupling). As a result of this population difference, the probability of the ^{13}C resonance is increased, which will be reflected in the intensity enhancement of carbon signals.

What is found is that NOEs are positive (W_2 is dominating) for small molecules in a nonviscous solution, but negative (W_0) dominates macromolecules in a very viscous solution. In cases where W_2 and W_0 take place at the same time and balance, NOE disappears. It seems reasonable that dipolar coupling may be able to cause cross-relaxation (W_0 and W_2 processes) as required by NOE. Relaxation through dipolar coupling was covered in Chapter 10.

Consider both nuclei, which have been discussed in the Solomon scheme (Figure 136), to be protons. If we go back to Figure 137 and look at the final population after double quantum relaxation, we notice that the population excess was increased from 10 to 15. This means that the signal intensity was increased up to 50% by the NOE. For any homonuclear decoupling experiment, the maximum NOE is 0.5, i.e. a 50% enhancement in the peak intensity. The maximum NOE enhancement factor η for any heteronuclear decoupling experiment mainly depends on the gyromagnetic ratios of the elements A and B and is expressed by the following equation:

$$\eta = 0.5 \frac{\gamma_A}{\gamma_B} = \frac{\gamma_A}{2\gamma_B} \tag{56}$$

If ^{13}C relaxation in a broadband decoupling experiment proceeds exclusively by the dipole–dipole interaction, the NOE factor η_C indicates the enhancement according to eq. 56 is 1.988:

$$\eta = \frac{\gamma_A}{2\gamma_B} = 1.988$$

The signal increase will then be $1 + 1.988 = 2.988$. NOE factors smaller than 1.9888 indicate the participation of other mechanisms in the spin–lattice relaxation.

We have shown that the NOE leads to more intense signals and therefore decreases the amount of time necessary to obtain a spectrum. NOE sensitivity enhancement is attained in the heteronuclear systems when the nucleus with low $\gamma(C)$ is observed while the nucleus with high $\gamma(H)$ is decoupled. In the proton decoupled ^{13}C-NMR measurements, the peak intensity can be increased threefold. The full NOE is only obtained for nuclei which are relaxed exclusively by dipolar relaxation. The presence of a contribution from a mechanism other than dipolar relaxation leads to a corres-ponding reduction in the observed NOE. Since the quaternary carbon atoms do not directly bear attached protons, they cannot undergo dipolar relaxation. Therefore, the contribution of NOE enhancement is reduced. Furthermore, we have previously seen that the contribution of the dipole–dipole interaction to the longitudinal relaxation time of two nuclei that are separated by a distance r is proportional to the factor $1/r^6$. An NOE can be observed only if the nuclei are in relatively close steric proximity, since it is only for this case that a dipole–dipole interaction represents a significant relaxation mechanism.

Summary: Broadband decoupling applied during the measurements causes a change in the population of the ^{13}C nuclei, which results in the formation of NOE. The intensity of ^{13}C-NMR are increased up to 300%. In the ^{1}H–NMR spectroscopy, the NOE enhancement is applied to assign the stereochemistry in a variety of compounds because of its strong dependence on nuclear distance.

In order to see the NOE effect, one has to record the normal spectrum and then the spectrum with NOE first, and then subtract the normal spectrum from the NOE enhanced spectrum. The *NOE difference spectrum* (NOE-Dif) will then only contain those signals for which there is a difference in intensity between the two spectra so that one can easily see which protons produce NOE upon irradiation at the resonance frequency of a given proton.

For better comprehension of homonuclear NOE experiments let us discuss two examples. Figure 138 shows the normal ^{1}H and NOE-Dif NMR spectra of the dibromo compound [80].

In this particular case the following questions have to be answered:

(1) the configuration of the cyclopropane ring (*syn* or *anti*);
(2) the configuration of the nitril group (*cis* or *trans*).

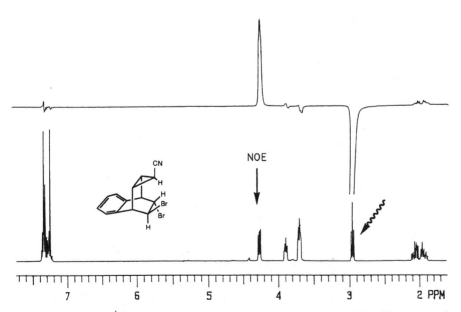

Figure 138 200 MHz ^{1}H-NMR spectrum of dibromo compound and its NOE-Dif spectrum after irradiation at the resonance frequency of cyclopropyl proton.

The *trans* configuration of the bromine atoms has been established on the basis of the asymmetrical ^1H, as well as the ^{13}C-NMR spectra. Irradiation at the resonance frequency of the cyclopropyl proton (2.95 ppm) next to the cyano group produced a positive NOE for the doublet of doublets at $\delta = 4.25$, which was previously assigned to one of the CHBr protons. The fact that only one of the CHBr protons gives positive NOE clearly indicates (i) that the configuration of the bromine atoms is *trans*, (ii) the cyclopropyl proton is in close proximity with one of the CHBr protons which in turn establishes the *anti* configuration of the cyclopropane ring, (iii) furthermore, the *trans* configuration of the cyclopropane protons is determined. The other cyclopropane protons resonating around 2 ppm produce a small NOE. This means that these protons are *trans* to the cyclopropane proton which is irradiated. Furthermore, the peak assignment to the CHBr proton resonances can be performed correctly with this experiment. This example demonstrates that through simple NOE experiments, one can determine the exact configuration of a compound having more than one isomeric center.

The second example (Figure 139) shows the correct configurational assignment of a diester having a five isomeric center. Theoretically, this diester can have a total of 32 isomers. By way of two NOE experiments we can eliminate 31 structures and obtain the correct configuration of the molecule [80].

Irradiation at the resonance signal of the methine protons H_2 and H_8 (br. singlet at $\delta = 3.0$) resulted in positive NOE for the H_{10} and H_{12} ($\delta = 1.6$) and H_1 and H_9 ($\delta = 3.6$) signals. This can be clearly seen in the NOE difference spectrum in Figure 138. The fact that the cyclopropane protons H_{10} and H_{12} produce NOE enhancement clearly

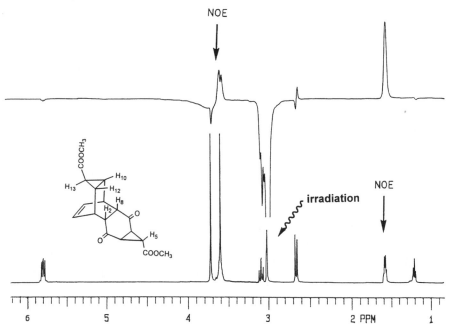

Figure 139   200 MHz ^1H-NMR spectrum of the pentacyclic compound and its NOE-Dif spectrum after irradiation at the resonance frequency of H_2 and H_8 protons.

indicates that these protons are in close proximity to the methine protons H_2 and H_8. This observation establishes the *endo* configuration of the cyclopropane protons. Furthermore, the fact that the other cyclopropane protons H_4 and H_6 do not produce any NOE indicates that these protons are located on the other side of the molecule. An NOE observation at 3.6 ppm shows that the bridgehead protons H_1 and H_9 signals are overlapped with one of the methyl resonances.

These two examples demonstrate that the exact configurations of the compounds having complicated structures can be easily determined. Modern instruments provide us with the possibility to obtain all kinds of NOE experiments in a single spectrum, which is called the NOESY spectra. We will discuss this technique in Chapter 16 in detail.

11.8   MEASUREMENTS OF NOE ENHANCED COUPLED ^{13}C-NMR SPECTRA: GATED DECOUPLING

We have already discussed in Section 11.6 that the broadband decoupling technique removes all of the possible couplings between the ^{13}C and ^{1}H nuclei, and ^{13}C signals are converted into singlets. Signal intensities are enhanced by NOE and the collapse of the multiplets lines into singlets. Although decoupling simplifies the spectrum and permits it to be obtained in substantially less time, valuable information on proton–carbon coupling values and the multiplicity of the signals are lost. Unfortunately, a price is paid for this increase in sensitivity. For example, the splitting pattern of the signals shows the substitution pattern (primary, secondary, tertiary, and quaternary carbons). Couplings between the ^{13}C and ^{1}H nuclei provide information about the hybridization of the corresponding C–H bonds. Of course, it is possible to obtain the coupled ^{13}C-NMR spectra by turning the decoupler off. Then, we can observe ^{13}C–^{1}H coupled spectra and observe the multiplicity of the peaks and determine the coupling constants. However, in this case we cannot retain the benefit of the NOE. Furthermore, the intensities of the signals are divided into multiplets. Therefore, we have to increase the number of the pulses and that requires longer recording times. It is particularly useful to obtain coupled spectra which retain the NOE. The *gated decoupled* technique provides us with this possibility. Pulse sequences of the gated decoupling techniques are given in Figure 140.

Firstly, the decoupling field B_2 will be applied to the sample, which covers the range of all proton Larmor frequencies. Proton decoupling begins immediately after the decoupler is switched on. The changes of the spin populations responsible for NOE will occur so that the number of the carbon nuclei populating the lower energy level will increase. Since the NOEs associated with proton decoupling are controlled by the relaxation times, they persist for several seconds. On the other hand, proton decoupling stops instantly if the decoupler is turned off. After turning the decoupler off, the carbon pulse will follow. During FID acquisition the proton decoupler will also be off. By way of these pulse sequences we will record ^{13}C–^{1}H coupled spectra and NOE will subsequently be effective. After the recording of the first FID signal, the same pulse sequences have to be applied from the beginning (Figure 140). In this way it is possible to observe ^{13}C–^{1}H couplings while retaining the NOE.

Figure 141 shows the normal (broadband decoupled) and gated decoupled ^{13}C-NMR spectra of biphenylene [81]. In the gated decoupled spectrum of biphenylene,

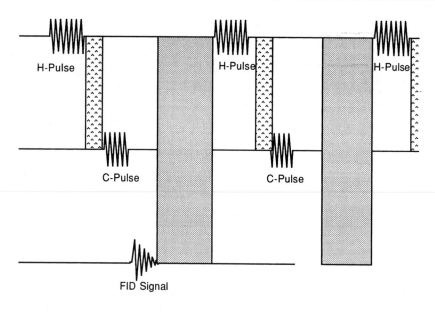

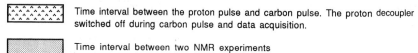

Time interval between the proton pulse and carbon pulse. The proton decoupler switched off during carbon pulse and data acquisition.

Time interval between two NMR experiments

Figure 140   Schematic presentation of carbon and proton pulses applied in a gated decoupling experiment.

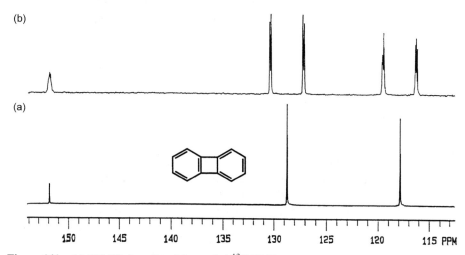

Figure 141   (a) 50 MHz broadband decoupled ^{13}C-NMR spectrum of biphenylene and (b) gated decoupled spectrum.

the α- and β-carbons are split into doublets due to the coupling with the directly bonded protons. Further fine splitting in the doublet lines arises from the coupling over two and three bonds. The peak around 152 ppm can be easily assigned to the quaternary carbon atoms in view of the splitting. It does not show any one-bond coupling. The line broadening arises from the couplings with the α- and β-protons.

Next we will discuss two additional measurement techniques, *off resonance decoupling* and *selective decoupling*, which have been entirely replaced by the modern techniques APT, DEPT and HETCOR.

11.9 OFF RESONANCE ^{1}H DECOUPLED AND SELECTIVE DECOUPLED EXPERIMENTS

In the case of broadband decoupled ^{13}C-NMR spectra, the bandwidth ν_i was set broad enough to cover all of the protons in a sample. However, in the proton off resonance decoupling, as the name indicates, the sample is irradiated with a strong radiofrequency field ν_2, which is a few hundred hertz away from the protons to be irradiated. This strong field, which is 'off' the resonance, will not completely remove the C–H couplings. The observed couplings are not equal to the actual one-bond coupling constants, but they are reduced or residual couplings. The degree of decoupling decreases as the separation $\Delta\nu = \nu_i - \nu_2$ increases. By the application of this measurement technique, the normal one-bond ^{13}C–^{1}H couplings around 120 Hz will be reduced down to 10–50 Hz depending on ν_2. Furthermore, all smaller couplings (geminal, vicinal, and long-range) will be removed (Figure 142).

What is the advantage of this technique? Firstly, the multiplicity of the carbon atoms can be easily determined. There is always peak overlapping in the coupled carbon spectra because of the large one-bond carbon–proton couplings that make it difficult for the interpretation of the spectra. In off resonance ^{1}H decoupled ^{13}C spectra, peak assignments are more easily made since the overlap between adjacent carbon resonance bands is less likely. Furthermore, in accordance with the number of directly bonded protons, carbon nuclei will resonate as singlet, doublet, triplet and quartets, respectively. However, this method does not allow one to determine the exact values of the corresponding coupling constants. This experiment shows high sensitivity since NOE will be retained during the measurements.

Much simpler techniques are applied in modern instruments, providing information about the multiplicity of the resonance signals. As we will see later, APT and DEPT are two of these methods that perform an exact assignment to the primary, secondary, tertiary and quaternary carbon atoms.

The next method of proton decoupling, selective decoupling (or single frequency proton decoupling), is useful for unequivocal assignments of the carbon signals. Before application of this method, the signal assignment of the protons in the ^{1}H-NMR spectrum has to be made. If a proton resonance can be identified in the ^{1}H-NMR spectrum, then it is possible to irradiate at the resonance frequency of this proton at a low radiofrequency power. As a consequence of this experiment, the signal of the carbon atom attached to that proton which is irradiated collapses into a singlet while the other protonated carbon atoms retain their C–H couplings. With this technique one can easily find the relation

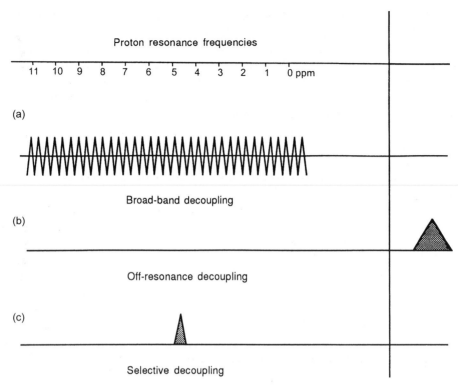

Figure 142 Schematic presentation of (a) broad-band, (b) off-resonance, and (c) selective decouplings.

between ^{1}H and ^{13}C lines. It is possible to obtain a set of single-frequency decoupled spectra by sequencing through the proton resonance frequencies. Irradiation at every proton resonance frequency will of course take more time. This outdated application has been replaced by the HETCOR experiments which allows us to perform an unequivocal peak assignment in a single spectrum.

11.10   INVERSE GATED DECOUPLING

In the ^{1}H-NMR section we discussed the integration of the area under the signal curve. In ^{1}H-NMR spectra, comparison of the integral of the signals directly provides the ratios of the protons in the molecule. For example, the integration value of a signal belonging to three equal protons is three times more intense than the integration value of a single proton. On the basis of the integration values, the peak assignments can be performed and the relative number of the protons can be determined easily. However, in the routine broadband decoupled ^{13}C-NMR spectra, the ^{13}C nuclei, whose relaxation times (T_1) vary over a wide range, are not equally relaxed between pulses, i.e. the relative number of the carbon atoms cannot be determined by the integration of the peak areas. The peak intensities of the carbon atoms will be different and will not correlate with the number of carbon atoms, as in proton NMR spectroscopy.

How can we work around these problems? The effect of unequal relaxation times can be overcome in different ways. The first one is the shortening of the relaxation times of different carbon atoms. Addition of some paramagnetic ions, chelate complexes of chromium or iron to the sample solution shortens the relaxation times and enables the quantitative evaluation of the carbon signals.

The second process of long delays between the pulses can be used to obtain quantitative results. The delay time has to be chosen carefully. In most cases, a 2 min delay may be sufficient. Generally, a waiting time of $3T_{1max}$ will be ideal in order to permit relaxation to the slowest carbon (with T_{1max}) of the sample molecule. However, NOE should be used during this measurement. On the other hand, we know that that the NOE response is not the same for all ^{13}C nuclei, resulting in a further loss of quantitation. This negative effect of NOE has to be removed. The following pulse sequences are subsequently applied. Carbon and proton pulses are applied at the same time on the sample in order to suppress the effect of NOE (Figure 143). This technique will eliminate the C–H couplings and will not allow the buildup of NOE. The intensities of the ^{13}C signals will not be distorted. This technique is known as *inverse gated decoupling*.

The broadband decoupled and inverse gated decoupled ^{13}C-NMR spectra of methyl homobenzobarrelene carboxylate are shown in Figure 144. In the broadband decoupled spectrum (lower spectrum), we can assign the small peaks around 146 and 172 ppm to the equivalent quaternary aromatic carbon atoms and the carbonyl carbon atom. These

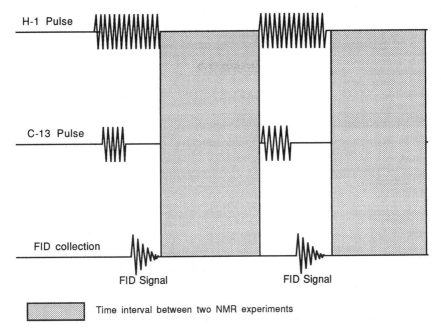

Figure 143 Presentation of pulse sequences in a inverse-gated decoupling experiment (quantitative carbon determination).

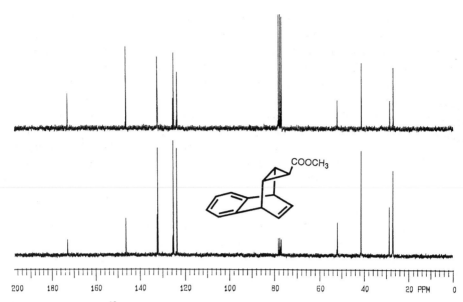

Figure 144   50 MHz ^{13}C-NMR spectrum of homobenzobarrelene derivative and its inverse-gated decoupled spectrum.

assignments can be made on the assumption that the quaternary carbon atoms are responsible for the weak peaks. In the inverse gated decoupled ^{13}C-NMR spectrum (upper spectrum) the intensity of the quaternary carbon atoms is increased and the intensities of all peaks are nearly equalized except for those (carbonyl carbon, methyl carbon and cyclopropyl carbon) representing single carbon resonances.

11.11   SAMPLE PREPARATION AND SOLVENTS

The sample solution is prepared as described for ^{1}H-NMR. In FT-NMR spectrometers, many FID signals are collected over a long period (a few hours or even days). During this time, very precise field-frequency stability is required. Any variation will cause line broadening—as a consequence of this, the fine splitting in the signal will be lost. The deuterium lock is the means by which a long-term stability of the magnetic field is achieved. An intense deuterium signal is often used for the stabilization of the magnetic field. Therefore, deuterated solvents have to be used in order to obtain a lock signal. Of course, the main purpose of using the deuterated solvents is to eliminate any signal arising from the solvent molecule. In the CW instruments it is possible to use nondeuterated solvents, such as carbon tetrachloride (CCl_4). However, in the FT instruments it is not possible to use a solvent which does not contain any deuterium atoms. A deuterium atom has to be present for the locking on of the magnetic field.

For ^{13}C-NMR spectra, owing to the low sensitivity and natural abundance of ^{13}C nuclei, the major requirement is to obtain a concentrated solution. A routine sample for carbon NMR should consist of about 10–100 mg of the compound, depending on the molecular size. Under favorable conditions, it is possible to record NMR spectra of samples that are less than 1 mg in a micro tube. It should be always kept in mind that the high concentration and the viscosity of the chosen solvent can have a considerable effect on the observed line width and loss of fine splitting. As for temperature measurements, the melting and boiling point of the solvent must be taken into account, and they of course have to be deuterated.

A large number of fully deuterated compounds are commercially available. The cheapest solvents are D_2O and $CDCl_3$. In carbon NMR, ^{13}C nuclei will couple with the deuterium atoms and give rise to multiplets. In general, the signal intensities arising from the solvent peak are smaller than expected since the deuterium atoms increase the relaxation time of the directly bonded carbon atoms. Nowadays, it is possible to remove the peak of unwanted solvents from the spectrum by a technique called *solvent suppression*. The sample is irradiated at the Larmor frequency of the solvent peak with greater amplitude during the recording of the spectrum. The resulting saturation greatly reduces the solvent peak. If there is any signal under the solvent peak, it will be revealed and the interpretation of the NMR spectra will be easier. Table 3.1 (see ^1H-NMR section) summarizes the most frequently used solvents and their properties.

– 12 –

Chemical Shift

For the interpretation of the carbon NMR spectra we have to discuss two important parameters in detail: chemical shift and coupling constant. At first we will deal with chemical shift and the second part we will reserve for spin–spin coupling.

As was discussed in the proton NMR section, there are many effects which influence chemical shift. The chemical shift of the carbon nuclei is mainly determined by the electron density surrounding the nucleus. All of the basic theories discussed concerning the proton NMR as pertains to the chemical shift are also valid for carbon NMR spectroscopy.

12.1 DIAMAGNETIC AND PARAMAGNETIC SHIELDING ON PROTON AND CARBON ATOMS

Induced magnetic fields generated by the electrons will cause either shielding or deshielding which will determine the chemical shift. Since the electrons carry a charge, they will circulate under the influence of the external magnetic field and generate their own magnetic fields. Following the Lenz law, this induced magnetic field will oppose the external magnetic field (Figure 13). The strength of the local magnetic field at the nucleus is then reduced and is dependent upon the external magnetic field.

$$H_{sec} = \sigma H_0 \tag{18}$$

The shielding constant, σ, is proportional to the electron density around the nucleus. Because of this diamagnetic shielding the local field at the carbon nucleus is smaller than the applied external field (see Chapter 3). Carbon nuclei are not only under the influence of the external and the induced magnetic field, they are also under the influence of the neighboring groups which generate their own magnetic fields. Therefore, we have to also consider the electronic circulation within the entire molecule.

The shielding constant σ in molecules corresponds to the sum of diamagnetic and paramagnetic components of the induced electronic motion. Additional terms are needed in order to take into account the effects of the neighboring groups. The most important of these terms is the magnetic anisotropy of the neighboring group σ_{neig} and solvent effects σ_{sol}.

$$\sigma = \sigma_{dia} + \sigma_{para} + \sigma_{neig} + \sigma_{sol} \tag{21}$$

However, the influence of these shielding parameters on protons and carbon is different. Diamagnetic shielding arises from the circulation of the electrons at the nucleus and it reduces the strength of the external magnetic field. As is known from the Lamb equation (from which one can calculate the size of the diamagnetic shielding)

$$\sigma_{\text{dia}} \propto r^{-1} \tag{57}$$

the diamagnetic term decreases with the distance r between the nucleus and circulating electrons. Therefore, the s-electrons will cause a stronger diamagnetic shielding than the p-electrons because of the longer distance between the electrons and the nucleus in the p-orbitals. Since a hydrogen atom has an unperturbed spherical electron distribution, it leads to a pure diamagnetic shielding effect. For atoms with s-electrons only, such as hydrogen, the diamagnetic shielding effect σ_{dia} is the dominant shielding. The lack of spherical symmetry for electron density surrounding the carbon nucleus results in the diminished circulation of electron density so that the diamagnetic shielding effect does not have as much influence on the carbon atom as in the case of the proton.

The paramagnetic term is considerably more difficult to deal with on both a qualitative and quantitative basis. Paramagnetic shielding is observed in the nonspherical molecules. Since the electrons in p-orbitals do not have a spherical symmetry, these electrons produce large magnetic fields at the nucleus. Although carbon atoms have four s-electrons which are spherically symmetric in their distribution, there are also two p-electrons in the ground state, which do not have spherically symmetric electron distribution. Nonspherical distribution of the electrons diminishes the electron circulation. Consequently, according to Lenz law, the strength of the induced magnetic field, i.e. the strength of the shielding, is reduced. For the paramagnetic shielding there must be unpaired electrons. Therefore, the molecule has to be excited for the participation of the paramagnetic term. For nuclei other than hydrogen, as in the carbon nucleus, the paramagnetic shielding term σ_{para} predominates. We have mentioned above that the contribution of the paramagnetic shielding in the proton is not observed because of the high excitation energy of the s-electrons. In conclusion, strong paramagnetic effect arises only for heavier nuclei where energetic low-lying atomic orbitals are available. Since carbon nuclei have p-electrons, which can be easily excited, the paramagnetic shielding term plays an important role in determining the chemical shift of the carbon nuclei. According to the Karplus–Pople equation [82],

$$\sigma_{\text{para}} = \frac{e^2}{m^2} \frac{h^2}{c^2} \Delta E^{-1} r_{2\text{p}}^{-3} \left[Q_{\text{AA}} + \sum Q_{\text{AX}} \right] \tag{58}$$

where ΔE is the average electronic excitation energy, $r_{2\text{p}}$ is the distance between a 2p electron and the nucleus, and $[Q_{\text{AA}} + \sum Q_{\text{AX}}]$ is the charge density around carbon nucleus. The paramagnetic term increases with a decreasing electronic excitation energy ΔE and with an inverse cube of the distance r between a 2p electron and the nucleus. It is useful to qualitatively discuss eq. 58. The average electronic excitation energy directly influences the paramagnetic shielding. Alkanes only consist of σ-bonds. For an alkane, the energy gap between the HOMO and LUMO ($\sigma \rightarrow \sigma^*$ transition) is very high ($\Delta \approx 10$ eV). Therefore, the paramagnetic shielding does not play an important role in the case of alkanes and the resonances of alkanes are observed at high field. However, alkenes have a higher energy π molecular orbital (HOMO) and the magnetically allowed $\pi \rightarrow \pi^*$ transition ($\Delta E \approx 8$ eV) is of reasonably low energy. The carbonyl groups also have low-lying excited states (n $\rightarrow \pi^*$ transition) so that the ΔE is smaller ($\Delta E \approx 7$ eV). Consequently, σ_{para} becomes an important factor. This means that the smaller the ΔE is, the greater the contribution of σ_{para} to the shielding. Increased paramagnetic shielding results in a low-field shift of the resonances. For example, alkenes and aromatic compounds resonate at lower field ($\delta_C \approx 100-150$ ppm). The carbonyl groups are the most deshielded functional groups ($\delta_{CO} > 170$ ppm). The different electronic excitation energies of the C=C and C=O correspond well with these different chemical shifts.

Similar examples are also frequently encountered by the other systems. An example is provided by a comparison of the ^{13}C-NMR chemical shifts for the C=O and C=S groups in acetone and thioacetone with the n $\rightarrow \pi^*$ transition in each of these groups. Carbonyl carbon of acetone resonates at 206.7 ppm, whereas thioacetone resonates at 252.7 ppm. This chemical shift difference can be attributed to the fact that the excitation energy of the nonbonding electrons located on sulfur is lower than that of acetone.

Another factor that affects the chemical shift of the carbon atoms is the distance (r_{2p}) of the 2p electron density from the nucleus. This distance depends on the electron density at the nucleus. An increased electron density at the carbons causes electronic repulsion. As a result, the bonding orbitals expand and r increases. This will reduce the paramagnetic shielding because of the reverse proportionality of the distance with the paramagnetic shielding. As it turns out, the chemical shifts move upfield.

The third term in eq. 58, which describes the electron density around the resonating nucleus, has an important effect on chemical shift. There is a linear correlation between the π-electron density and the chemical shift of the carbon atoms as we have seen earlier in the case of the proton chemical shifts. If the electron density around the carbons is increased, they resonate at higher fields, whereas in the case of decreasing the electron density, the resonances move to the lower field. Figure 145 shows the existence of a linear correlation between the carbon resonance frequencies and the electron density of some aromatic ions.

In the proton NMR we have shown that the resonance frequencies of the aromatic protons are shifted around 1–2 ppm to the lower field when compared to the olefines. This effect was explained in terms of an induced ring current. This paramagnetic shift (downfield shift) is nowadays the best criteria for aromaticity. Downfield shift of the aromatic protons around 1–2 ppm is a remarkable shift in the proton NMR spectroscopy when compared to the resonance region of 10 ppm for all kinds of protons. In the ^{13}C-NMR spectroscopy the ring current effect is less important, as it contributes only a little to the total shielding (2 ppm), which is generally hidden by other larger effects.

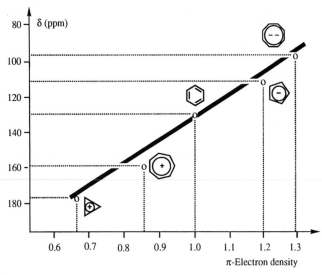

Figure 145 ^{13}C chemical shifts of carbon atoms in some selected aromatic compounds versus π-electron density.

7.27 ppm   H 5.80 ppm   5.6 ppm   −0.5 ppm δ_C=34.8 ppm

H   H   H   H   H   H

                                                                                                                 6.95–7.20 ppm

δ_C=128.5 ppm δ_C=131.5 ppm δ_C=125.5 and 26.0 ppm δ_C=128.7, 126.1 and 114.6 ppm

15   **135**   **136**   **20**

For example, when we compare the proton resonance frequencies of benzene with that of cyclooctatetraene, which is an olefinic compound, it can be seen that benzene resonates at about 1.5 ppm at the lower field. This 1.5 ppm chemical shift difference in a range of 10 ppm is a remarkable shift and it arises from the diamagnetic ring current. However, the carbon chemical shift difference between those molecules is only 3 ppm. In a resonance range of 250 ppm, this value is less at 1.5%. On the other hand, cyclooctatetraene carbons resonate contrary to the expectation at the lower field. Similar chemical shifts are observed in 1,6-methano[10]annulene (**20**). Methylene protons at the bridge resonate at high field (−0.5 ppm) due to the location of the protons in the middle of the diamagnetic ring current. However, the resonance frequency of the bridge carbon atom is not affected by the diamagnetic ring current. When compared with cyclohexadi-1,4-ene, the bridge carbon in **20** resonates at an even lower field than the methylene carbon in **136**. The use of ^{13}C-NMR data as criteria for aromaticity in potentially aromatic hydrocarbons is out of the question.

12.2   FACTORS WHICH INFLUENCE THE CHEMICAL SHIFTS

In this section we will discuss the other factors that influence the chemical shift. Generally, ^{13}C chemical shifts correlate with some typical properties of carbon within its molecular environment such as: carbon hybridization, electronic effects, steric effects, etc.

12.2.1   Influence of the hybridization

The hybridization of carbon atoms determines to a great extent the range of the ^{13}C resonances. According to hybridization, the following general statements can be made.

$$\delta_{sp^2} > \delta_{sp} > \delta_{sp^3}$$

Generally, sp^3 carbon atoms resonate between -10 and 80 ppm relative to TMS and sp-hybridized carbon atom resonances, which are shielded relative to olefinic carbons that are observed in a range of 60 and 95 ppm. In the case of sp^2 carbons we have to distinguish between the olefinic carbons and carbonyl carbons. Alkene resonances are found between 100 and 150 ppm while carbonyl carbons of aldehydes, ketones, and carboxylic acids, including all of the derivatives, occur between 160 and 220 ppm.

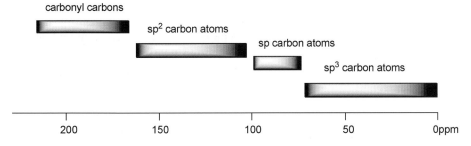

These ranges for the chemical shifts of the different hybridized carbon atoms are not exact ranges. Outside of these ranges, one can also encounter chemical shifts depending on the number of the attached substituents as we have seen in the ^{1}H-NMR spectroscopy.

12.2.2   Electronic effects

Electronic effects will be treated under two different sections.

Inductive effect

At first we want to show how the inductive effect can influence the chemical shifts. This topic has been discussed in detail in the ^{1}H-NMR section. As the electronegativity of the substituents ($+I$, $-I$ effect) that are attached at the carbon atom changes, the resonance frequencies will also change. Inductive and mesomeric substituent effects are transmitted through chemical bonds. The inductive effect of the substituents is also observed at the

Table 12.1

The effect of halogen substituents on the chemical shifts in pentane derivatives (in ppm)

X	CH_2	CH_2	CH_2	CH_2	CH_3
H	13.7	22.6	34.5	22.6	13.7
I	−7.4	10.5	−2.1	−1.1	−0.1
Br	19.3	10.1	−4.1	−0.7	0.0
Cl	30.6	10.0	−5.3	−0.5	−0.1
F	70.1	8.0	−6.7	0.1	0.0

γ-carbon atom. For example, the influence of the inductive effect in halopentanes is illustrated in Table 12.1 [74b].

A correlation between the ^{13}C chemical shifts and substituent electronegativities is observed for the α-carbons, since the electronegative substituents reduce the shielding owing to their −I effect. The bond polarization due to electronegative substituents should propagate along the carbon chain. Careful examination of the chemical shifts in Table 12.1 clearly indicates that there is no correlation between the substituent electronegativities and the observed chemical shifts of β- and γ-carbon resonances. The expected low-field trend is not observed in the case of β- and γ-carbon atoms. The α-effect increases with increasing electronegativity of the substituent (F, Cl, Br) while the effect of iodine is an exception to the pattern. The α-carbon of iodopentane resonates at −7.4 ppm. Table 12.2 illustrates the upfield shift of the substituted carbon atoms when the number of the attached bromine and iodine atoms is increased.

The upfield trend in the chemical shifts produced by bromine and iodine atoms is called the 'heavy atom effect', which increases where there is a multiple substitution. This is attributed to increased diamagnetic shielding caused by the large number of electrons introduced by heavy atoms. This diamagnetic shielding is given by the following equation:

$$\sigma_{dia} = \frac{e^2 \mu_0}{4\pi 3 m_e} \sum Z \cdot r^{-1} \qquad (59)$$

where Z is the atom number, r is the internuclear distance, m_e is the electron mass, and e is the electron charge.

The atomic number of the substituent attached to the carbon atom directly influences the chemical shift. This contribution is of particular importance, where increased shielding is observed with an increasing atomic number.

Table 12.2

^{13}C-NMR chemical shifts value in halosubstituted methane derivatives

	CH_3X	CH_2X_2	CHX_3	CX_4
X = I	−24.0	−53.8	−139.7	−292.5
X = Br	9.6	21.6	12.3	−28.5
X = Cl	25.6	54.4	77.7	96.7
X = F	75.0	109.0	116.4	118.6

Since the diamagnetic shielding is inversely proportional with the distance, the heavy atom effect is not observed at the β-carbon atoms. The β-effect is always a deshielding effect. Since the inductive effect plays a dominating role, all of the β-carbon resonances in the halopentanes are shifted to around 8–11 ppm downfield (Table 12.1). However, there appears to be no direct relationship to the electronegativity.

Generally, the α-effects of halogen atoms follow the electronegativities of the elements in series as F, Cl, Br (except I). The β-effect does not show such a correlation, but regardless the β-carbon resonances are shifted to lower field. Since the inductive effect propagate through the σ-bonds and decrease with the inverse cube of the distance, much smaller effects are expected in β- and γ-positions. However, contrary to this expectation, the γ-carbon atoms are shielded more and their resonances appear at a higher field. The effect, which causes the upfield shift of the carbon resonance frequencies, is called the 'γ-gauche effect', which will be treated more in detail later under the steric effects.

Similar effects are also observed in the cyclic systems (Table 12.3). Substituents effects have been studied in particular detail for the cyclohexane system. One of the striking effects in the cyclohexane ring is the shift of $C_{3,5}$ and C_4 carbon resonances to upfield, which is caused by γ-gauche steric compression.

Mesomeric effect

The mesomeric effect (M) produces, as a result of an interaction through the π-electrons, an electron excess or deficiency depending on the nature of the substituents. If a substituent having double bond or nonbonding electrons is directly attached to a conjugated system, the electron density and consequently the chemical shift will change. As we have seen in the ^{1}H-NMR spectroscopy, introduced substituents will extend their range of chemical shifts. Let us analyze two substituted aromatic compounds: anisole and nitrobenzene (see page 53). The electron-releasing substituent, the methoxyl group (+M-effect), increases the π-electron density in the aromatic ring, while the

Table 12.3

^{13}C-NMR chemical shift values in monosubstituted cyclohexane derivatives

Substituent, X	Chemical shifts of the carbon atoms			
	C_1	$C_{2,6}$	$C_{3,5}$	C_4
–H	27.6	27.6	27.6	27.6
–F	90.5	33.1	23.5	26.0
–Cl	59.8	37.2	25.2	25.6
–Br	52.6	37.9	26.1	25.6
–I	31.8	39.8	27.4	25.5
–OH	70.0	36.0	25.0	26.4
–NH$_2$	51.1	38.7	25.8	26.5
–NO$_2$	84.6	31.4	24.7	25.5
–CN	28.4	30.1	27.3	25.8
–COOCH$_3$	43.4	29.6	26.0	26.4

Table 12.4

^{13}C-NMR chemical shift values in substituted benzene derivatives

Substituent, X	Chemical shifts of the carbon atoms				
	C_1	$C_{2,6}$	$C_{3,5}$	C_4	Substituent
–H	128.5	128.5	128.5	128.5	
–OCH$_3$	159.9	121.4	128.1	125.3	54.8
–NO$_2$	149.1	124.2	129.8	134.7	
–CHO	136.7	129.7	129.0	134.3	192.0

electron withdrawing group, the nitro group ($-$M-effect), decreases π-electron density in the ring. Examples given in Table 12.4 show that the electron-releasing substituents cause a shielding in the *ortho* and *para* positions and shift the carbon resonances to high field, whereas the electron-withdrawing substituents cause a deshielding in *ortho* and *para* positions. Consequently, the corresponding carbons resonate at lower field compared to benzene. *Meta* positions are the least affected by substituents, in accordance with the mesomeric forms involved. The resonances of the aromatic compounds will be treated in more detail in Section 13.5.

In α,β-unsaturated carbonyl compounds, the chemical shifts are also largely affected by the conjugation. Carbonyl carbons are shielded relative to those in saturated carbonyl compounds. The olefinic carbons in cyclopentene resonate at 130.8 ppm, while the β-olefinic carbon in cyclopent-2-enone resonates at 30 ppm at the lower field (165.3 ppm) because of the conjugation. However, α-olefinic carbon atom resonance is observed at 134.2 ppm.

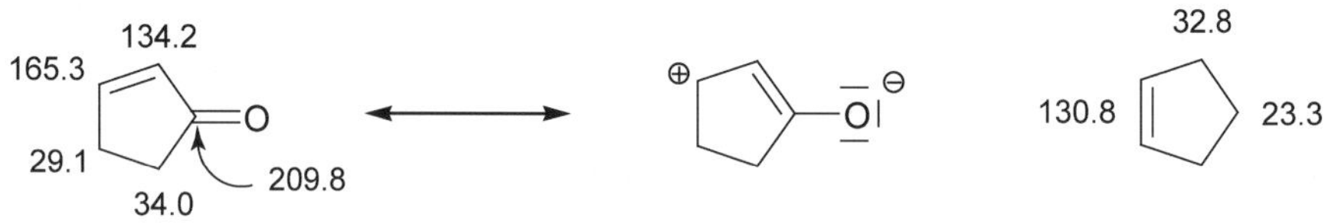

Those examples discussed under the electronic effects demonstrate that the ^{13}C resonances are very sensitive against any kind of electronic effects (inductive and mesomeric) that can change the electron density.

Steric effect (γ-gauche effect)

Steric interactions arise from an overlapping of van der Waals radii of the protons or substituents, which are closely spaced. The carbon atom is always shielded if substituents are introduced in the γ-position. It is not restricted to the alkyl groups, and it has also been observed for other substituents. In acyclic and cyclic systems, the steric effect can be observed if the protonated carbons have the gauche conformation as shown below. Because of the necessity of this conformation, this interaction is called the γ-gauche steric effect [83].

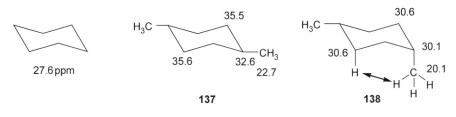

Due to this steric interaction, the chemical shift of the γ-carbon atom will move upfield. This nonbonding interaction causes a polarization of the C–H bonds accompanied by an increase of the electron densities at the two carbon atoms involved. Increased electron density is responsible for this diamagnetic shielding, upfield resonance.

γ-Gauche steric effect can be up to −2 ppm in the acyclic systems when the substituent is a methyl group, since the gauche rotamers population is approximately 30%. It is also noteworthy that the γ-effect shows a linear dependence on the electronegativity of the substituent X. For example, halogens cause the γ-effect of up to −7 ppm.

27.6 ppm

35.5
H₃C
35.6   32.6
CH₃
22.7

137

trans-1,4-dimethylhexane

30.6
H₃C
30.6   30.1
H   C 20.1
H   H
H

138

cis-1,4-dimethylhexane

The γ-gauche effects are better observed in conformationally rigid systems and it can be applied to the configurational assignments of the molecules. For example; in the *cis*- and *trans*-1,4-dimethylcyclohexane (**137**, **138**), the γ-gauche effect can be observed depending on the configuration of the methyl groups. In the *trans*-isomer (**137**), the γ-gauche effect is not observed because of the equatorial positions of the methyl groups. However, in the *cis*-isomer (**138**), the γ-gauche effect is observed. The –CH₃ carbon atom and the ring carbon atoms C₃ and C₅ show upfield shifts of 2.6 and 5 ppm relative to **137**, respectively.

47.63 ppm

Br
Br
Br
Br

139

43.49 ppm

Br
Br
Br
Br

140

37.34 ppm

Br
Br
Br
Br

141

In the bicyclic rigid systems **139**–**141**, the γ-effect arising from the bromines can be observed [61]. Comparison of the bridge carbon resonances reveals that the *exo*-orientation of one bromine atom in **140** causes an upfield shift of the bridge carbon resonance about 4 ppm, and the second *exo*-bromine causes an additional upfield shift of 6 ppm. Furthermore, this example demonstrates the additivity of the γ-effect.

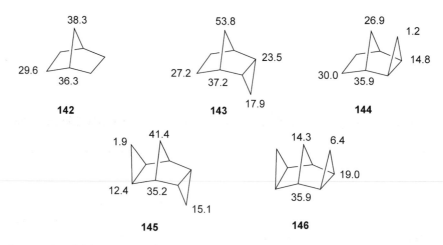

142   **143**   **144**

145   **146**

In certain rigid systems, the γ-gauche effect can have unusual values. This is encountered in the substituted norbornane systems **142–146** [84, 85]. The chemical shifts of the bridge carbons in **143–146** are drastically affected by the configuration (*exo* or *endo*) of the fused cyclopropane ring. When the cyclopropane ring is in the *exo*-position, the bridge methylene carbon is strongly shielded. The methano-bridge resonance in **142** is shifted from 38.3 up to 26.9 ppm in **144** ($\Delta\delta = 11.4$ ppm). This can be readily explained by means of the *syn*-γ-gauche effect. If the cyclopropane ring is in the *endo*-position, the bridge carbon is deshielded and its resonance is shifted down to 53.8 ppm while the resonance of the carbon atoms of the ethano bridge is shielded by 2.4 ppm. The chemical shift difference of the bridge carbon in **143** compared to **142** is 15.5 ppm. This effect is called the *anti*-γ-gauche effect, which is explained by the electron-accepting ability of the cyclopropane ring. The fact that the *syn*-γ-gauche effects and the *anti*-γ-gauche effects are additive, is demonstrated well in the structures **145** and **146**. The methano-bridge carbon resonance in the *exo,exo*-isomer **146** is strongly shielded and it resonates at 14.3 ppm. However, in the *endo,exo*-isomer **145**, the *syn*-γ-gauche effect and *anti*-γ-gauche effect cancel each other out, so that the methylene carbon resonates at 41.4 ppm which is a comparable value with those of **142**.

Generally, γ-gauche effects cause the ^{13}C nuclei concerned to be more shielded as well as deshielded. If the interacting groups have *syn*-periplanar conformation, the corresponding carbon atoms are shielded and the resonances are shifted upfield. In the case of *anti*-periplanar conformation, the carbon nuclei are deshielded and resonate at the lower field.

– 13 –

^{13}C Chemical Shifts of Organic Compounds

We have already discussed in detail that ^{13}C resonances mainly depend on the electronic effects (inductive and mesomeric) and steric effects. In this section, we will discuss under the headings of the common functional groups of organic compounds with consideration of the above-mentioned effects. As noted earlier, ^{13}C resonances extend over a range of 200 ppm, while protons resonate in a range of 10 ppm. Therefore, knowledge of the relationships between chemical shifts and molecular structure is even more important in ^{13}C-NMR spectroscopy than in ^{1}H-NMR spectroscopy. In order to determine the ^{13}C chemical shifts in a given molecule, a set of simple rules has been proposed. Empirical substituent constants or increments have been calculated for the different substituents which allow us in general to have a good prediction of resonance frequencies by very small deviations. These are based on the assumption that the substituents' effects are additive. A selection of these is described in the following sections. Let us start with the chemical shift of hydrocarbons because of their simplicity.

13.1 ALKANES

The chemical shifts of the carbon nuclei in linear and branched saturated hydrocarbons extend over a range of about 60 ppm. Generally, as the number of the carbon substituent is increased, the chemical shifts move down to low field:

$$\delta_C > \delta_{CH} > \delta_{CH_2} > \delta_{CH_3}$$

A large number of pure hydrocarbons have been completely analyzed and from these obtained values an incremental system has been developed which can be used to predict the chemical shifts of linear and branched hydrocarbons. The chemical shifts of some alkanes are given in Table 13.1 [86].

The simplest relationship describing the effect of the substituent on the shift of a given carbon, k, in alkanes is given by the following equation.

$$\delta_k = -2.3 + \sum A_l n_{kl} + \sum S_{kl} \tag{60}$$

where δ_k is the chemical shift of the carbon nucleus of interest, n_{kl} is the number of the carbon atoms in position l with respect to C atom k, A are the additive shift parameters, and S_{kl} are the steric correction parameters.

Table 13.1

^{13}C-NMR chemical shift values of some hydrocarbons

Compound	Chemical shifts of the carbon atoms				
	C_1	C_2	C_3	C_4	C_5
Methane	− 2.3				
Ethane	5.7				
Propane	15.8	16.3			
n-Butane	13.4	25.2			
n-Pentane	13.9	22.8	34.7		
n-Hexane	14.1	23.1	32.2		
n-Heptane	14.1	23.2	32.6	29.7	
n-Octane	14.2	23.2	32.6	29.9	
n-Nonane	14.2	23.3	32.6	30.0	30.3
n-Decane	14.2	23.2	32.6	31.1	30.5
i-Butane	24.5	25.4			
2-Methylbutane	22.2	31.1	32.0	11.7	
2,2-Dimethylbutane	29.1	30.6	36.9	8.9	
2,3-Dimethylbutane	19.5	34.3			
2,2,3-Trimethylbutane	27.4	33.1	38.3	16.1	
2-Methylpentane	22.7	28.0	42.0	20.9	14.3
3-Methylpentane	11.5	29.5	36.9	18.8 (C_6)	
3,3-Dimethylpentane	7.7	33.4	32.2	25.6 (C_6)	

The constant − 2.3 ppm is the chemical shift of methane. Eq. 60 shows that the shift and steric correction parameters have to be added to the chemical shift value of methane. The corresponding increment values (shift parameters) are listed in Tables 13.2 and 13.3. For example, for the α position the additive shift parameter A is 9.1 ppm. This means that any carbon atom in the α position in respect to the resonating carbon will shift the resonance of the carbon atom of interest 9.1 ppm downfield. If the number of the α-carbon atoms is n, the shift parameter A_l will be multiplied by the number n and the resulting value will be added to the constant − 2.3 ppm. For example, in the case of two α-carbon atoms $n = 2 \times 9.1 = 18.2$ ppm has to be added to − 2.3 ppm.

Besides the shift parameters, steric correction parameters are also required in order to calculate the chemical shift parameters. First, we have to determine the substitution of the carbon atoms. Primary, secondary, tertiary, and quaternary carbon atoms are symbolized

Table 13.2

Additive shift parameters [86]

Position of the carbon atom l	A_l values (ppm)
α	9.1
β	9.4
γ	− 2.5
δ	0.3
ε	0.1

Table 13.3

Additive correction parameters

Observed carbon atom	Neighbor carbon atom l			
	1°	2°	3°	4°
1° (primary)	0.0	0.0	− 1.1	− 3.4
2° (secondary)	0.0	0.0	− 2.5	− 7.5
3° (tertiary)	0.0	− 3.7	− 9.5	− 15.0
4° (quaternary)	− 1.5	− 8.4	− 15.0	− 25.0

by 1°, 2°, 3° and 4°, respectively. The next step we have to determine the number and substitution of the directly attached carbon atoms. The connectivity symbol 2°(4°) indicates the observed carbon atom k to be secondary and attached to a quaternary carbon l. For example, let us assume that two primary, one secondary and one quaternary carbon atoms are attached to a tertiary carbon atom. In this case, the steric correction parameter for the primary carbon atom is 0 ppm, for the secondary − 3.7 ppm and for the quaternary − 15.0 ppm (Table 13.3).

For a better understanding and illustration of the shift parameters, we calculate the chemical shift values of two different hydrocarbons by using the shift and steric correction parameters given in Tables 13.2 and 13.3 and compare the calculated values with those of the experiment.

For example, we calculate the ^{13}C chemical shifts of n-hexane. Because of the symmetry in the molecule, there are only three types of different carbon atoms. We have to calculate the chemical shifts of all three carbons one by one using eq. 60 and the given parameters in Tables 13.2 and 13.3.

$$CH_3 \overline{\quad} CH_2 \overline{\quad} CH_2 \overline{\quad} CH_2 \overline{\quad} CH_2 \overline{\quad} CH_3$$
$$\quad 1 \qquad 2 \qquad 3 \qquad 4 \qquad 5 \qquad 6$$

$$\delta_{C1} = -2.3 + \sum A_\alpha + A_\beta + A_\gamma + A_\delta + A_\varepsilon + \sum S_{kl}$$
$$\delta_{C1} = -2.3 + \sum 9.1 + 9.4 - 2.5 + 0.3 + 0.1 + \sum 0.0$$
$$= 14.1 \text{ ppm (calculated)}; \quad 13.7 \text{ ppm (experimental)}$$
$$\delta_{C2} = -2.3 + \sum 2A_\alpha + A_\beta + A_\gamma + A_\delta + \sum S_{kl}$$
$$\delta_{C2} = -2.3 + \sum 2 \times 9.1 + 9.4 - 2.5 + 0.3 + \sum 0.0$$
$$= 23.1 \text{ ppm (calculated)}; \quad 22.8 \text{ ppm (experimental)}$$
$$\delta_{C3} = -2.3 + \sum 2A_\alpha + 2A_\beta + A_\gamma + \sum S_{kl}$$
$$\delta_{C3} = -2.3 + \sum 2 \times 9.1 + 2 \times 9.4 - 2.5 + \sum 0.0$$
$$= 32.2 \text{ ppm (calculated)}; \quad 31.9 \text{ ppm (experimental)}$$

This example demonstrates that the agreement with the determined values for such calculations is very good. n-Hexane has a linear structure and does not have any branched carbon atoms; therefore, the value of steric correction parameters is zero. After this simple and unbranched hydrocarbon, we calculate the chemical shifts of the carbon atoms in 2,2-dimethylpentane and compare the values with those of the experiment.

$$
\begin{array}{c}
\text{CH}_3 \\
| \\
\text{CH}_3 \;\text{---}\; \underset{1}{\overset{}{\text{C}}} \;\text{---}\; \text{CH}_2 \;\text{---}\; \text{CH}_2 \;\text{---}\; \text{CH}_3 \\
| \\
\text{CH}_3
\end{array}
$$

CH$_3$ ---C--- CH$_2$ --- CH$_2$ ---CH$_3$

147

$$\delta_{C1} = -2.3 + \sum A_\alpha + 3A_\beta + A_\gamma + A_\delta + \sum S_{kl}$$
$$\delta_{C1} = -2.3 + \sum 9.1 + 28.2 - 2.5 + 0.3 + \sum -3.4$$
$$= 29.4 \text{ ppm (calculated);   29.5 ppm (experimental)}$$

$$\delta_{C2} = -2.3 + \sum 4A_\alpha + A_\beta + A_\gamma + \sum S_{kl}$$
$$\delta_{C2} = -2.3 + \sum 36.4 + 9.4 - 2.5 + \sum 3(-1.5) - 8.4$$
$$= 28.1 \text{ ppm (calculated);   30.1 ppm (experimental)}$$

$$\delta_{C3} = -2.3 + \sum 2A_\alpha + 4A_\beta + \sum S_{kl}$$
$$\delta_{C3} = -2.3 + \sum 18.2 + 37.6 + \sum -7.5$$
$$= 48.8 \text{ ppm (calculated);   47.3 ppm (experimental)}$$

$$\delta_{C4} = -2.3 + \sum 2A_\alpha + A_\beta + 3A_\gamma + \sum S_{kl}$$
$$\delta_{C4} = -2.3 + \sum 18.2 + 9.4 - 7.5 + \sum 0.0$$
$$= 17.8 \text{ ppm (calculated);   18.1 ppm (experimental)}$$

$$\delta_{C5} = -2.3 + \sum A_\alpha + A_\beta + A_\gamma + 3A_\delta + \sum S_{kl}$$
$$\delta_{C5} = -2.3 + \sum 9.1 + 9.4 - 2.5 + 0.9 + \sum 0.0$$
$$= 14.6 \text{ ppm (calculated);   15.1 ppm (experimental)}$$

Carbon atom 1 has 1α-, 3β-, 1γ-, and 1δ-carbon atoms and is attached to a quaternary 1°(4°) carbon atom. Steric correction parameter is -3.4 ppm. Carbon atom 2 has 4α-, 1β-, and 1γ-carbon atoms and is attached to three primary 4°(1°) carbon atoms and one secondary carbon atom 4°(2°). Total steric correction parameter is -12.9 ppm. The other carbon chemical shifts are calculated in the same manner. Again, the agreement between the calculated and experimental values is very good. Thus, such additive rules provide a reasonable assignment of the ^{13}C resonances.

Figure 146 illustrates the ^{13}C-NMR spectrum of heptane. The signal with the small intensity belongs to the carbon C$_4$ which resonates at the higher field compared to the neighboring carbon atoms C$_3$ and C$_5$. High-field resonance of this carbon atom is attributed to the steric compression of a γ-gauche effect arising from the methyl groups.

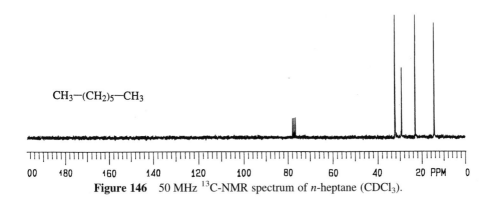

CH₃—(CH₂)₅—CH₃

Figure 146   50 MHz ^{13}C-NMR spectrum of *n*-heptane (CDCl₃).

13.2   SUBSTITUTED ALKANES

As we have seen earlier in the ^{1}H-NMR spectroscopy section, substituents have a considerable influence on the chemical shifts of carbon atoms. The electronic nature of the substituents determines the chemical shift of the attached carbon atom. Generally, the α-effect correlates with substituent electronegativity (inductive effect), while the β-effect is much smaller and is always a deshielding effect. However, there appears to be no direct relationship to the electronegativity. The upfield shift of the γ-carbon atoms results from the steric compression of a gauche interaction. Table 13.4 provides the shift values of different substituents.

Thus, for example, if we want to calculate the ^{13}C chemical shifts of a substituted hydrocarbon, we first use eq. 60 to calculate the chemical shifts of the corresponding parent hydrocarbon. These values can also be taken from the tables. Later on, the shifts parameter and steric correction parameter has to be added to the appropriate shift values of the alkanes given in Table 13.4 [87].

$$\delta_C = \delta_{(RH)} + Z_i + \sum S_{kl} \tag{61}$$

where Z_i is the shift parameter of the substituent and S_{kl} are the steric correction parameters.

By way of two examples we will attempt to demonstrate that the agreement between the calculated values and experimental values is sufficiently good. For example, we can calculate the ^{13}C shifts for 1-octanol. Table 13.4 provides the substituent shift parameters. These values must be added to the appropriate shift values of octane.

	CH₃	—CH₂	—CH₂	—CH₂	—CH₂	—CH₂	—CH₂	—CH₂—OH
	8	7	6	5	4	3	2	1
Octane	13.9	22.9	32.2	29.5	29.5	32.2	22.9	13.9
Subst. Effect	–	–	–	–	0.0	−6.2	10.1	49.0
Calculated	13.9	22.9	32.2	29.5	29.5	26.0	33.0	62.9
Experimental	13.9	22.8	32.1	29.6	29.7	26.1	32.9	61.9

Table 13.4

Shift parameters of the substituents [87]

Substituent	Position of the substituent			
	α	β	γ	δ
–F	70.0	7.8	−6.8	0.0
–Cl	31.0	10.0	−5.1	−0.5
–Br	18.9	11.0	−3.8	−0.7
–I	−7.2	10.9	−1.5	0.9
–OH	49.0	10.1	−6.2	0.0
–OR	58.0	7.2	−5.8	0.0
–OCOR	54.0	6.5	−6.0	0.0
–COOH	20.1	2.0	−2.8	0.0
–COOR	22.6	2.0	−2.8	0.0
–COCl	33.1	2.3	−3.6	0.0
–CN	3.1	2.4	−3.3	−0.5
–CHO	29.9	−0.6	−2.7	0.0
–COR	22.5	3.0	−3.0	0.0
–CH=CHR	20.0	6.9	−2.1	0.4
–C≡C–	4.4	5.6	−3.4	−0.6
–Ph	22.1	9.3	−2.6	0.3
–NH$_2$	28.3	11.2	−5.1	0.1
–NO$_2$	61.6	3.1	−4.6	−1.0
–SH	10.6	11.4	−3.6	−0.4
–SCH$_3$	20.4	6.2	−2.7	0.0

The agreement is sufficiently good in order to immediately give the assignment of the spectrum (Figure 147).

If more than one substituent is attached to a carbon atom, the chemical shift of that carbon atom can be calculated by using the additivity principle. Let us calculate the chemical shift values of the amino acid leucine by using substituent shift parameters [88]. First of all we replace the substituents by hydrogen atoms and then we obtain the parent

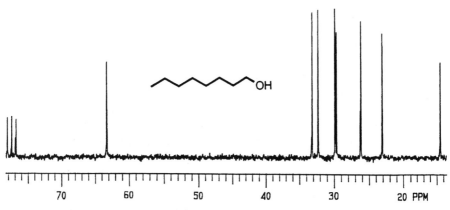

Figure 147   50 MHz ^{13}C-NMR spectrum of *n*-octanol (CDCl$_3$).

hydrocarbon 2-methylbutane. The carbon chemical shifts of 2-methylbutane are known. We add the substituents shift parameters and steric correction parameter to the appropriate chemical shift values of 2-methylbutane and calculate the chemical shift values of leucine. All of the calculated shifts are in reasonable agreement with the determined values.

$$\underset{NH_2}{\overset{1\qquad 2\qquad 3\qquad 4\qquad 5}{HOOC-CH-CH_2-CH-CH_3}} \qquad Leucine$$

with CH₃ on position 4; NH₂ on position 2.

$$\underset{CH_3}{\overset{1\qquad 2\qquad 3\qquad 4}{H_3C-CH_2-CH-CH_3}} \qquad 2\text{-Methylbutane}$$

	1	2	3	4
2-Methylbutane (calculated)	11.2	32.2	30.7	22.0 ppm
2-Metylbutane (experimental)	11.4	31.7	29.7	21.9
Z–COOH	20.1	2.0	−2.8	0.0
Z–NH₂	28.3	11.2	−5.1	0.0
Steric correction	−3.7	−5.0	−3.7	−1.1
Leucine (calculated)	56.1	40.0	18.1	20.8
Leucine (experimental)	54.8	41.0	22.1	25.4

13.3 CYCLOALKANES

The chemical shifts of unsubstituted cycloalkanes are given in Table 13.5. Common cycloalkanes absorb within a very narrow range (22–27 ppm) [89]. However, the remarkable high-field shift of the cyclopropane carbon atoms, analogous to the shift of its proton absorptions, deserves some explanation. This high-field shift is explained in terms of a ring current effect. Some ring strain is also responsible for this unusual chemical shift.

Another striking feature in the chemical shifts of cycloalkanes is low-field shift (28.20 ppm) of the carbon resonances with increasing ring size (up to a seven-membered ring). With further increasing ring size, the chemical shift starts to drop to 23.2 ppm by a 12-membered ring. Then, it starts again to increase and with higher cycloalkanes approaches the value of 29.7 ppm for the central methylene carbon in long-chain unbranched alkanes. In the medium-sized cycloalkanes, the steric effect caused by the internal hydrogen atoms is responsible for the high-field shift of the carbon resonances.

As we have discussed in the case of acyclic alkanes, the replacement of a proton by any substituent in cyclic alkanes gives rise to changes of the chemical shifts, particularly at the carbon bearing the substituent. The α-, β-, and γ-effects are also observed in cycloalkanes. Although each ring skeleton has its one set of shift parameters, substituents have a very similar effect on the cyclohexane ring as in the case of acyclic systems. Chemical shifts of substituted cyclohexanes can be calculated by the following equation:

$$\delta_C = 27.6 + \sum Z_i \tag{62}$$

Table 13.5

^{13}C-NMR chemical shifts of some cycloalkanes

Number of the ring carbon atoms n	δ
3	− 3.8
4	22.10
5	26.30
6	27.60
7	28.20
8	26.60
9	25.80
10	25.00
11	25.40
12	23.20
13	25.20
14	24.60
15	26.60
16	26.50
17	26.70
18	27.50

The shift parameters of various substituents on the cyclohexane ring are given in Table 13.6 [90].

131   **132**

Cyclohexane ring rapidly flips back and forth between two *equivalent* chair conformations (**131** and **132**) at room temperature. By the replacement of a hydrogen atom by a substituent, the ring inversion process will still take place. However, the conformers are not identical any more. The monosubstituted cyclohexane ring has two different interconvertible chair conformations. The chemical shifts of these conformational isomers are also different. In one conformation, the substituent occupies an equatorial position; in the other one the substituent occupies the axial position. Therefore, the shift parameters of the substituents are changing in connection with their positions (axial or equatorial). The shift parameters of some substituents depending on their positions are given in Table 13.7.

The ring inversion process which causes the substituent to exchange between equatorial and axial positions can be frozen out at lower temperatures. Then, two different conformers can be observed separately by NMR spectroscopy. This provides us with the ability to determine the shift parameters of the substituents in equatorial and axial positions separately. As one can see from Table 13.7, there are striking differences between the shift parameters of the substituents in connection with their positions.

Table 13.6

Shift parameters of the substituents attached to cyclohexane ring

Substituent	Shift parameters (cyclohexane $\delta_C = 27.6$ ppm)			
$\langle\ \rangle$—X	C_1	$C_{2,6}$	$C_{3,5}$	C_4
–F	62.9	5.5	–4.1	–1.6
–Cl	32.2	9.6	–2.4	–2.0
–Br	25.0	10.3	–1.5	–2.0
–I	4.2	12.2	–0.2	–2.1
–OH	42.4	8.4	–2.6	–1.2
–OCH$_3$	51.0	4.7	–3.3	–0.9
–OCOCH$_3$	44.7	4.6	–3.2	–1.5
–COOH	16.1	2.0	–1.4	–1.0
–CHO	19.6	–1.0	–2.4	–1.4
–COCH$_3$	15.8	2.0	–1.6	–1.2
–COCl	27.8	2.1	–2.1	–1.7
–CN	0.7	2.5	–3.0	–1.8
–NH$_2$	23.5	10.1	–1.8	–1.1
–NH$_3^+$	23.9	5.8	–2.0	–1.6
–NO$_2$	57.0	3.8	–2.9	–2.1
–SH	10.9	10.9	–0.8	–1.7
–Ph	17.5	7.3	–0.2	–0.9
–CH$_3$	5.8	8.4	–0.5	–0.6
–C$_2$H$_5$	12.6	6.1	–0.5	–0.2
–C$_4$H$_9$	10.8	6.5	–0.5	–0.3
–C(CH$_3$)$_3$	21.2	0.5	0.1	–0.5

Notable features of this are the γ-gauche steric effect of about -6 ppm of an axially positioned methyl group. The substituents bonded to cyclopentane or cycloheptane rings have comparable effects on the chemical shifts of the ring carbon atoms. The α- and β-effect can be seen much more clearly. Since cyclopentane has a half-chair conformation, the γ-effect (1,3-diaxial steric interaction) is small.

One of the most encountered molecules in organic chemistry is the norbornane skeleton. We can observe similar substituents effect in this compound. Substituents attached to the C_2 carbon atom can have two different configurations: *exo* or *endo* (**147** and **148**).

147   148

The upfield shift of the methylene bridge carbon atom when there is a substituent in the

Table 13.7

Shift parameters of the substituents attached to cyclohexane ring in axial and equatorial position

Substituted cyclohexanes	Shift parameters Z_i			
	C_1	$C_{2,6}$	$C_{3,5}$	C_4
CH$_3$ (axial)	1.5	5.4	-6.4	-0.1
CH$_3$ (equatorial)	5.9	9.0	0.0	0.2
COOCH$_3$ (axial)	12.1	0.7	-3.9	-0.3
COOCH$_3$ (equatorial)	16.3	2.5	-1.1	-0.7
Br (axial)	28.4	7.9	-5.5	-0.6
Br (equatorial)	25.1	11.8	1.4	-1.4

exo configuration is readily explained in terms of the γ-gauche effect. If the substituent is positioned in the *endo* position, then the γ-gauche effect is observed at the C_6 carbon atom. By comparison of the chemical shifts of the C_2 substituted norbornane derivatives, it is possible to distinguish between the *exo* and *endo* isomers. Values given in Table 13.8 demonstrate this finding clearly.

13.4   ALKENES

In the introduction we have discussed that the diamagnetic shielding is not so effective in determining the ^{13}C chemical shifts of alkenes. However, alkenes have a higher energy π molecular orbital (HOMO) and the magnetically allowed $\pi \rightarrow \pi^*$ transition ($\Delta E \approx 8$ eV) is of reasonably low energy. Therefore, paramagnetic shielding is responsible in the shifting of the resonance frequencies of alkenes downfield. The alkenes substituted only by alkyl groups absorb in the range of 110–150 ppm. However, the chemical shift range for substituted alkenes is 75–175 ppm relative to TMS. The electronic nature and the number of the substituents attached to the alkene carbon can shift the resonance frequencies down- or upfield. Before analyzing the electronic effects of the substituents let us first discuss the chemical shifts of alkyl-substituted olefines. Again, shifts parameter of the alkyl groups have to be considered in order to determine the alkene carbon

Table 13.8

^{13}C-NMR chemical shift values in *endo*- and *exo*-2-substituted norbornanes

Chemical shifts (ppm)

X =	C$_1$	C$_2$	C$_3$	C$_4$	C$_5$	C$_6$	C$_7$	X
H	36.8	30.1	30.1	36.8	30.1	30.1	38.7	
exo-OH	44.5	74.2	42.4	35.8	28.8	24.9	26.6	–
endo-OH	43.1	72.2	39.6	37.7	30.3	20.4	37.8	–
exo-NH$_2$	45.7	55.4	42.5	36.4	28.9	27.0	34.3	–
endo-NH$_2$	43.6	53.4	40.6	38.0	30.7	20.6	39.0	–
exo-CN	42.3	31.1	36.4	36.5	28.6	28.5	37.4	123.4
endo-CN	40.2	30.2	35.6	37.0	29.4	25.2	38.7	122.6
exo-COOCH$_3$	41.9	46.5	34.3	36.4	29.0	28.7	36.6	175.7
endo-COOCH$_3$	40.8	46.0	32.3	37.5	29.4	25.1	40.4	174.6
exo-CH$_3$	43.5	36.8	40.2	37.3	30.3	29.0	35.0	22.3
endo-CH$_3$	42.2	34.6	40.7	38.2	30.6	25.1	38.9	17.4

chemical shifts.

$$-C_\gamma-C_\beta-C_\alpha-\mathbf{C}_k\!=\!\mathbf{C}-C_{\alpha'}-C_{\beta'}-C_{\gamma'}-$$

To calculate the chemical shifts of alkene carbon atoms, the substituent shift parameters of the substituent attached to two alkene carbon atoms (C_k and C) have to be considered at the same time. Furthermore, steric correction parameters have to be added. With a minimal deviation, the chemical shifts of alkene carbon atoms can be calculated by eq. 63. The constant value 122.1 is the chemical shift of ethylene.

$$\delta_k = 122.1 + \sum A_l n_{kl} + \sum S_{kl} \tag{63}$$

where δ_k is the chemical shift of the carbon nucleus of interest, n_{kl} is the number of the carbon atoms in position l with respect to C atom k, A are the additive shift parameters, and S_{kl} are the steric correction parameters.

Table 13.9 shows the shift parameters, α, β, and γ represent substituent shift parameters on the same end of the double bond as the alkene carbon to be determined, whereas α', β', and γ' represent substituents on the other olefinic carbon atom. For better comprehension, let us calculate the chemical shifts of sp^2 carbon atoms of an alkyl-substituted alkene, 2-ethyl-pentene (**149**).

Table 13.9

Alkene shifts and correction parameters

Position of the carbon atom l	Shift parameters	Type of substitution	Steric shift parameters
α	11.0	*trans*	0.0
β	6.0	α,α' (*cis*)	−1.2
γ	−1.0	α,α (*geminal*)	−4.9
α'	−7.1	α',α' (*geminal*)	1.2
β'	−1.9	β',β' (*geminal*)	1.3
γ'	1.1		

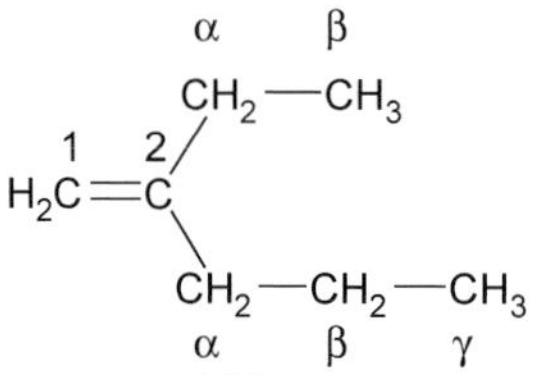

$$\begin{array}{cc}
\alpha & \beta \\
CH_2 & \!\!\!-CH_3
\end{array}$$

$$\begin{array}{cc}
1 & 2 \\
H_2C = C
\end{array}$$

$$\begin{array}{ccc}
CH_2 & \!\!\!-CH_2\!\!\! & -CH_3 \\
\alpha & \beta & \gamma
\end{array}$$

149

$$\delta_{C2} = 122.1 + 2\alpha + 2\beta + \gamma + \alpha, \alpha \text{ (gem)}$$
$$\delta_{C2} = 122.1 + 2(11.0) + 2(6.0) - 1.0 - 4.9$$
$$= 150.2 \text{ ppm (calculated)};\quad 150.9 \text{ ppm (experimental)}$$
$$\delta_{C1} = 122.1 + 2\alpha' + 2\beta' + \gamma' + \alpha', \alpha' \text{ (gem)}$$
$$\delta_{C1} = 122.1 + 2(-7.1) + 2(-1.9) + 1.1 + 1.2$$
$$= 106.4 \text{ ppm (calculated)};\quad 108.1 \text{ ppm (experimental)}$$

Comparison of the calculated values with those of measured values indicates that the chemical shifts of alkyl-substituted olefines can be estimated with a small deviation. Representative alkene chemical shifts are presented in Table 13.10 and cyclic alkenes in Table 13.11. The ¹³C-NMR spectra of two hydrocarbons are given in Figure 148.

Some anomalous chemical shifts in some cyclic alkenes can be observed. For example, the exocyclic double bond carbon atoms in 7-methylenenorbornadiene (**150**) show a chemical shift difference of about 100 ppm. The terminal carbon atom in **150** resonates at 78.5 ppm, whereas the bridge carbon atom appears at 177.1 ppm. This unusual chemical shift difference cannot be explained with the different substitution pattern of the double bond carbon atoms. It arises from the homoconjugative interaction between the endocyclic and the exocyclic double bond electrons as shown below.

Table 13.10

^{13}C-NMR chemical shift in some selected alkenes

Compounds	Chemical shifts (ppm)					
	C_1	C_2	C_3	C_4	C_5	C_6
Ethene	123.5					
Propene	115.9	133.4	19.4			
1-Butene	113.5	140.5	27.4	13.4		
1-Pentene	114.5	139.0	36.2	22.4	13.6	
1-Hexene	114.2	139.2	33.8	31.5	22.5	14.0
cis-2-Butene	11.4	124.2				
trans-2-Butene	16.8	125.4				
trans-2-Pentene	17.3	123.5	133.2	25.8	13.6	
cis-3-Hexene	14.1	25.9	131.2			
1,3-Butadiene	116.6	137.2				
2,4-Hexadiene	12.9	124.9	125.3			
2,3-Dimethyl-1,3-butadiene	113.0	143.8				

Table 13.11

^{13}C-NMR chemical shift in some selected cyclic alkenes

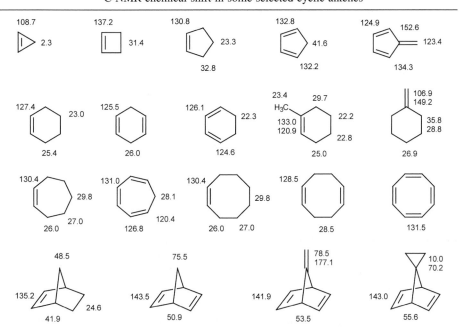

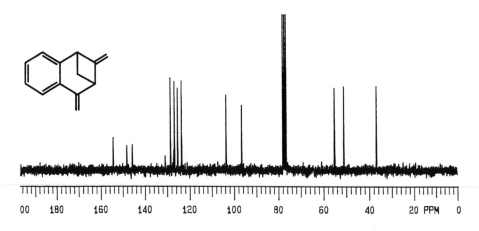

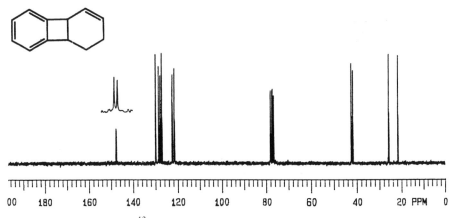

Figure 148   50 MHz ^{13}C-NMR spectrum of two different hydrocarbons (CDCl$_3$).

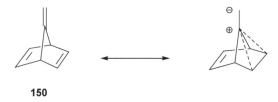

150

This homoconjugative interaction pushes the electrons toward the *exo* methylene carbon that resonates at high field. The low-field resonance of the bridge carbon atom is attributed to the decreased electron density. This example demonstrates how the ^{13}C chemical shifts are sensitive against the electron density.

By the attachment of a substituent on an alkene carbon, the chemical shift of the alkene carbon atoms can be easily estimated by using the corresponding shift parameters (Table 13.12). Substituents with different mesomeric effects ($+$M or $-$M effect) of course have a different influence on the α- and β-carbon atoms. Changes in the electron

Table 13.12

Shift parameters in substituted alkenes

Substituent	Shift parameters Z_i	
$\overset{1}{X}\overset{}{-}\overset{2}{CH}\text{=}CH_2$	C_1	C_2
$-H$	0.0	0.0
$-CH_3$	10.6	-8.0
$-C_2H_5$	15.5	-9.7
$-CH_2-CH_2-CH_3$	14.0	-8.2
$-CH(CH_3)_2$	20.3	-11.5
$-C(CH_3)_3$	25.3	-13.3
$-CH\text{=}CH_2$	13.6	-7.0
$-Ph$	12.5	-11.0
$-CH_2OR$	13.0	-8.6
$-CHO$	13.1	12.7
$-COCH_3$	15.0	5.9
$-COOH$	4.2	8.9
$-COOR$	6.0	7.0
$-CN$	-15.1	14.2
$-OR$	28.8	-39.5
$-OCOR$	18.0	-27.0
$-NR_2$	16.0	-29.0
$-NO_2$	22.3	-0.9
$-F$	24.9	-34.3
$-Cl$	2.6	-6.1
$-Br$	-7.9	-1.4
$-I$	-38.1	7.0

densities of a double bond caused by the mesomeric substituents are shown below.

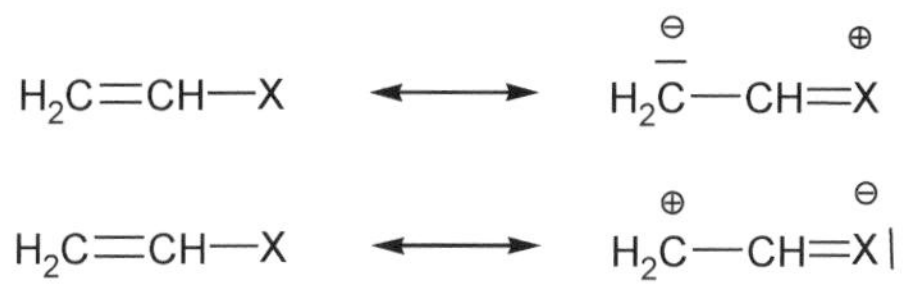

$$H_2C\text{=}CH-X \longleftrightarrow \overset{\ominus}{H_2C}-CH\text{=}\overset{\oplus}{X}$$

$$H_2C\text{=}CH-X \longleftrightarrow \overset{\oplus}{H_2C}-CH\text{=}\overset{\ominus}{X}|$$

Generally, mesomeric electron-donating groups ($+M$ effect) increase the electron density of the double bond and shift the resonances to higher field. On the other hand, electron-withdrawing substituents decrease the electron density and cause low-field resonance of the alkene carbon atoms. Shifts for polysubstituted alkene carbon atoms can be approximated by applying the principle of substituent shift additivity. However, it should be always kept in mind that the electronic and steric interactions can always cause some deviation in the chemical shifts by the increased number of the substituents. Therefore, one has to be careful by using the shift parameters presented in Table 13.12 for the highly substituted alkenes. These values always cause deviation for the tri- and tetrasubstituted alkenes.

Table 13.13

^{13}C-NMR chemical shift in some selected alkynes and allenes

CH≡CH
71.9

H₃C—C≡C—C≡CH
3.9   74.4 65.4 68.8 64.7

H₂C=C=CH₂
73.5 212.6

H₃C   200.0   CH₃
    C=C=C
H₃C   91.6   CH₃
20.0

Ring structures with values: 20.8, 34.7, 94.4, 29.7; 20.2, 95.8; 92.7, 206.5, 27.2, 25.8, 27.9

We may include alkynes and allenes in this chapter and discuss the chemical shifts of them here. Carbon-13 chemical shifts of alkynes, which contain sp-hybridized carbon atoms, are found in the range of approximately 60–95 ppm. High-field shift of acetylenic carbon atoms can be explained on the basis of higher electronic excitation energy of the triple bond which decreases the contribution of the paramagnetic shielding. Therefore, the chemical shifts of the alkyne carbon atoms are observed at the higher field compared with alkenes.

Allenes form a unique class of compounds because of the extremely low-field shift of the central carbon atom (195–215 ppm). The sp carbon atom in allene shows a low-field shift of more than 100 ppm compared with alkynes (Table 13.13). On the other hand, the terminal sp²-hybridized carbon resonances appear at high field. This anomalous shift difference observed in allenes can be attributed to the following resonance structure.

13.5   AROMATIC COMPOUNDS

The fact that the aromatic protons resonate 1–2 ppm at the lower field in the ^{1}H-NMR spectroscopy compared with the olefinic protons is the best criteria for the aromaticity. Although proton shielding is markedly influenced by the ring current effect, the same influence is less significant in ^{13}C-NMR spectroscopy. Therefore, there is no low-field shift of aromatic carbon atom resonances compared with those of olefinic carbon atoms. The resonances for aromatic carbons appear in the same region as those for olefinic carbons. Generally, aromatic carbon resonances occur between 100 and 150 ppm. On the attachment of more than one electron-withdrawing and electron-releasing substituents, this shift range may expand to 90–180 ppm.

There is a direct relationship between the aromatic carbon resonances and electron density in the aromatic ring. By the attachment of a substituent to the benzene ring, the electron density, consequently the chemical shifts of the ring carbon atoms will either increase or decrease depending on the electronic nature of the substituents and on their position (*ipso*, *ortho*, *meta* and *para*).

Generally, alkyl substituents have practically no influence on the chemical shifts of the *meta*-carbon atoms. For the *para* position, a high-field shift of 3.0 ppm is found in toluene compared with benzene. However, for the *ipso* position a low-field shift of 9.3 ppm is found, going from benzene to toluene. Increasing the branching of the alkyl chain will shift the resonance of the *ipso*-carbon atom further down field. For example, benzene carbons resonate at 128.5 ppm, the *ipso*-carbon atom of toluene appears at 137.8 ppm, whereas the *ipso*-carbon atom in *t*-butylbenzene resonates at 150.9 ppm. Increasing the branching of the alkyl chain (γ effect) causes an upfield shift in the resonances of *ortho*-carbon atoms. Carbon chemical shifts of some selected aromatic compounds are given in Table 13.14.

Substituents with ±M effect will affect the electron density in the *ortho* and *para* positions. *meta*-Carbon atoms remain almost unaffected by all kinds of substituents. However, the inductive effect, operating through σ-electrons will mainly influence the chemical shift of the *ipso*- and *ortho*-carbon atoms. Substituents with an electron-donating ability will increase the electron density at the benzene ring. As a result of this effect, aromatic carbon resonances will appear at the higher field as discussed in the aromatic proton resonances. By decreasing the electron density, the opposite effect will be observed. On the other hand, some polar substituents such as a nitro or carbonyl group can shift the resonance frequency of the *ortho*-carbon atoms upfield in spite of the fact that they are strongly electron-withdrawing substituents by the mesomeric effect. Such groups generate an intramolecular *electric field* which influences the electron density in the molecule. As a consequence of this effect, C–H can be distorted and the bonding electrons are shifted towards the carbon atom so that σ-electron densities in the *o*- and *o′*-carbon atoms are increased. This effect overcompensates for the electron-withdrawal effect of the nitro group. As a result of this effect, the *ortho*-carbon atoms in nitrobenzene resonate 4.3 ppm upfield (124.3 ppm).

Steric effect is clearly seen in the chemical shifts of *ortho*-carbon atoms of mono- and *ortho*-disubstituted benzene derivatives. Shielding of *ipso*-carbon atoms in benzonitrile and acetylbenzene arises from the *anisotropy effect* of the C≡C and C≡N triple bond. The *ipso*-carbon in acetylene resonates around 5.8 ppm upfield, whereas the corresponding carbon atom in benzonitrile appears 15.5 ppm upfield. The large upfield shift (32.3 ppm) of *ipso*-carbon resonance in iodobenzene is attributed to the *heavy atom effect*.

Chemical shifts for polysubstituted aromatic compounds can be easily predicted on the basis of a simple additive relationship for the induced shifts. Empirical substituent shift

Table 13.14

^{13}C-NMR chemical shift in some selected aromatic compounds

parameters are given in Table 13.15, which permit us to predict the carbon chemical shifts in multisubstituted benzenes according to the following equation:

$$\delta_C = 128.5 + \sum Z_i \tag{64}$$

Table 13.15

Shift parameters in substituted benzenes [87]

X =	ipso	ortho	meta	para
−H	0.0	0.0	0.0	0.0
−CH$_3$	9.3	−0.1	0.7	−3.0
−CH(CH$_3$)$_2$	20.3	−0.2	−0.1	−2.6
−CH=CH$_2$	9.1	−2.4	−0.2	−0.9
−C≡CH	−5.8	3.9	0.1	0.4
−Ph	13.0	−1.1	0.5	−1.0
−F	35.0	−14.4	0.9	−4.4
−Cl	6.4	0.2	1.0	−2.0
−Br	−5.9	3.0	1.5	−1.5
−I	−32.3	9.9	2.6	−0.4
−OH	26.6	−12.8	1.6	−7.1
−OCH$_3$	31.4	−14.4	1.0	−7.8
−OCOCH$_3$	23.0	−6.4	1.3	−2.3
−SH	2.0	0.6	0.2	−3.3
−NH$_2$	20.0	−14.1	0.6	−9.6
−N(CH$_3$)$_2$	22.2	−15.8	0.5	−11.8
−NO$_2$	20.6	−4.3	1.3	6.2
−COCl	4.8	2.9	0.6	6.9
−COOH	2.9	1.3	0.4	4.6
−COOCH$_3$	2.1	1.2	0.0	4.4
−CHO	8.2	0.5	0.5	5.8
−COCH$_3$	8.9	−0.1	−0.1	4.5
−CN	−15.5	1.4	1.4	5.0

Benzene resonates at 128.5 ppm. As we have discussed for the other systems, the effects of the substituents on chemical shifts are additive. In most cases, there is a good agreement between the calculated and experimental values. For example, let us calculate the carbon resonances of methyl 3,5-dinitrobenzoate.

$$\delta_{C1} = 128.5 + \sum Z_{ipsoCOOR} + 2 \times Z_{metaNO_2}$$

$$\delta_{C1} = 128.5 + 2.1 + 2 \times 1.3 = 133.2 \text{ ppm (calculated)}; \quad 134.2 \text{ ppm (experimental)}$$

$$\delta_{C2,6} = 128.5 + \sum Z_{orthoCOOR} + Z_{orthoNO_2} + Z_{paraNO_2}$$

$$\delta_{C2,6} = 128.5 + 1.2 - 4.3 + 6.2$$

$$= 131.6 \text{ ppm (calculated)}; \quad 129.4 \text{ ppm (experimental)}$$

$$\delta_{C3,5} = 128.5 + \sum Z_{ipsoNO_2} + Z_{metaCOOR} + Z_{metaNO_2}$$
$$\delta_{C3,5} = 128.5 + 20.6 + 0.0 + 1.3$$
$$= 149.4 \text{ ppm (calculated); } 148.5 \text{ ppm (experimental)}$$
$$\delta_{C4} = 128.5 + \sum Z_{paraCOOR} + 2 \times Z_{orthoNO_2}$$
$$\delta_{C4} = 128.5 + 4.4 - 2 \times 4.3$$
$$= 124.3 \text{ ppm (calculated); } 122.2 \text{ ppm (experimental)}$$

It has to be kept in mind always that some deviations from the calculated chemical shifts can be encountered, especially in the case of *ortho* di-, tri- and polysubstituted benzene derivatives. Increased crowding of the substituent around the benzene ring can cause steric interactions between the substituents with π-bonds, which in turn may hinder the conjugation of the substituents with the benzene ring. For example, the carbonyl carbon in methyl benzoate (**153**) resonates at 166.9 ppm. However, by introducing two methyl and two *tert*-butyl groups in the *ortho* positions (**154, 155**), the chemical shift of the carbonyl group moves successively to the lower field. The bulky groups prevent the conjugation of the carbonyl group with the benzene ring because of the twisting of the planes of both π-systems.

153	**154**	**155**
δ_{co} = 166.9 ppm	δ_{co} = 170.4 ppm	δ_{co} = 173.1 ppm

The ^{13}C-NMR spectra of three different monosubstituted benzene derivatives are illustrated in Figure 149. By the careful analysis of the chemical shifts in these spectra, the effect of the substituents on the ring carbon atoms can be clearly seen. Since the methoxyl group in anisol donates the electrons to the benzene ring (+M effect) the *ortho*- and *para*-carbon resonances shifted to higher field, whereas the *meta* position is not much different from benzene. The *ipso*-carbon atom resonates at 159.9 ppm. In the case of benzaldehyde, the most affected carbon atom is the *para*-carbon atom. The *ortho*-carbon atom is under the influence of two opposite effects: (i) mesomeric effect and (ii) electric field effect of the carbonyl group which are compensated. The *meta*-carbon atom remains unaffected.

The carbon chemical shifts of some selected substituted aromatic compounds are presented in Table 13.16.

In substituted naphthalenes the effect of the substituents on the chemical shift values depends very strongly on their position. For example, a substituent attached to the α-carbon atom has a different influence on the neighboring carbon atoms C_2 and C_9. For

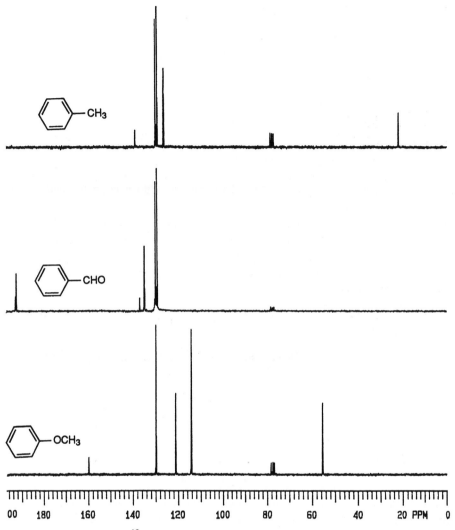

Figure 149 50 MHz ^{13}C-NMR spectra of toluene, benzaldehyde and anisol (CDCl$_3$).

example, an electron-donating substituent will increase the electron density on the carbon atoms C$_2$ and C$_9$ according to the following resonance structures of naphthalene:

loss of aromaticity

However, the canonical formula C will not contribute much to the resonance, due to the disrupted aromatic sextets. In the case of β-substituted naphthalene derivatives, the aromatic sextet of the second ring is disrupted by the canonical formula C′. Therefore, the C_1 carbon atom will be influenced more than the C_3 carbon atom. Generally, the C_2 carbon atom in α-substituted naphthalenes and C_1 carbon atom resonances in β-substituted naphthalenes are mostly affected by

Table 13.16

^{13}C-NMR shift parameters in some mono-, di- and trisubstituted benzenes

Compounds	Chemical shifts						
	C_1	C_2	C_3	C_4	C_5	C_6	Substituent
Toluene	137.8	129.2	128.4	125.5			21.3
Ethylbenzene	144.3	128.1	128.6	125.9			29.7, 15.8
t-Butylbenzene	150.9	125.4	128.3	125.7			34.6, 31.4
Styrene	137.6	126.1	128.3	127.6			137.1, 113.3
Phenol	155.1	115.7	130.1	121.4			
Anisol	159.9	114.1	129.5	120.7			
Aniline	148.7	114.4	129.1	116.3			
Nitrobenzene	149.1	124.2	129.8	134.7			
Benzonitrile	112.8	132.1	129.2	132.8			119.5
Benzaldehyde	136.7	129.7	129.0	134.3			192.0
Benzophenone	137.6	130.0	128.3	132.3			196.4
Benzoic acid	131.4	129.8	128.9	133.1			168.1
o-Dichlorobenzene	132.6		130.5	127.7			
m-Dichlorobenzene	134.9	128.5		126.9	130.8		
Catechol	147.1		119.3	124.1			
Hydroquinone	151.5	118.5					
Phthalic acid	120.7		133.6	138.6			181.0
Phthalaldehyde	137.2		133.1	135.8			196.1
o-Chlorophenol	151.8	120.3	129.3	121.5	128.5	116.3	
p-Chlorophenol	154.1	116.8	129.7	125.9			
o-Nitrophenol	155.3	133.8	125.2	120.4	137.7	120.1	
p-Nitrophenol	161.5	115.9	126.4	141.7			
o-Nitroaniline	144.5	131.7	125.6	118.5	135.2	116.5	
m-Nitroaniline	151.4	109.0	150.4	117.7	131.1	121.6	
o-Aminobenzoic acid	111.5	152.9	116.4	135.3	118.0	132.8	169.1
p-Aminobenzoic acid	118.7	132.8	114.3	154.6			171.3
2,3-Dichlorophenol	152.0	118.5	129.7	121.8	133.9	116.8	
2,4,6-Trinitrophenol	153.2	139.2	126.2	138.2			

Table 13.17

^{13}C-NMR shift parameters in some substituted naphthalenes [91]

Compounds	C_1	C_2	C_3	C_4	C_5	C_6	C_7	C_8	C_9	C_{10}
1-Bromonaphthalene	129.9	130.7	127.1	128.9	129.2	127.5	128.2	127.3	132.6	135.5
1-Hydroxynaphthalene	151.5	108.7	125.8	120.7	127.6	126.4	125.2	121.4	124.3	134.6
1-Aminonaphthalene	142.0	109.4	126.2	118.7	128.3	125.6	124.6	120.7	123.4	134.3
1-Nitronaphthalene	146.5	123.8	123.9	134.5	128.5	127.2	129.3	122.9	124.9	134.2
1-Carboxynaphthalene	132.2	137.4	125.9	135.8	129.3	127.6	129.6	125.3	131.0	134.5
2-Bromonaphthalene	130.6	120.1	129.8	130.7	128.6	127.2	127.8	127.8	135.4	132.8
2-Hydroxynaphthalene	109.4	153.2	117.6	129.8	127.7	123.5	126.4	126.3	134.5	128.5
2-Aminonaphthalene	108.4	144.0	118.1	129.0	127.6	122.3	126.2	125.6	134.8	127.8
2-Nitronaphthalene	125.1	146.3	119.7	130.4	128.7	130.5	128.7	130.8	132.7	136.6
2-Carboxynaphthalene	135.1	135.2	123.1	129.8	128.8	129.8	127.9	130.3	133.5	137.1

the substituents. ^{13}C chemical shifts of some monosubstituted naphthalene derivatives are given in Table 13.17.

13.6   CARBONYL COMPOUNDS

The carbonyl carbon has been the most investigated of all carbons by NMR. Carbonyl carbon atoms absorb in a characteristic region of 150–220 ppm [92]. In this characteristic region ranging over 70 ppm, the carbon resonances of some heteroaromatic compounds and allenes are also encountered. Substituents directly attached to the carbonyl carbon atom influence the carbonyl carbon resonances. Generally, carbonyl carbon resonances are observed in two distinct regions. The carbonyl groups of carboxylic acids and derivatives appear in the range of 160–180 ppm, whereas the aldehyde and ketone resonances appear at the lower field of 195–220 ppm. The fact that the carbonyl carbon signals have a smaller intensity and appear in a characteristic field makes it easier to recognize the carbonyl absorptions from the other resonances. Furthermore, recording of the proton-coupled carbon spectra helps us to easily identify the ketone and aldehyde carbonyl carbons. A simple APT spectrum can also distinguish between these resonances since the aldehyde carbonyl carbon is directly attached to a proton.

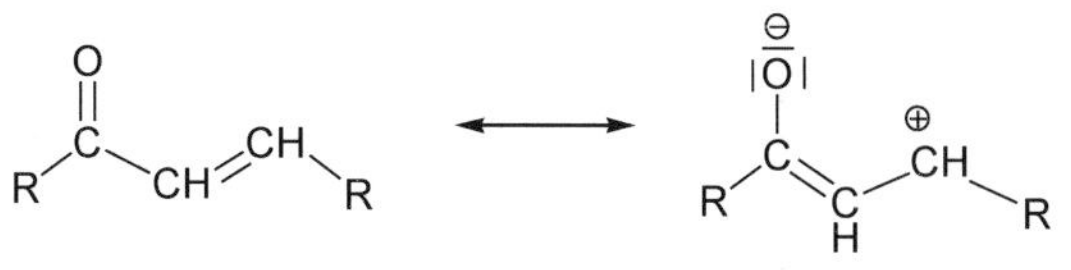

The low-field shift of carbonyl carbon resonances is due to the polarization of carbonyl groups as shown above. The formed positive charge is located on the carbonyl carbon atom. However, in acids and their derivatives this positive charge is compensated by a relatively greater contribution from the mesomeric form B. The carbonyl carbon resonances in α,β-unsaturated carbonyl compounds are shifted upfield relative to those in the corresponding saturated compounds. The mesomeric structure B shows that the electron density at the carbonyl carbon atom is increased due to the conjugation of the carbon–carbon double bond electrons. For example, cyclohexanone carbonyl carbon (156) resonates at 209.7 ppm. The carbonyl carbon resonance in 157 is shifted to the higher field due to the conjugation. The introduction of a second double bond forming a cross-conjugated system as in the case of 158 causes a further shift of the carbonyl carbon resonance to the higher field. p-Quinone (159) resonates at 187.0 ppm.

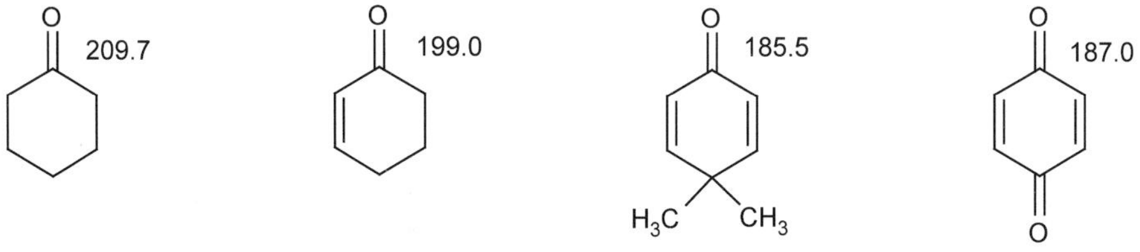

Some selected carbonyl compounds having different carbon resonances are given below. Furthermore, full ^{13}C-NMR spectra of these compounds are presented in Figure 150.

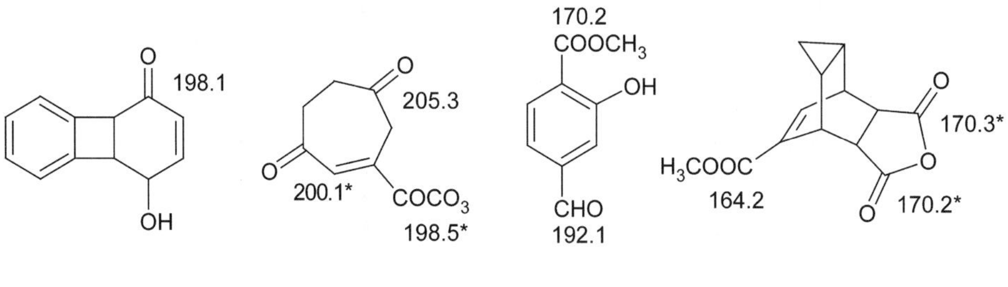

* interchangable

Tables 13.18 and 13.19 show the carbonyl carbon resonances of some selected aldehydes and ketones.

Replacement of a proton on the α-carbon atom by alkyl substituents causes a shift to the lower field. There is a direct relationship between the number of alkyl groups and the chemical shift of the carbonyl carbon atom. Electronegative substituents directly

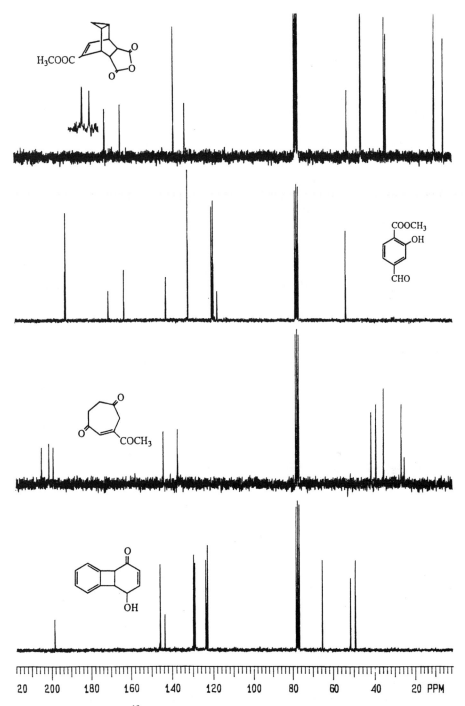

Figure 150   50 MHz ^{13}C-NMR spectra of some selected carbonyl compounds (CDCl$_3$).

Table 13.18

^{13}C-NMR shift parameters in selected saturated and unsaturated aldehydes [93]

Compounds	C_1	C_2	C_3	C_4	C_5	$C_{6/8}$	C_7
Methanal	197.0						
Ethanal	200.5	31.2					
Propanal	202.7	36.7	5.2				
Pentanal	201.3	43.6	24.3	22.4	13.8		
2-Methylpropanal	204.6	41.1	15.5				
2,2-Dimethylpropanal	205.6	42.4	23.4				
Propenal	193.3	136.0	136.4				
(E)-2-Butenal	193.4	134.8	153.9	18.5			
(E)-3-Phenylpropenal	193.5	128.2	152.5	134.1	129.0	128.5	131.0
Trichloroethanal	176.9	95.3					
Formylcyclohexane	204.7	50.1	26.1	25.2	25.2		

attached to α-carbon atom shift the carbonyl carbon resonances to the higher field since they partly hinder the polarization of the carbonyl group.

Since ^{13}C-NMR spectroscopy is very sensitive to any change in the molecular geometry of the compounds, it has been found to be a useful application in the determination of keto–enol tautomerism. In the case of 2,4-pentadienone and dimedone the existence of such equilibrium has been clearly demonstrated by ^{13}C-NMR spectroscopy.

Table 13.19

^{13}C-NMR shift parameters in some saturated and unsaturated ketones [94]

Compounds	C_1	C_2	C_3	C_4	C_5	$C_{6/8}$	C_7
Acetone	30.7	206.7					
Butanone	27.5	206.3	35.2	7.0			
2-Pentanone	29.3	206.6	45.2	17.5	13.5		
3-Pentanone	18.0	35.5	210.7				
3-Methylbutanone	27.5	212.5	41.6	18.2			
3,3-Dimethylbutanone	24.5	212.8	44.3	26.5			
Butenone	25.7	197.5	137.1	128.0			
(E)-3-penten-2-one	18.0	142.7	133.2	197.0	26.6		
3,5-Heptadien-2-one	26.7	198.5	128.3	143.3	130.0	139.7	18.3
Chloropropanone	49.4	200.1	27.2				
Cyclobutanone	208.9	47.8	9.9				
Cyclopentanone	219.6	37.9	22.7				
Cyclohexanone	209.7	41.5	26.6	24.6			
2-Cyclopentenone	209.8	134.2	165.3	29.1	34.0		
2-Cyclohexenone	199.0	129.9	150.6	25.8	22.9	38.2	
2-Norbornanone	49.3	216.8	44.7	34.8	26.7	23.7	37.1
7-Norbornanone	37.9	24.3					216.2

The methylene carbon atoms located between two carbonyl groups in these compounds resonate at 58.7 and 57.0 ppm due to the sp^3 hybridization of the corresponding carbon atoms. However, the hybridization of these carbon atoms in the enol-form changes from sp^3 to sp^2. Consequently, they resonate at 101.1 and 102.5 ppm.

Carbocyclic acid carbonyl resonances can be found in the range 165–185 ppm, depending on the electronic nature of the attached substituents. There is some parallelism between the substituent effects observed in the ketones and acids. Alkyl carbon atoms attached in a position α, β, or γ to a carboxyl carbonyl group change the resonance by about 10, 4 and − 1 ppm, respectively. The chemical shifts data which show the effect of branched-chain alkyl groups on the chemical shifts of the carbonyl carbon atom in carboxylic acids are given in Table 13.20. Solvent effect can also influence the chemical shifts of carboxylic acids up to 4 ppm. Equilibration of carboxylic acid dimers and monomers will be affected by the electronic nature of the solvent molecule. On going from carboxylic acids to the corresponding anhydrides, esters, amides and halides, carbonyl carbon atoms generally display an upfield shift due to the reduction or lack of hydrogen bonding. For example, the carbonyl carbon atom in formic acid resonates at 166.5 ppm. Upon methyl esterification, the carbonyl carbon atom resonance is shifted 6 ppm upfield to 160.9 ppm.

13.7 HETEROCYCLIC COMPOUNDS

It is useful to divide heterocyclic compound into saturated and unsaturated heterocyclic compounds.

13.7.1 Saturated heterocyclic compounds

Oxiranes prepared by the treatment of an alkene with peroxy acids are the smallest member of heterocyclic compounds and they are important intermediates in organic chemistry. Oxirane carbon atoms absorb in a characteristic region of 40–75 ppm depending on the electronic nature of the substituents attached to the three-membered ring. The other three-membered heterocycles: aziridine and thiirane carbon atoms resonate at high field such as their carbocyclic analogues cyclopropane. The chemical

Table 13.20

^{13}C-NMR shift parameters in some acid and acid derivatives [95]

Compounds	C_1	C_2	C_3	C_4	C_5	C_6
Formic acid	166.7					
Ethanoic acid	175.7	20.3				
Propanoic acid	179.8	27.6	9.0			
Butanoic acid	179.3	36.2	18.7	13.7		
Pentanoic acid	180.6	34.8	27.7	22.7	14.2	
2-Methylpropanoic acid	185.9	38.7	27.1			
2,2-Dimethylpropanoic acid	185.9	38.7	27.1			
Propenoic acid	168.9	129.2	130.8			
(E)-2-Butenoic acid	169.3	122.8	146.0	17.3		
(Z)-2-Butenoic acid	169.8	121.0	146.2	15.0		
Chloroethanoic acid	170.4	63.7				
Trichloroethanoic acid	167.0	88.9				
2-Hydroxypropanoic acid	176.8	66.0	19.9			
Oxalic acid	160.1					
Malonic acid	169.2	40.9				
Adipic acid	174.9	33.5	24.9			
Ethanoic anhydride	167.4	21.8				
Propanoic anhydride	170.9	27.4	8.5			
Butendioic anhydride	165.9	137.4				
Butandioic anhydride	172.5	28.2				
Methyl methanoate	160.9	49.1				
Ethyl methanoate	160.3	58.8	13.0			
Ethyl ethanoate	21.0	170.6	60.3	14.2		
Propyl ethanoate	19.6	169.0	64.6	21.6	9.3	
Methyl 2-methylpropanoate	16.2	31.0	173.5	49.8		
Ethanoyl chloride	170.5	33.6				
2-Methylpropanoyl chloride	177.8	46.5	19.0			
Ethanedioic acid dichloride	170.5					
Methanamide	167.6					
N,N-Dimethylamide	162.6	31.5 (s)	36.5 (a)			
Ethanamide	178.1	22.3				
N,N-Dimethylethanamide	170.4	21.5	35.0 (s)	38.0 (a)		
N,N-Diethylethanamide	21.4	169.6	40.0 (s)	13.1 (s)	42.9 (a)	14.2 (a)

shifts of the ring carbon atoms in heterocyclic compounds are determined by the heteroatom electronegativity. Representative data for some saturated heterocyclic compounds are given in Table 13.21.

13.7.2   Unsaturated heterocyclic compounds

Unsaturated heterocyclic compounds should be treated under two different groups:

(1)   nonaromatic heterocycles;
(2)   aromatic heterocycles.

Table 13.21

^{13}C-NMR shifts of representative saturated heterocyclic compounds [96]

Table 13.21 (continued)

In the first group, the compound can have an endo- or exocyclic double bond. For example, in coumarin, β-carbon atom is deshielded due to $-$M effect of the carbonyl group. On the other hand, β-carbon atom in indol is shielded due to the electron release $+$M effect of a nitrogen atom. Generally, the effects observed in unsaturated heterocyclic compounds are similar to the effects described for the open-chain analogues.

Heterocyclic aromatic compounds are generally five- and six-membered ring compounds containing one or more heteroatoms in the ring. Carbon atom resonances of aromatic heteroatoms are found between 105 and 170 ppm (Table 13.22). Heteroatoms in the heterocyclic aromatic compounds contributes one π-electron to complete the aromatic sextet as in the case of pyridine. The lone pair of electrons on the pyridine nitrogen atom is not involved in bonding. However, the pyrrole nitrogen atom uses all five of its valence electrons in bonding. The two lone-pair electrons are involved in aromatic π-bonding, as shown by the following resonance structures. Because of the decreased electron density in pyridine, the resonances of the ring carbon atoms are shifted downfield (123.8, 136.0 and 149.9 ppm). Comparison of these values with that of benzene clearly indicates that the chemical shifts of C_2 and C_4 carbon atoms, where the positive charge is located, resonate at lower field rather than the benzene carbon atoms. However, the carbon resonances in pyrrole are shifted upfield due to the increased electron density in the aromatic ring (107.6 and 117.3 ppm). Theoretical calculations also support the existence of a linear correlation between carbon chemical shifts and electron density at the individual carbon atoms.

Table 13.22

^{13}C-NMR shifts of representative saturated heterocyclic compounds [74b]

electron deficient heteroaromatic compound

electron reach heteroaromatic compound

One of the most important reactions involving the pyridine ring is the reaction of the lone-pair located on the nitrogen atom with different electrophiles. Pyridine can easily react with acid protons to form pyridinium salts. Protonation of pyridine influences the carbon chemical shifts. For example, the α-carbon atoms are shielded due to the change of the C–N bond order. The observed deshielding at C_3, C_4 and C_5 carbon atoms is attributed to the increased electron-withdrawing effect of the positively charged nitrogen atom. The α-carbon atoms in protonated pyridine derivatives are shifted by 5–12 ppm upfield whereas the other carbon atoms downfield by 7–8 ppm.

A second important reaction of pyridine ring is the formation of pyridine oxide derivative upon N-oxidation. Since the electron density in the ring increases by the oxidation, the chemical shift values of the ring carbon atoms move to the higher field, as expected.

In the case where the heteroaromatic ring contains more than one heteroatom, the effects of the heteroatoms on the chemical shifts are additive. Starting from this point it is possible to estimate the chemical shifts of the ring carbon atoms by using the chemical shift values. The chemical shift values of some representative aromatic heterocycles are given in Table 13.22.

– 14 –

Spin–Spin Coupling

In the ^{1}H-NMR section we discussed the spin–spin coupling between the protons and therein we have shown the importance of the coupling constants by the elucidation of the molecular structures. In the case of ^{13}C-NMR spectra, the coupling constants also provide valuable information concerning the chemical structure. In ^{13}C-NMR spectroscopy, the carbon nuclei couple with protons, as well as with the carbon nuclei. The natural abundance of ^{13}C nuclei is only 1.1%. ^{13}C–^{13}C coupling can arise only in molecules which contain at least two adjacent ^{13}C nuclei. The probability for having two adjacent ^{13}C nuclei is approximately 0.01%. Therefore, the direct observation of ^{13}C–^{13}C couplings is very difficult in normal ^{13}C spectra, usually; homonuclear carbon couplings are lost in the noise. For that reason, we prefer to utilize the heteronuclear couplings between ^{13}C and ^{1}H nuclei and we will therefore discuss them first. Since the spin quantum number of the carbon nucleus has the same value as the proton,

$$I_{^{13}C} = \tfrac{1}{2}$$

the peak splitting and peak intensities in the split ^{13}C-NMR spectra can be determined by the Pascal triangle as we did in the case of the proton–proton couplings (see ^{1}H-NMR section). Appreciable ^{13}C–^{1}H couplings also extend over two or more bonds, $^1J_{CH}$, $^2J_{CH}$, and $^3J_{CH}$, in which we will discuss each of them separately.

14.1 COUPLINGS OVER ONE BOND ($^1J_{CH}$)

There are important factors affecting coupling constants between proton and carbon nuclei. These are

(1) dependence on the hybridization of the carbon atom;
(2) electronic nature of the substituents attached to the carbon atom;
(3) strain in the molecule.

There is an empirical correlation between the carbon–proton coupling constant J_{CH} and fractional s-character (denoted by s) in the hybrid carbon orbitals of the carbon–hydrogen bond:

$$J_{CH} = 500s \qquad (65)$$

s is the carbon *s*-character

$s = 0.25$ for sp^3-hybridized carbon atom

$s = 0.33$ for sp^2-hybridized carbon atom

$s = 0.50$ for sp-hybridized carbon atom

As shown in eq. 65, when the *s* ratio of the hybrid orbitals increases, the corresponding coupling constant $^1J_{CH}$ also increases. One-bond spin–spin coupling constants ($^1J_{CH}$) range from 100 to 320 Hz. According to eq. 65, the size of the coupling for a saturated system (sp^3-hybridized carbon) is 125 Hz, for unsaturated systems (sp^2-hybridized carbon) 150–170 Hz and for acetylenic/allenic systems (sp-hybridized carbon) 250 Hz. These values are observed only in cases in which one electronegative substituent is attached to the coupled carbon atom. The range of one-bond coupling ($^1J_{CH}$) expands considerably if more than one electronegative substituent is attached at the coupled carbon atom. The coupling constant values ($^1J_{CH}$) increase with the increasing electronegativity of the substituents. This trend is exemplified in a series of mono-, di- and trisubstituted methanes. For example, monosubstituted methane derivatives (CH$_3$F, CH$_3$Cl, CH$_3$Br, and CH$_3$I) show an almost equal one-bond coupling of about 150 Hz. The $^1J_{CH}$ coupling constant in methane is 125.0 Hz. Replacement of one proton by a chlorine atom increases the $^1J_{CH}$ coupling from 125.0 to 150 Hz. Replacements of the second and third proton by chlorine atoms change the coupling constants to 178 and 209 Hz, respectively (Table 14.1). The effects of the electron-withdrawing groups on coupling constants are largely additive.

Some representative one-bond coupling constants are given in Table 14.2. The coupling constants in nonstrained cyclic systems have approximately the same size as those observed in acyclic systems. In strongly strained systems, very large coupling constants J_{CH} are observed. For example, a one-bond coupling in cyclohexane, $J_{CH} = 127.0$ Hz, is found to be close to the value for ethane $J_{CH} = 124.9$ Hz. However, the coupling in cyclopropane is $J_{CH} = 161.0$ Hz. This is attributed to the change of the hybridization in the C–H bond. It is well established that with increasing ring strain the carbon hybridization also changes in a manner that the exocyclic orbitals (C–H bond) take on an increased *s*-character. The exceptional chemical reactivity of cyclopropane, such as undergoing bromination, hydrogenation, etc., supports this finding. Considerable variations are also observed in cyclopropene. The observed coupling for the double bond carbon atom of 228.2 Hz in cyclopropene is more related to an alkyne. The one-bond coupling J_{CH} for methylene carbon is 167.0 Hz. Similarly, the three-membered

Table 14.1

$^1J_{CH}$ coupling constants (Hz) in methane halides [84]

X	H	F	Cl	Br	I
CH$_4$	125.0				
CH$_3$X	125.0	149.1	150.0	151.5	151.1
CH$_2$X$_2$	125.0	184.5	178.0		
CHX$_3$	125.0	239.1	209.0		

Table 14.2

$^1J_{CH}$ coupling constants (Hz) in some selected hydrocarbons [84]

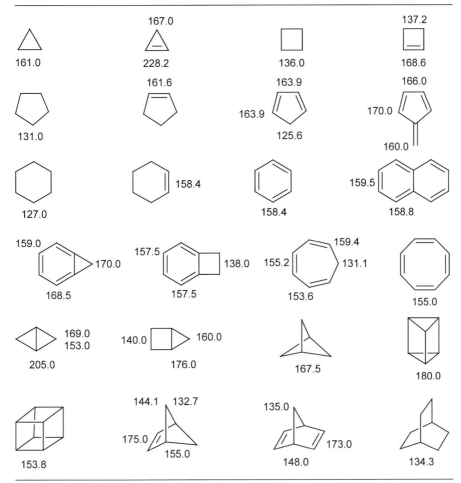

heterocyclic compounds: oxirane $J_{CH} = 176.0$ Hz, thiirane $J_{CH} = 170.6$ Hz and aziridine $J_{CH} = 168.0$ Hz show much larger values than cyclopropane. An additional increase in the couplings in the small heterocyclic compounds arises from the inductive effect of the ring heteroatom (Table 14.3).

It is chemically well established that strong bases can easily abstract the protons attached to the bridgehead carbon atoms in highly strained systems. The increased acidity of bridge protons is attributed to the high s-character of the bridgehead carbon hybrid orbitals. When the s-character in the exocyclic C–H bonds is increased, the p-character in the endocyclic orbitals is increased in order to decrease the strain in the ring. As a consequence of this rehybridization, the coupling constants in the exocyclic orbitals are increased. It is unusual when large coupling constants are found in polycyclic strained

Table 14.3

$^1J_{CH}$ coupling constants (Hz) in some selected heterocycles [84]

hydrocarbons. For example, in bicyclo[1.1.0]butane, the observed coupling constant J_{CH} is 205.0 Hz. Therefore, the one-bond coupling constants can be used to evaluate the bonding situation (hybridization) in hydrocarbons.

We have already mentioned that substituent effects on $^1J_{CH}$ are approximately additive. For a molecule of the type CHXYZ, one-bond coupling $^1J_{CH}$ can be determined by adding the empirical increment values i using the following equation (Table 14.4):

$$^1J_{CHXYZ} = i_X + i_Y + i_Z \qquad (66)$$

where i_X, i_Y, i_Z are the increment values of the substituent attached to the carbon atom.

For example, let us estimate the one-bond carbon–proton coupling J_{CH} in benzyl alcohol, Ph–CH$_2$OH. There are three substituents besides the coupled proton. They are the –H, –OH, and –Ph groups.

$$^1J_{CH(H)(OH)(Ph)} = 41.7 + 42.6 + 58.6 = 142.9 \text{ Hz (calculated);} \quad 140.0 \text{ Hz (experimental)}$$

The one-bond coupling constants can be easily estimated in nonstrained cyclic hydrocarbons and in heterocycles. In a strained system, there are always large deviations from that of calculated values.

On the basis of the one-bond coupling constants, it is possible to determine the position of a dynamic equilibrium and make signal assignments.

Table 14.4

Increment values of some substituent in $^1J_{CH}$ coupling constants [97]

Substituent	Increment values
−H	41.7
−CH$_3$	42.6
−Phenyl	42.6
−C=O	43.6
−CH=CH$_2$	38.6
−COOH	45.6
−OH	58.6
−OCH$_3$	56.6
−NH$_2$	49.6
−NO$_2$	63.6
−CN	52.6
−F	65.6
−Cl	66.6
−Br	67.6
−I	67.6

δ = 15.7ppm
$^1J_{CH}$ = 161.0Hz

δ = 19.7ppm
$^1J_{CH}$ = 129.0Hz

In the above-shown system there are some questions which have to be answered. The first one is: Is this system in a dynamic equilibrium or is the equilibrium shifted towards the side of one valance isomer? If the equilibrium is shifted to one side, the question is to which side? Furthermore, how far can we use the one-bond coupling constants by peak assignments? A ^{13}C-NMR spectrum of this system shows the presence of two absorption peaks at 15.7 and 19.7 ppm. By recording the proton-coupled ^{13}C-NMR spectrum of this system, we see that the peaks mentioned above are split into triplets with a coupling of 161.0 and 129.0 Hz, respectively. A one-bond C−H coupling of 129.0 Hz is an expected value for the nonstrained hydrocarbons. However, the coupling constant of 161.0 Hz clearly indicates the presence of a cyclopropane ring. On the basis of these measured coupling constants, we can predict that this equilibrium is static and shifted towards the side of the tricyclic compound. Furthermore, one can easily distinguish between the resonance signals appearing at 15.7 and 19.7 ppm.

We have shown that one-bond coupling between carbon and proton nuclei can be determined by measuring proton-coupled ^{13}C-NMR spectra. Those coupling constants J_{CH} can also be determined from the ^{1}H-NMR spectra. In the ^{1}H-NMR spectra we normally see proton−proton couplings. But, we have to consider the fact that all of the compounds contain 1.1% ^{13}C nuclei (natural distribution). Just as the spin of a ^{13}C nucleus couples with the spins of neighboring protons, so the spin of one proton also couples with the spins of neighboring carbon atoms. Therefore, proton–carbon splitting has to be observed in the ^{1}H-NMR spectra. Because of the low natural abundance of ^{13}C nuclei, peaks arising from the proton–carbon couplings are superimposed by the main peaks. It is possible to observe these signals with lower intensity by amplification of the spectrum. These signals, called satellites, are located on the right and left sides of the main peak.

Let us discuss the ^{13}C satellites appearing in the ^{1}H-NMR spectra using maleic anhydride as an example. It is expected that the olefinic protons in maleic anhydride resonate as a singlet at 7.05 ppm. By the amplification of that signal we observe additional signals on the left and right side of the main signal (doublets) with lower intensities (Figure 151).

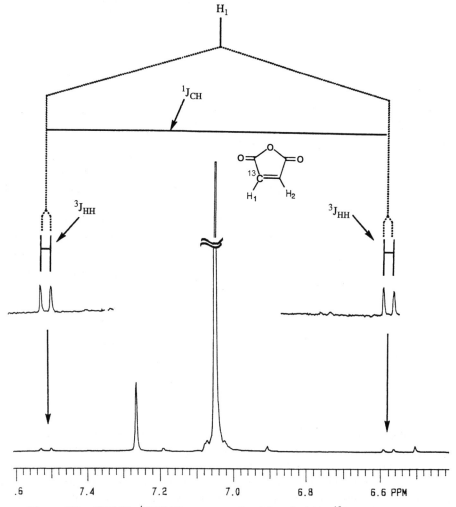

Figure 151   200 MHz ^{1}H-NMR spectrum of maleic anhydride: ^{13}C satellite signals.

nonequal protons

160

The main peak at 7.05 ppm arises from the protons that are attached to ^{12}C isotopes. In 1.1% of the sample, the protons are attached to ^{13}C isotopes. This proton attached to ^{13}C nuclei will couple with the bonded carbon atom (^{13}C isotope) and give rise to a doublet $J = 187.5$ Hz. Since the coupling between a proton and a sp^2-hybridized carbon atom is large, the proton signal attached to the ^{13}C nucleus will be far from the main signal depending on the size of the proton–carbon coupling constants. Careful examination of these lines further indicates doublet splitting by 5.3 Hz. The probability that both of the olefinic protons in maleic anhydride are attached at the same time to two ^{13}C nuclei is very low (0.01%). Therefore, it is likely that the other proton will be attached to the ^{12}C nucleus. These different bonding modes will make these protons chemically nonequivalent, so that they will couple with each other. The observed doublet splitting in the satellite spectra arises from the proton–proton coupling in maleic anhydride. In summary, we can determine the carbon–proton coupling constants from the proton-coupled ^{13}C-NMR spectra, as well as from the ^1H-NMR spectra. In the case of ^1H-NMR spectra, it is not possible every time to observe the satellite spectra, since they can be superimposed by the other signals. Furthermore, analysis of the ^{13}C satellites can provide the coupling constants between the equivalent protons, which cannot be determined from the main signal.

The recording of the proton-coupled ^{13}C-NMR spectra requires more time because of the decrease of the signal intensities due to the signal splitting. Peak multiplicity in the proton-coupled ^{13}C spectra provides us with valuable information about the substitution pattern of the carbon nucleus and structure of the molecule. Let us discuss the ^{13}C-NMR spectra of the following compounds **161** and **162**.

161   **162**

The proton-decoupled ^{13}C-NMR spectrum of **161** is interesting in several ways. Note that only six distinct resonance signals are observed even though the molecule contains 12 carbon atoms [98]. The molecule has a plane symmetry that makes two carbons equivalent. Another interesting feature of this spectrum is that the peaks are not uniform in size due to the different relaxation times of the carbon atoms. Four of these signals appear in the range of 120–150 ppm and belong to the aromatic and olefinic carbon atoms. As expected, three of these signals are split into doublets in the coupled ^{13}C-NMR spectrum of **161** (Figure 152).

Careful inspection of these doublet lines reveal the existence of further splitting that arise from $^2J_{CH}$ and $^3J_{CH}$ couplings. The measured coupling constants between the aromatic carbon atoms and adjacent protons are $^1J_{CH} = 161.2$ and 162.2 Hz, respectively, and those are in agreement with the expected values. The one-bond coupling $^1J_{CH}$ measured for olefinic carbons is 170.3 Hz. This coupling is larger than the coupling for the aromatic carbons. It is well established that the one-bond couplings for α,β-unsaturated carbonyl compounds are always greater than the aromatic carbon–proton couplings. The remaining signals appear as broad singlets (over two- and three-bond couplings) and belong to the quaternary carbon atoms.

The proton-decoupled ^{13}C-NMR spectrum of **162** shows 13 carbon signals because of its unsymmetrical structure [25b]. In the coupled ^{13}C-NMR spectrum, nine of these signals are split by the adjacent protons. Thus, the aromatic ring carbon atoms (CH) show four absorptions in the range of 120–130 ppm and appear as doublets. Careful examination of the doublets lines indicates the presence of fine splitting arising from the coupling with neighboring protons ($^2J_{CH}$, $^3J_{CH}$). The extracted coupling constants from the doublet lines are $^1J_{CH} = 161.0$, 158.0, 157.1, and 156.4 Hz, respectively. These values are in agreement with those determined in benzenoid aromatic compounds. High-field resonances at 103.8 and 96.9 ppm belong to the exocyclic double bond carbon atoms and they are split into triplets by the adjacent methylene protons. The corresponding coupling constants are $^1J_{CH} = 157.1$ and 157.9 Hz. Since the resonance signals at 55.2 and 51.2 ppm are split into doublets, they can arise only from the bridgehead carbon atoms C_1 and C_5. The measured coupling constants, $^1J_{CH} = 147.0$ and 147.3, vary considerably from that reported for acyclic and strain-free cyclic alkanes ($^1J_{CH} = 125.0$ Hz). This is again a hybridization effect, resulting from the strain introduced by the cyclobutane ring. The s-character in the exocyclic C–H orbitals is increased. The resonance signal at 37.0 ppm can be assigned to the methylene carbon atom on the basis of the fact that it resonates as a triplet in the proton-coupled ^{13}C-NMR spectrum. The extracted coupling constant is $^1J_{CH} = 142.0$ Hz and it is smaller than the coupling constants obtained from the bridgehead carbon atoms. These values show that the strain in the four-membered ring is more localized at the bridgehead carbons.

These two examples demonstrate that the proton-coupled ^{13}C-NMR spectra of the compounds make it possible to determine how many protons are bonded to each carbon atoms. Furthermore, they can provide us with very useful information about the number of protons that are directly attached to the resonating carbon atoms. Furthermore, measured coupling constants will help in the evaluation of the bonding situation in hydrocarbons.

14.2 COUPLINGS OVER TWO BONDS ($^2J_{CH}$) (GEMINAL COUPLING)

As in the case of ^{1}H-NMR spectroscopy, we observe carbon–proton couplings over two bonds and those couplings are less important in ^{13}C-NMR than in ^{1}H-NMR. Table 14.5 shows some representative $^2J_{CH}$ values, which range from -10 to 65 Hz. In aliphatic systems, typical values range from -6 to -4 Hz. The couplings of the aldehyde protons

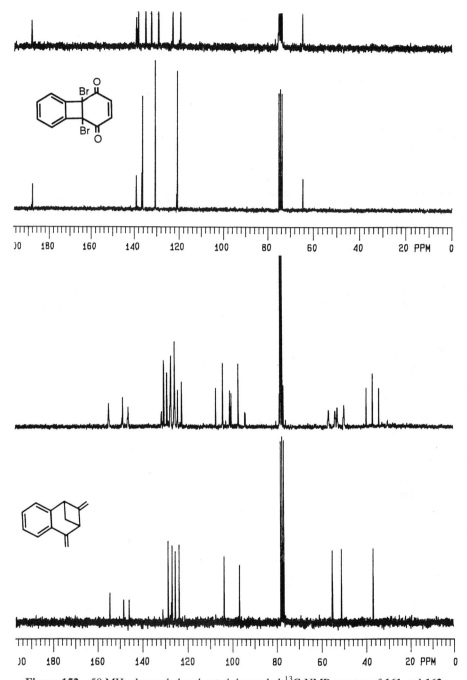

Figure 152 50 MHz decoupled and gated decoupled ^{13}C-NMR spectra of **161** and **162**.

Table 14.5

$^2J_{CH}$ geminal coupling constants in some selected compounds [99, 100]

H_3C-CH_3	−4.5 Hz	▷	−2.6 Hz
$H_3C-CH_2-CH_3$	−4.3 Hz	☐	−3.5 Hz
$H_3C-CH_2-CH_3$	−4.4 Hz	(pentagon)	−3.0 Hz
$CH_2=CH_2$	−2.4 Hz	(hexagon)	−3.7 Hz
$CH_2=CH-CH_3$	−6.8 Hz	(benzene)	+1.1 Hz
$CH\equiv CH$	+49.6 Hz	(cyclopentadienone)	+4.5 Hz
$H_3C-CH=O$	+26.7 Hz	(pyrrole, positions 4 3 5 2 N1 H)	C_2-H_3 8.7 Hz C_3-H_2 8.3 Hz C_3-H_4 4.6 Hz
$Cl_3C-CH=O$	+46.3 Hz	(furan)	C_2-H_3 11.0 Hz C_3-H_2 13.8 Hz C_3-H_4 4.1 Hz
(bicyclobutane, 1 2)	C_1-H_2 3.3 Hz C_2-H_1 5.3 Hz	(pyridine)	C_2-H_3 3.1 Hz C_3-H_2 8.5 Hz C_3-H_4 0.9 Hz C_4-H_3 0.7 Hz
(cyclobutene =CH)	+1.2 Hz		
(cyclopentene =CH)	+4.7 Hz		
(cyclohexene =CH)	−0.3 Hz		

are between $+20$ and $+50\,\mathrm{Hz}$, and those for acetylenic protons are higher ($+40$ to $+65\,\mathrm{Hz}$).

Since we do not observe any proton–proton coupling over one bond in ^{1}H-NMR (except the one-bond coupling in the H_2 molecule), it is not possible to compare one-bond couplings in ^{13}C-NMR with ^{1}H-NMR. However, it is possible to compare the coupling constants over two bonds in ^{13}C-NMR with those in ^{1}H-NMR. A proportionality relationship between **C–C–H** and **H–C–H** has been determined, which is given by the following equation:

$$^nJ_{CH} = a\,^nJ_{HH} \tag{67}$$

where n is the number of bonds between coupled nuclei.

$a = 0.25$ for sp^3-hybridized carbon atom
$a = 0.4$ for sp^2-hybridized carbon atom
$a = 0.6$ for sp-hybridized carbon atom

The factor 'a' in this equation depends on the hybridization of the carbon atom.

As in the case of geminal proton–proton couplings, two-bond carbon–proton couplings $^2J_{CH}$ strongly depend on the C–C–H valence angle [99]. The corresponding coupling constants become more positive when the C–C–H valence angle increases. Geminal couplings vary with substitution on the carbon atom with different electronegativity and C–C bond length.

14.3   COUPLINGS OVER THREE BONDS ($^3J_{CH}$) (VICINAL COUPLING) [101]

Vicinal couplings $^3J_{CH}$ are distributed in a range of 0–$16\,\mathrm{Hz}$ and have positive values. There is also a relationship between the vicinal proton–proton couplings and vicinal carbon–proton couplings as was discussed earlier. All of the factors influencing vicinal proton–proton couplings (see Section 4.3.3) can also be applied to vicinal carbon–proton couplings. Those factors are

(1)   the dihedral angle ϕ;
(2)   the bond length R_{CC};
(3)   the valance angle θ;
(4)   the electronegative effect of the substituents.

Vicinal carbon–proton couplings mainly depend on the dihedral angle ϕ. As we have discussed in the ^{1}H-NMR section (see Section 4.3.2.1), the magnitude of the vicinal coupling approaches zero when the dihedral angle is $90°$, whereas the coupling constants are largest for $\phi = 0°$ or $180°$. Generally, *trans* coupling is larger than *cis* coupling.

It is not common to use vicinal carbon–proton coupling constants to elucidate the configuration of a given molecule. However, in some specific cases, these couplings can be useful. For example, imagine a trisubstituted double bond that can have two different configurations (*cis* or *trans*).

$^3J_{CHtrans}$ coupling $^3J_{CHcis}$ coupling   propene

With ^{1}H-NMR spectroscopy, the exact configuration of the above compounds cannot be determined. However, the configurational assignment can be performed by measuring the $^3J_{CH}$ coupling over three bonds. For example, the *trans* coupling between the methyl carbon and methylene proton is $^3J_{CHtrans} = 12.7$ Hz, whereas the corresponding *cis* coupling is $^3J_{CHcis} = 7.6$ Hz.

A similar relationship between vicinal carbon–proton couplings and vicinal proton–proton couplings is also observed. This relation is given by the following equation:

$$^3J_{CH} = 0.6\,^3J_{HH} \tag{68}$$

Table 14.6 shows some representative $^3J_{CH}$ values, which range from 4 to 15 Hz [102].

Table 14.6

$^3J_{CH}$ vicinal coupling constants in some selected compounds [99, 100]

$^3J_{CHcis} = 12.7$ Hz $^3J_{CHtrans} = 7.6$ Hz	$^3J_{C_2H_4} = 7.0$ Hz $^3J_{C_2H_5} = 6.9$ Hz $^3J_{C_3H_5} = 6.1$ Hz
$^3J_{CHcis} = 15.2$ Hz $^3J_{CHtrans} = 9.4$ Hz	$^3J_{CH} = 7.6$ Hz
$^3J_{C_2H_4} = 7.5$ Hz $^3J_{C_2H_5} = 6.6$ Hz $^3J_{C_3H_5} = 7.4$ Hz	$^3J_{C_2H_4} = 6.8$ Hz $^3J_{C_2H_5} = 11.1$ Hz $^3J_{C_3H_5} = 6.6$ Hz
$^3J_{C_2H_4} = 10.0$ Hz $^3J_{C_2H_5} = 5.0$ Hz $^3J_{C_3H_5} = 9.8$ Hz	$^3J_{C_1H_3} = 10.0$ Hz $^3J_{C_2H_6} = 4.4$ Hz

14.4   CARBON–DEUTERIUM COUPLING ($^1J_{CD}$)

Since the NMR spectra are recorded in deuterated solvents, the couplings between the proton and carbon, as well as between the carbon and deuterium, are always encountered due to the spin quantum number of D ($I_D = 1$). The most important factor influencing the magnitude of the couplings between two elements is the gyromagnetic constant. For example, carbon–proton and carbon–deuterium coupling constants are related to each other by the magnitude of the gyromagnetic constants.

$$\frac{J_{CH}}{J_{CD}} = \frac{\gamma_H}{\gamma_D} = 6.51 \tag{69}$$

Attachment of a deuterium atom to carbon will split carbon resonances in triplets with equal intensities. Spin quantum number for deuterium is equal to $I_D = 1$. It means the magnetic quantum number of deuterium will be $m = 2I + 1 = 3$. This indicates that the deuterium atom will have three different alignments in a magnetic field which in turn will split the line of carbon in triplets. For example, the most used solvent molecule $CDCl_3$ resonates as a triplet at 77.0 ppm (Figure 153).

The distances between the triplet lines directly give the coupling constant ($^1J_{CD} = 32.0$ Hz) between the carbon and deuterium atom. In normal chloroform $CHCl_3$ the measured coupling constant is $^1J_{CH} = 210.5$ Hz. The ratio between those experimental values is 6.58, which is in good agreement with eq. 69. Signal multiplicities of carbon resonances will change upon the number of attached deuterium atoms. Triplets for CD, quintet for CD_2 and septet for CD_3 will be observed.

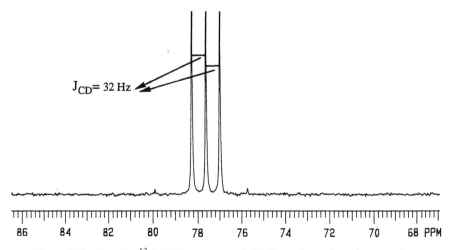

Figure 153   50 MHz ^{13}C-NMR spectrum of $CDCl_3$; carbon–deuterium coupling.

– 15 –

Multiple-Pulse NMR Experiments

In the previous chapters we described certain NMR experiments which involve the application of a single radiofrequency field. Firstly, in the case of one-pulse experiments a pulse is applied on the sample. After the pulse, the FID signals representing the difference between the applied frequency and Larmor frequency are collected. Later on, the FID signals are Fourier transformed by computer into the conventional NMR spectrum. However, in the multiple-pulse experiments, more than one pulse is applied on the sample before collecting the FID signals. Furthermore, an evolution period between the pulses is inserted. Although two important parameters such as chemical shifts and coupling constants are obtained from the spectra recorded with the single-pulse method, much more information can be obtained from multiple-pulse experiments. For example, measurements of the relaxation times, relation between the signals from spectrum correlation and identification of the peaks in ^{13}C-NMR spectra can be performed with multiple-pulse experiments. Furthermore, multiple-pulse experiments can be applied in order to increase the sensitivity of some nuclei with low natural abundances. The basic principles of the multiple-pulse method should be investigated for better comprehension. The next two sections are set out in order to provide this knowledge. However, for the interpretation of the NMR spectra, obtained by the application of multiple-pulse experiments on a sample, there is no need to know the detailed mechanism of the different multiple-pulse experiments in detail. Therefore, those who may not be interested in the multiple-pulse sequences can go directly to the application and interpretation of the spectra.

One of the most frequently applied multiple-pulse experiments is the recording of APT (attached proton test) spectra, which provide information about the number of the attached protons to a ^{13}C atom. In APT spectra, carbon atoms having an odd number of protons such as CH and CH_3 appear as negative signals, whereas the carbon atoms with an even number (0 or 2) of protons such as C or CH_2 appear as positive signals. For the interpretation of an APT spectrum, there is no need to know the complex multiple-pulse sequences of the APT spectra.

The APT spectrum of 1-acetyl-cycloheptatriene is shown in Figure 154. There are nine different kinds of carbons in the APT spectrum. The molecule possesses two quaternary, five tertiary, one secondary and one primary carbon atoms. The fact that the positive signals which are observed at 200 and 133 ppm belong to quaternary carbon atoms is easily defined; the negative signals between 120 and 140 ppm belong

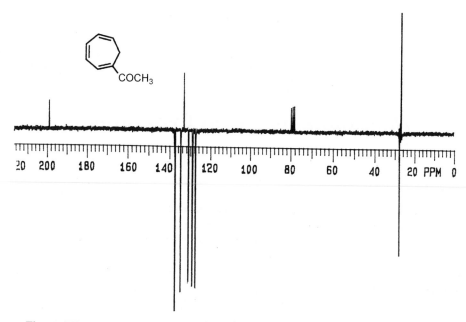

Figure 154   50 MHz APT (attached proton test) spectrum of 1-acetyl cycloheptatriene.

to tertiary olefinic carbon atoms (C=CH). Moreover, the positive signal at 28 ppm can be easily assigned to the methylene carbon atom and the negative one to the methyl carbon atom. This example shows us that one can interpret the APT spectrum without having knowledge about the pulse sequences involved during measurements.

To understand the mechanism better why some signal resonances appear to be negative and some as positive signals, we will start to discuss the simplest multiple-pulse experiments. At first, we will discuss the application of multiple-pulse experiments on the measurements of relaxation times which will be frequently encountered in the preceding parts.

15.1 MEASUREMENTS OF RELAXATION TIMES

While investigating the relaxation mechanisms of the carbon nuclei, we have pointed out that there are two kinds of relaxation mechanisms:

1. spin–lattice relaxation, T_1;
2. spin–spin relaxation, T_2.

In the following section, we will explain how to measure these relaxation times.

15.1.1 Measurement of the spin–lattice relaxation time T₁: inversion-recovery method (longitudinal magnetization)

The most popular method for the measurement of the spin–lattice relaxation time is the inversion-recovery method [103]. Before going into detail, two terms should be explained:

1. 180° pulse or π pulse;
2. 90° pulse or $\pi/2$ pulse.

As we have described in the resonance concept, when a macroscopic sample is placed in a magnetic field, the net magnetization cone is formed on the z-axis, which is called longitudinal magnetization. Interaction of the radiofrequency field with the net magnetization vector M_z will tip the magnetization vector on the y-axis. The magnetization can be tipped in desired angles by changing the time for which the pulse is applied. From now on, we will call the pulse that is needed to turn the magnetization vector through 180° on the $-z$-axis the π pulse, and the one that turns the magnetization vector 90° on the y-axis the $\pi/2$ pulse (Figure 155).

When a $\pi/2$ pulse is applied to a macroscopic sample, the magnetization vector will be turned onto the y-axis and a maximum signal will be obtained in the receiver (since maximum induction is formed on the y-axis). A different situation exists in a 180° pulse.

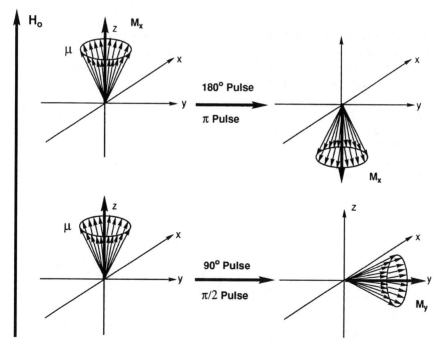

Figure 155 180° and 90° pulses: (a) 180° pulse magnetization on $-z$-axis; (b) 90° pulse magnetization on $+y$-axis.

Since the magnetization is turned onto the $-z$-axis (no induction on the y-axis), we will not observe any signal in the receiver. In order to observe an NMR signal, there should be a magnetization vector or a component of this vector on the xy-plane. Think about the 180° pulse on the sample that we have sent. From now on, we will call the period that the pulse is acting on the sample the 'preparation time'. At first, the magnetization vector M_0 will be turned to the $-z$-axis. We can ask the question 'what will happen when we will leave the system?'. The system will now begin to relax back to the previous position according to the spin–lattice relaxation mechanism by emitting all of its energy to the lattice. We will call this period the 'evolution time'. However, it is impossible to directly observe the effect of this pulse by NMR spectroscopy, because all of the induced magnetization is formed along the z-axis, as can be seen in Figure 156. Only the induction of a magnetization at the xy-plane can be observed by NMR spectroscopy, since we detect the component in the y-direction of any magnetization that has occurred in the xy-plane as an NMR signal.

The time necessary to regain the original magnetization after a 180° pulse is called the spin–lattice relaxation time. The evolution period should be observed in some way in order to measure this time. If we wish to follow the relaxation process, we must generate a component of the magnetization in the xy-plane. This can be easily done by the application of a second pulse, this time a 90° pulse, which will tip any z-magnetization into the xy-plane, producing a signal. This applied pulse is called the 'read pulse', which transfers the information from the z-axis into the xy-plane. The xy-plane can be called the read-plane. Let us consider the case where at some time after the 180° pulse (this time must be shorter than the relaxation time T_1) the magnetization vector M_z has changed to the value M_{-z}. We then apply a 90° pulse and rotate the magnetization vector onto the $-y$-axis as shown in Figure 157.

When the receiver is turned on after a 90° pulse, the magnetization along the $-y$-axis will be recorded as a FID signal. Fourier transformation of that signal will result in the formation of an inverted signal (a negative signal). If the magnetization M_z is in

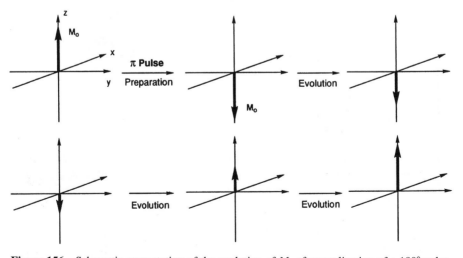

Figure 156   Schematic presentation of the evolution of M_0 after application of a 180° pulse.

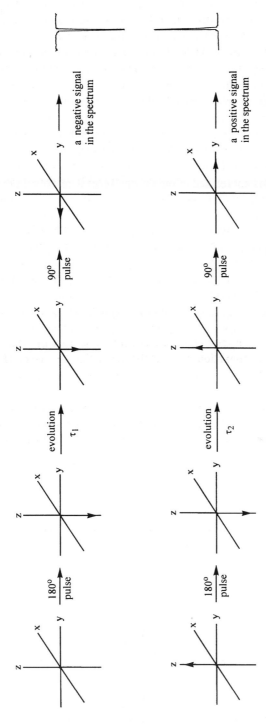

Figure 157 Motion of the magnetization vector during the inversion-recovery experiment. The effect of different delay times on the magnetization and signal intensity.

the $+z$ direction, we apply a 90° pulse, which will rotate the magnetization partially towards the $+y$-axis. The receiver is then turned on and the obtained signal will be a positive signal (Figure 157). Let us demonstrate these processes in an example.

Figure 158 shows a series of ^{1}H-NMR signals of the chloroform molecule that was recorded in the following way. At first a 180° pulse was sent to the sample, then different delay times were given in the evolution period. The operator can set up a series of experiments with different delay times. In that experiment the delay time between two pulses is increased by a constant time interval of 0.5 s. After different delay times, a 90° pulse is applied and the spectra are recorded. Depending on the position of the magnetization, positive or negative NMR signals are recorded. At the first spectrum, a very short delay time is given so that a maximum negative signal is recorded, which means that the system is not relaxed or only a small part is relaxed. In the case of a very long delay time, the system will have returned to equilibrium (to its original position) by the time the 90° pulse is applied. In this case we will record the most intense positive signal. If the magnetization M_z reaches a value of zero, a 90° pulse will not generate a component in the xy-plane, so that no signal will be detected (Figure 158, $t = 4.25$ s). This time t helps us to calculate the relaxation time. Let us demonstrate the magnetization vector as $+M_0$ when the system is in equilibrium. After the 180° pulse, the magnetization vector just changes its sign and is shown as $-M_0$. The magnetization will change between $-M_0$ and $+M_0$ in the course of the relaxation processes. Let us consider the magnetization measured at any time with t as M_z. Magnetization M_z is changed with respect to the delay times after the 180° pulse. It is well known from the Bloch equations

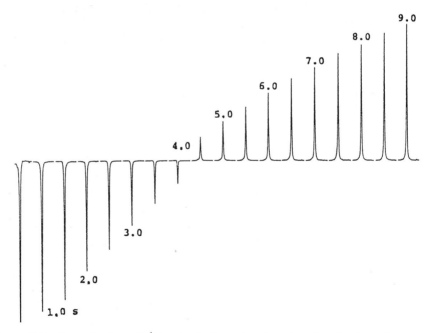

Figure 158 Determination of ^{1}H spin–lattice relaxation time by the inversion-recovery experiment in chloroform.

that the decay of the magnetization in the z-direction is given by the following equation:

$$\frac{dM_z}{dt} = -\frac{M_z - M_0}{t} \tag{70}$$

Integration of this equation gives

$$\ln\frac{M_0 - M_\tau}{2M_0} = \frac{\tau}{t} \tag{71}$$

Since the magnetization is proportional to the peak intensity, this equation may be expanded to

$$\ln\frac{I_0 - I_\tau}{2I_0} = \frac{\tau}{t} \tag{72}$$

where M_0 is the magnetization at equilibrium ($t = 0$), M_τ is the magnetization in the z-direction at the time $t = \tau$, I_0 is the peak intensity at $t = 0$, and I_τ is the peak intensity at a given time $t = \tau$.

The spin–lattice relaxation time T_1 can be easily determined from the zero relaxation time τ_0 where the signal intensity is equal to zero. Eq. 72 simplifies to

$$T_1 = \frac{\tau_0}{\ln 2} = \frac{\tau_0}{0.69} \tag{73}$$

Relaxation times of all the nuclei in a sample can be determined with the inversion-recovery technique in a single experiment. Figure 159 shows the ^{1}H-NMR spectra of acetylcycloheptatriene containing different protons by inversion-recovery measurements [104]. The differences at the zero transition times show that the protons in this molecule have different relaxation times.

It can be seen qualitatively from the spectra that the methylene protons are relaxed faster than the olefinic and methyl protons.

15.1.2   Measurement of spin–spin relaxation time T_2: transverse magnetization

Let us turn back to the single-pulse experiment and think of an experiment where we apply a 90° pulse to a macroscopic sample having one nucleus, either hydrogen or a carbon. This pulse will invert the magnetization from its equilibrium position towards the y'-axis. We assume that the frequency of the rotating frame is equal to the Larmor frequency of the nuclei. In a system like this, the magnetization vector will be stationary in the xy-plane. There are many factors which can influence the decay of the magnetization in the xy-plane. Some of the nuclei will transfer their energy to the lattice and relax. Of course, this spin–lattice relaxation process will contribute to the transverse relaxation. Suppose there is no longitudinal relaxation just for a moment. Consider whether the transverse magnetization can protect its position and value for a long time or not. The magnetization in the xy-plane will in fact decay. There are two mechanisms which are responsible for the decay of the magnetization in the xy-plane. Let us explain the first one. The magnetization M_z consists of many individual magnetic vectors. In an ideal case, all of the individual magnetic moments should experience the same magnetic

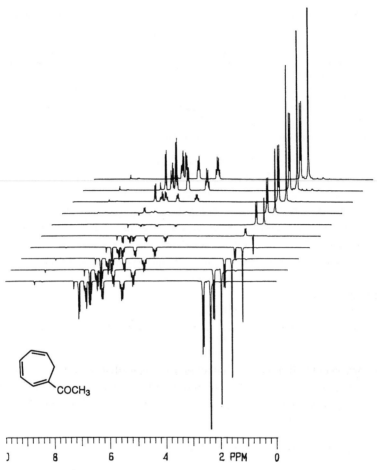

Figure 159   200 MHz ^{1}H-NMR spectra of acetylcycloheptatriene recorded by inversion-recovery technique.

field and they should have the same Larmor frequencies. In fact, this is not the case. Since the inhomogeneity of the magnetic field cannot be avoided, the Larmor frequencies of the nuclei will be altered. Some of them will precess faster than the frame, and some a little slower (Figure 160).

As a consequence of this fact, the spins will fan out and lose their coherence. It is easy to appreciate that this process causes the transverse magnetization to finally disappear. Actually, this decay of the magnetization arising from the inhomogeneity of the magnetic field is not the real mechanism for the transverse relaxation. However, the main contribution to the spin–spin relaxation is of a quite different nature. The energy is transferred between the nuclei. For example, one nucleus in the xy-plane (higher energy state) transfers its energy to a spin aligned in the z-direction (lower energy state). One nucleus changes its position from a higher energy level to a lower level, whereas another one simultaneously changes from a lower energy level to a higher level. By this process

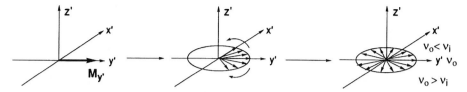

Figure 160   Decay of the transverse magnetization by fanning out of the spins in the xy-plane followed by a 90° pulse.

equal amounts of energy are absorbed and released. This process is not changing the energy of the sample, as the level populations are not affected. Only the phase coherence between the nuclei, and the amount of the order is lost. This type of relaxation is described as an entropy process, whereas the spin–lattice relaxation is an enthalpy process. As a consequence of this energy transfer process, some of the nuclei will change their phases and cause phase coherence, which will contribute to the spin–spin relaxation process. In summary, three different factors influence the spin–spin relaxation:

(i)   longitudinal relaxation;
(ii)   energy transfer between the spins;
(iii) phase coherence caused by the inhomogeneity of the magnetic field.

Is there any way to eliminate the inhomogeneity contribution of the magnetic field or cancel it to measure the real spin–spin relaxation time? To overcome these difficulties some pulse sequences, such as the *spin-echo* experiment, have been developed.

15.1.3   The spin-echo experiment [105]

Let us return again to the simple experiment which we discussed above. We apply a 90° pulse on the sample and the macroscopic magnetization vector M_z and rotate it towards the y'-axis and then the individual spins will begin to spread out. We wait for the time τ s (the time τ cannot be longer than the relaxation time) and then apply a 180° pulse. The spins which have already started to fan out will be inverted across to the other side of the xy-plane (Figure 161). Since the rotational sense of their motion is unchanged they will move towards the $-y$-axis. However, now the slower ones are in front and the faster ones are behind so that they will reach the $-y$-axis after an identical delay time τ s. At the end of this second evolution period, the magnetization will be aligned along the $-y$-axis. This experiment is called the spin-echo experiment. This experiment will cancel all of the effects that result from the inhomogeneity of the magnetic field.

After refocusing the spins along the $-y$-axis, they will continue their motion and then the spin will start to fan out again. After the time τ s, we again apply a second 180° pulse on the sample and the spins will be inverted again. After the period τ s the spin will refocus along the $+y$-axis. This experiment will be repeated with a time interval of 2τ s and will produce an echo at every 2τ s (Figure 162).

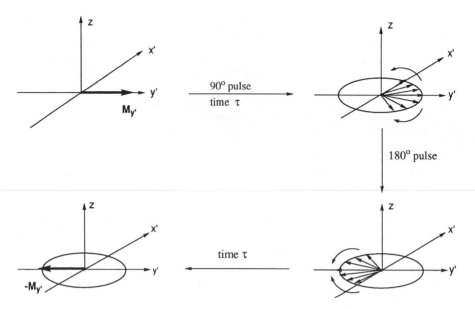

Figure 161   Schematic presentation of a spin-echo experiment.

The signal intensity will be measured at each echo and a plot of $(\ln M_{y'} - M_y^0)$ against t will yield a straight line whose gradient is $-1/T_2$. This method was developed by Carr, Purcell, Meiboom, and Gill and is referred to as the CPMG spin-echo experiment.

15.2   *J*-MODULATED SPIN-ECHO EXPERIMENTS

In the previous sections we have discussed the concept of single-pulse experiments, where we apply a 90° pulse (or a smaller pulse) on the sample and rotate the magnetization vector towards the y-axis. This induced magnetization along the y-axis is then recorded as a FID signal which can be easily transformed into the real spectrum by the Fourier transform technique. In the case of a multiple-pulse experiment, more than one pulse is applied on the sample before recording the FID signal. One of the most applied multiple-pulse experiments is the spin-echo experiment. As we will see later, the spin-echo experiment is an essential component of many multiple-pulse experiments. We have already discussed the application of the spin-echo experiment to measure the spin–spin relaxation time where the inhomogeneity of the magnetic field was eliminated. In

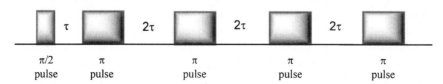

Figure 162   The CPMG spin-echo pulse sequences applied to measure the transverse relaxation T_2.

that experiment we did not consider the effect of the spin–spin couplings on the spin-echo experiment. We just took a sample giving a single line resonance. Now we will discuss the effect of the coupling on spin-echo experiments and show how the phases of the signals in a coupled system are affected by spin–spin coupling (*J*-modulation). The spin-echo experiment can provide some very useful information. For example, the *J*-modulated spin-echo experiment can provide us with information as to whether each carbon in a molecule has an odd or even number of protons directly attached to the carbon atoms. One of the simple applications of multiple-pulse experiments is the APT experiment. Qualitatively, we have already discussed an APT spectrum at the beginning of this chapter. Now we will discuss the mechanism of this technique.

15.2.1 Attached proton test experiment: assignment of ^{13}C signals [106]

In this section we will discuss some techniques which will provide information about the number of the protons attached to the resonating carbon atoms and attempt to comprehend the underlying theory. In Section 11.9 we have shown the traditional method of the *off-resonance decoupling* technique which can easily determine the multiplicity of the carbon atoms. In these complex molecules, there are always peaks overlapping in the coupled carbon spectra. Therefore, the *off-resonance decoupling* technique cannot be successfully applied in complex molecules. Nowadays, some multiple-pulse experiments such as APT (attached proton test), SEFT (spin-echo Fourier transform) or [107] DEPT (distortionless enhancement by polarization transfer) techniques are applied to make an exact assignment to the primary, secondary, tertiary and quaternary carbon atoms.

Let us start with a quaternary carbon resonating as a singlet (we ignore the long-range coupling constants). Application of a 90° pulse on the sample will rotate the magnetization vector towards the *y*-axis. In the case that the Larmor frequency of the carbon atom is equal to the frequency of the rotating frame, the tipped magnetization vector will no longer appear to precess at the *xy*-plane, but it will be stationary. Let us consider now a carbon atom bearing a single proton. Because of the heteronuclear coupling between the carbon and proton nuclei, the carbon atom will resonate as a doublet in ^{13}C-NMR spectroscopy. The resonance frequency of the resonating carbon atom is located in the middle of the doublet lines (Figure 163). The distance (in Hz)

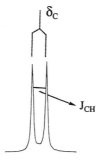

Figure 163 Presentation of the chemical shift and the coupling constant in a doublet.

between the doublet lines is referred to as the corresponding coupling constant between the carbon and proton.

In a system such as this, let us assume that the resonance frequency of the carbon atom is equal to the frequency of the rotating frame. Upon the application of a 90° pulse, both vectors (doublet lines) will rotate onto the y-axis. From the chemical shift, the Larmor frequency is exactly in the middle of these lines, one vector will rotate ahead with the frequency of $J_{CH}/2$ of the frame while the other will rotate in the opposite direction at the same rate $(-J_{CH}/2)$. These individual vectors will begin to spread out (Figure 164).

When we apply a 180° pulse after a certain delay time τ, the two vectors will be inverted about the x-axis back to the other side of the xy-plane. Since the rotational sense of their motion is unchanged they will move towards the $-y$-axis. They will reach the $-y$-axis after the delay time τ. At the end of this second evolution period, the magnetization will be aligned along the $-y$-axis. In that experiment we have applied exactly the same pulse sequences as in spin-echo experiment and shown that the vectors which are spread out as a result of a spin–spin coupling can also be refocused along the $-y$-axis after the time 2τ. In this specific example we have set the resonance frequency of the carbon atom equal to the frequency of the rotating frame. If we have a sample containing different sorts of carbon atoms with different Larmor frequencies and different coupling constants, the frequency of the rotating frame cannot be equal any more to the frequencies of the individual nuclei. Let us analyze a case where the resonance frequency of a doublet line is displayed by some arbitrary frequency (shown by the angle α), from the frequency of the rotating frame (Figure 165).

The application of a 90° pulse will rotate the magnetization vectors of the doublet lines into the y-direction. Now they will not be in phase with the frequency of the rotating frame. These vectors will move in the xy-plane at a rate of $\delta + J_{AB}/2$ and $\delta - J_{AB}/2$, respectively. After the delay time τ they will reach a certain position. A 180° pulse will invert the vectors to the other side of the xy-plane without changing the directions of rotation as we discussed above. Now the slower ones are in front and the faster ones are

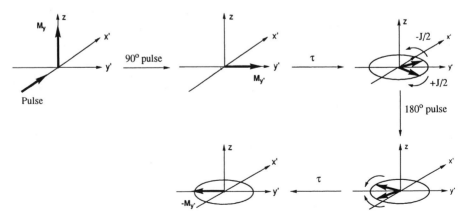

Figure 164 Schematic presentation of the magnetic vectors of a heteronuclear-coupled system during a spin-echo experiment. (The Larmor frequency is set equal to the frequency of the rotating system.)

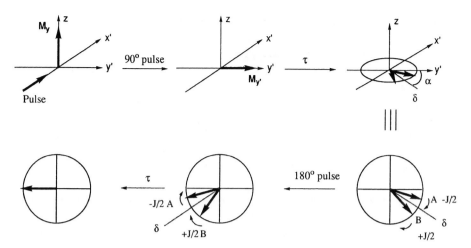

Figure 165   Schematic presentation of the magnetic vectors of a heteronuclear-coupled system during a spin-echo experiment. The sequence is applied off-resonance for the observed spins.

behind so that they will reach the $-y$-axis after the second delay time τ. At the end of this second evolution period, the magnetization will align along the $-y$-axis so that the spin-echo processes will be completed. This experiment demonstrates that the chemical shifts or heteronuclear couplings can also be refocused besides the phase coherence caused by the inhomogeneity of the magnetic field.

All of the experiments described above have been carried out with proton coupling throughout all of the pulses. Spin-echo experiments include all the three time periods: preparation, evolution and detection. If we carry out the above-mentioned spin-echo experiment with proton decoupling throughout all of the three periods, then the resulting spectrum, of course, will contain each carbon as a singlet. A spectrum of this kind will provide us with information about the chemical shifts, but not about the carbon multiplicities. However, we are interested in obtaining the valuable information about the multiplicity of the carbon atoms. How can we modify the spin-echo experiment in order to obtain the information about the carbon multiplicities? The spin-echo experiment starts with the initial 90° pulse generating transverse magnetization (Figure 166).

The proton decoupler should be on during the preparation period in order to benefit from NOE. Thereafter, the proton decoupler is switched off, so that the magnetization vectors evolve under the influence of proton–carbon coupling (*J*-modulation) during the first delay. During the second delay time and detection period, the proton decoupler must be turned on in order to obtain a proton-decoupled spectrum in the final period. During the second delay time, in the absence of a coupling, the vectors will collapse into a singlet. In an experiment such as that we can obtain information about the carbon multiplicities. Let us now explain how we can obtain information about the carbon multiplicities after this experiment.

Let us go back again to the spin-echo experiment of a tertiary carbon atom (C–H). We have discussed that the doublet lines (see Figure 164) will start to evolve after the 90° pulse. In a case where the Larmor frequency is equal to the frequency of the rotating

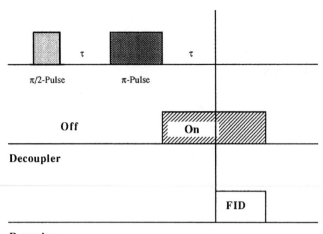

Figure 166   Pulse sequences in an APT experiment.

frame, the individual lines will move relative to the rotating frame at the same rate of $J_{CH}/2$ at the opposite direction. Let us calculate the time which is necessary to complete a single rotation. They rotate with a speed of $J_{CH}/2$ Hz. This means that they can rotate $J_{CH}/2$ times in a second. The time required for one rotation is therefore the inverse of this rate and given by the following equation:

$$\frac{1}{J_{CH}/2} = \frac{2}{J_{CH}} \tag{74}$$

Now, we know the exact time $(2/J_{CH})$ which is necessary for a single rotation. We can now calculate the time for a half rotation, as well as for a one-fourth rotation. At the time

$$\frac{2}{J_{CH}} \frac{1}{4} = \frac{1}{2J_{CH}}$$

the vectors will have made one-fourth of a rotation and will be opposite to each other along the x'-axis. At this point we cannot observe any signal in the detector because there is no component of the magnetization vector along the y'-axis. At the time of $1/J_{CH}$ the vectors will be focused along the $-y'$-axis and produce a negative signal if the signal was collected at this time. The position of the magnetization vectors of doublet lines during different evolution times is given in Figure 167.

After analyzing the position of the doublet lines at different evolution times, let us examine the fate of a triplet arising from methylenic carbons ($-CH_2-$). In a coupled spectrum, methylenic carbon will of course resonate as a triplet. Again, let us assume that the frequency of the rotating frame is equal to the chemical shift. We will follow the same pulse sequences and first apply a 90° pulse as was done before. After rotating the triplet lines on the y-axis, the central line (chemical shift) of the triplet signal will align along the y-axis and the outer lines will rotate relative to the rotating frame at the opposite direction. But, the most important point in this case is the rotation rate of the outer lines. Since the difference between the central line (chemical shift) and the outer lines is equal

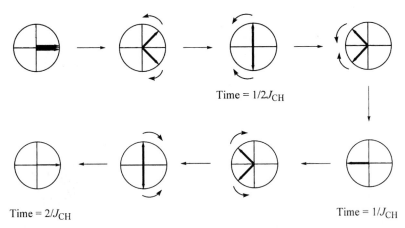

Figure 167   Vector diagrams for the magnetic vectors of a doublet and their dependence on the evolution period.

to the corresponding coupling constant J_{CH}, these lines will rotate in contrast to the doublet lines (see above), and with the frequencies $+J_{CH}$ and $-J_{CH}$, respectively. Let us now calculate the time for a single rotation. The time required for one rotation is again the inverse of this speed or $1/J_{CH}$. For a one-fourth rotation the time required is $\frac{1}{4}J_{CH}$ and for a half rotation $\frac{1}{2}J_{CH}$ seconds. At the time of $\frac{1}{2}J_{CH}$ the vectors will realign along the y'-axis and will generate a positive signal. The position of the magnetization vectors of triplet lines during different evolution times is given in Figure 168.

For CH$_3$ quartets, we have a similar situation as in the case of a CH doublet. There are two slower, but intense vectors that rotate with a speed of $+J_{CH}/2$ and $-J_{CH}/2$ while the outer lines of the quartets rotate at the speed of $3J_{CH}/2$. The position of the magnetization vectors of quartet lines during different evolution times is shown schematically in

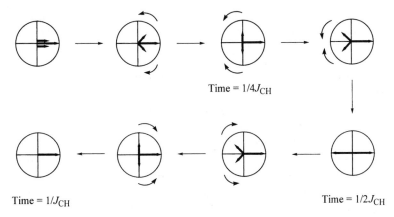

Figure 168   Vector diagrams for the magnetic vectors of a triplet and their dependence on the evolution period.

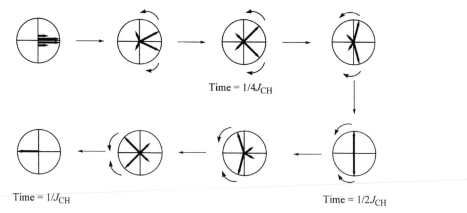

Time = 1/4J_{CH}

Time = 1/J_{CH} Time = 1/2J_{CH}

Figure 169   Vector diagrams for the magnetic vectors of a quartet and their dependence on the evolution period.

Figure 169. At the time $\tau = \frac{1}{2}J_{CH}$ the resultant magnetization is zero. Recording the spectrum at the time $\tau = 1/J_{CH}$ will produce a negative signal in the detector because the magnetic vectors of the inner as well as the outer lines will be refocused along the $-y'$-axis.

A similar analysis of the quaternary carbon atom shows that the magnetization vector will align along the $-y'$-axis and will be stationary and produce a positive signal at any time.

In all of the above-mentioned experiments we have set the Larmor frequency of each nucleus equal to the frequency of the rotating frame. But there must always be phase differences. In Figure 165 we have shown that these phase differences can be eliminated by applying a 180° pulse. Therefore, the phase differences will not prevent the focusing of the magnetic vectors.

After explaining the position of the magnetization vectors of different groups (C, CH, CH_2, CH_3) at different times, we may raise the question of how we can distinguish between the different substituted carbon atoms by way of spin-echo experiments. We have to analyze the positions of different carbon magnetization vectors at different times (Figures 167–169). At the time $\tau = \frac{1}{2}J_{CH}$, the resultant magnetization for all protonated carbon atoms (CH, CH_2, and CH_3) will be either zero or they will not have any component along the y-axis in order to be able to produce a signal in the detector. Only the quaternary carbon atoms can give a signal. On the other hand, at the time $\tau = 1/J_{CH}$ all of the carbon atoms will produce a signal in the detector, a positive signal for the C and CH_2 and a negative signal for the CH and CH_3 carbons (Figure 170).

In summary, in the APT experiments the delay time τ is set to be equal to $1/J_{CH}$ because this value produces the most notable effect in the spectrum. However, for a successful experiment one has to know the corresponding coupling constants in the molecule. In a molecule, the carbon–proton coupling constants can vary. It is suggested to use an average coupling constant. In that case, there will be some deviations from the ideal couplings. These small deviations will not affect the sign of the signal, but the intensity of the signal, which should not be important in the APT experiments, can vary.

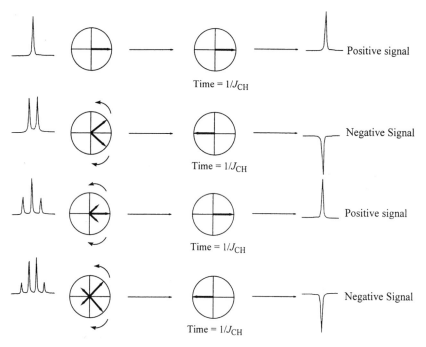

Figure 170   The position of the magnetic vectors of a singlet, doublet, triplet and quartet after the evolution period $1/J_{CH}$ in a *J*-modulated spin-echo experiment.

In an APT experiment one can also distinguish between the quaternary and methylenic carbon on the basis of the peak intensities. However, it will be difficult to distinguish between the methyl and methine carbon atoms. DEPT experiments can provide clear-cut information about the substitution pattern of the carbon atoms.

Figure 171 shows two APT spectra of two different compounds. The lower one, a normal decoupled ^{13}C-NMR spectrum belongs to menthol and shows 10 distinct resonances as expected. Menthol has three CH_3, three CH_2 and four CH carbon atoms. Since all of these carbon atoms are protonated, one cannot distinguish between them on the basis of the intensities. APT spectrum of the same molecule also consists of 10 signals, whereas seven of them are inverted, indicating that they belong to CH and CH_3 carbon atoms and three are positive signals that belong to CH_2 carbon atoms.

The second spectrum in Figure 171 shows the APT spectrum to be a ketoaldehyde [72a]. The total number of carbon resonances is nine. There are two carbonyl carbons in the molecule. One of them resonates as a positive signal at 210 ppm, whereas the other one appears as an inverted signal at 195 ppm. We can easily assign the positive signal to the ketone functionality and the other one to the aldehyde carbonyl carbon. This example demonstrates that the APT spectra can easily distinguish between the ketone and aldehyde functional groups. The negative signals that resonate between 133 and 155 ppm belong to the olefinic carbon atoms, those appearing between 60 and 80 ppm to the tertiary carbon atoms.

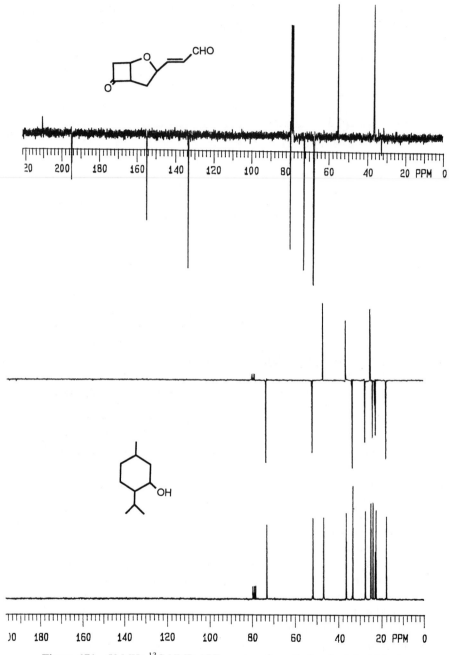

Figure 171　50 MHz ^{13}C-NMR APT spectra of menthol and a ketoaldehyde.

15.3   SIGNAL ENHANCEMENT BY POPULATION TRANSFER: SELECTIVE POPULATION TRANSFER AND SELECTIVE POPULATION INVERSION

It has already been mentioned that the low natural abundance of some nuclei has delayed the development of ^{13}C-NMR spectroscopy. On the other hand, poor sensitivity of these nuclei has forced the scientists to develop new techniques in order to increase the sensitivity of insensitive nuclei. Increasing of the magnetic field strength and fast signal accumulation are the most important ones.

One of the most important factors influencing the sensitivity of a given element is the energy difference ΔE between the energy levels formed in the presence of an external magnetic field. The gyromagnetic constant γ determines the energy difference ΔE between two spin states (eq. 10). The nuclei with greater gyromagnetic ratio γ are much more sensitive than the others. According to the Boltzmann distribution law, the larger the energy difference, the larger the population difference between the energy levels. The gyromagnetic ratio of a carbon nucleus is four times weaker than the proton ($\gamma_H = 4\gamma_C$). The population difference between the carbon nuclei in the presence of a static magnetic field will be correspondingly smaller than the proton. Therefore, carbon nuclei are much less polarized (less population excess) by a magnetic field. From that point we come to the following conclusion: The population difference between the energy levels is the most important factor influencing the sensitivity, i.e. the signal intensity. The question of how can we affect or increase the population difference between the different energy levels in a static magnetic field gains importance. In this chapter we will show how we can affect the population difference and as a consequence of that increase the sensitivity of elements with low-sensitivity. One should always keep in mind that any technique which can increase the sensitivity of elements has always been the most attractive area of NMR spectroscopy and it always will.

From now on, any process which can change the population difference between the energy levels will be called population transfer or polarization transfer [108]. We have already used the population transfer in some experiments without specific reference to the populations. For example, in the double resonance experiments, irradiation at the resonance frequency of a proton causes rapid transitions between the spin states so that the populations of the spin states will become equalized by keeping the irradiation on long enough. Secondly, a 180° pulse will rotate the bulk magnetization from the $+z$-axis to the $-z$-axis. This process is equivalent to an inversion of the populations of two different energy levels. Let us assume that there are 10 nuclei in the upper cone (parallel aligned) and 8 nuclei in the lower cone (antiparallel aligned). After the 180° pulse, the number of nuclei populating the different levels will be inverted, 8 nuclei in the upper cone and 10 nuclei in the lower cone. The basic mechanism of the NOE experiments is also based on the effects of manipulating the population difference between the spin states through the space dipolar interaction.

To understand the population transfer and its application, let us to clarify this phenomenon with some numbers. Let us consider a homonuclear system of two coupled protons. If the chemical shift difference $\Delta\delta$ between these protons is much larger than the corresponding coupling constant J_{AX}, these protons will form an AX-system. The intensity of the doublet lines will be equal. The energy level diagram of a homonuclear AX system is shown in Figure 172 together with a representation of the spectrum that we

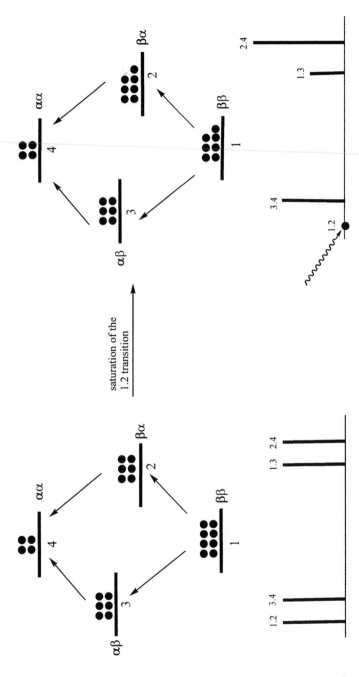

Figure 172 The energy levels and populations of a homonuclear AX system before and after selective saturation of the 1,2-transition and the corresponding signal intensities.

would obtain from each nucleus. It is there that four completely different energy levels exist:

(1)   both nuclei are parallel with the external magnetic field;
(2)   nucleus A parallel, nucleus B antiparallel;
(3)   nucleus B parallel, nucleus A antiparallel;
(4)   both nuclei antiparallel.

There are four different transitions which are shown by the arrows. Since the protons A and X are different, we have to distinguish between the energy levels 2 and 3. The number of the spins populating the energy levels are shown by dots.

Since the energy separation between the levels are almost equal, the population differences between the energy levels are also equal. The equal intensities of the doublet lines in an AX system also clearly indicate the equal population differences between the energy levels. As one can easily see from Figure 172 there are eight nuclei at the lower energy level, and four at the upper energy level, whereas there are six nuclei at the energy levels 2 and 3. These numbers show arbitrary units of excess population. The most important part of these populations is the difference between the energy levels. For example, the population differences between the energy levels that are responsible for the transitions ($1 \rightarrow 2$; $1 \rightarrow 3$; $2 \rightarrow 4$; $3 \rightarrow 4$) are 2 in all four cases. Therefore, the peak intensities are also equal. Now we selectively irradiate the resonance frequency of the line responsible for the transition $1 \rightarrow 2$ with a low-power soft pulse. If we keep the selective irradiation on long enough, the population levels 1 and 2 will become equalized, each having seven units of nuclei. When we subsequently record the ^{1}H-NMR spectrum of that system, the spectrum (AX system) will change. The line arising from the transition $1 \rightarrow 2$ will disappear due to the equal population of the energy levels. The intensity of the line from the transition $3 \rightarrow 4$ will not change. However, the intensity of the line from the $1 \rightarrow 3$ transition will be reduced to half of the original intensity because the difference between the nuclei at the energy levels has been reduced from 2 to 1. On the other hand, the intensity of the line $2 \rightarrow 4$ will have 1.5 times the original intensity, since the population difference has been changed from 2 to 3. In that experiment we have irradiated at one line of doublet and affected the intensities of the other coupled doublet lines. This process is called the *selective population transfer* (SPT).

Let us now conduct another experiment with the same system as described above. We apply a selective 180° pulse (rather long, low-power pulse) on the $1 \rightarrow 2$ transition. This proton pulse will induce a selective inversion of the population levels (Figure 173). In other words, the magnetization cones will be inverted. As a consequence of that, the number of the nuclei populating the energy levels 1 and 2 are now inverted. Recording of a spectrum immediately after the pulse will change the NMR spectrum. The signal from the transition $1 \rightarrow 2$ will appear as a negative signal. The intensity of the line from transition $3 \rightarrow 4$ will remain unchanged. However, the intensity of the line from transition $2 \rightarrow 4$ will increase by two times the original intensity. Since the energy levels 1 and 3 now have equal populations, the corresponding line from transition $1 \rightarrow 3$ will disappear. The total intensity of the unirradiated lines remains the same as before. This experiment is called *selective population inversion* (SPI).

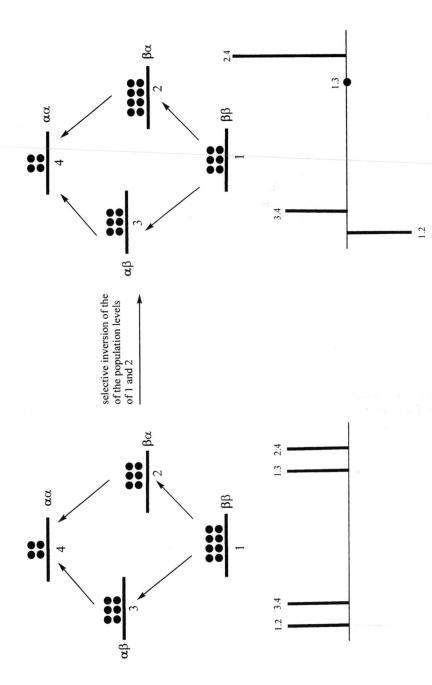

Figure 173 The energy levels and populations of a homonuclear AX system before and after selective inversion of the 1,2-transition and the corresponding signal intensities.

With the above-described experiments we have attempted to demonstrate how the populations at the different energy levels are changed upon application of a soft and hard pulse on a single line of a homonuclear system. Let us now describe similar SPT and SPI experiments with a heteronuclear system, chloroform. The ^{13}C and ^{1}H nuclei in the chloroform molecule will form a heteronuclear AX system and resonate as doublets (of course a proton doublet in the ^{1}H-NMR and carbon doublet in the ^{13}C-NMR spectra). Doublet lines in the proton NMR spectrum appear as satellites on the left and right side of the main signal arising from ^{12}CHCl$_3$.

In the homonuclear system, we have taken an equal population excess of the different energy levels. This will not be the case in a heteronuclear system since the gyromagnetic ratios determine the populations of the levels. Since the gyromagnetic ratio of a proton is four times larger than that of carbon, we have intentionally increased the numbers of the nuclei of the levels responsible for the proton resonances. Figure 174 shows the population differences when the nuclei are at equilibrium in the presence of an external magnetic field. Population differences are reflecting the gyromagnetic ratios of the corresponding elements. In that connection we will ignore the probability for the resonance. (We know the resonance probability of carbon to proton is 1:60 even in a case where the population difference ratio is 1:4).

Let us show the energy levels of a coupled C–H system as we did in the case of a homonuclear coupled system (Figure 174). The transitions $1 \rightarrow 3$ and $2 \rightarrow 4$ indicate the proton resonances and the transitions $1 \rightarrow 2$ and $3 \rightarrow 4$ to carbon resonances. The excess of the nuclei for the carbon resonances is 2, whereas for the proton resonances it is 8. This number reflects the gyromagnetic ratios, or in other words, the relationship between the populations and the relative intensities of the corresponding transitions. Selective irradiation with a soft pulse at the resonance frequency of the transition $1 \rightarrow 3$ will saturate the transition so that the number of the nuclei at levels 1 and 3 will be equalized. As a result of this experiment, ^{13}C transitions will be completely altered. The signal from transition $1 \rightarrow 2$ will be inverted. On the other hand, the carbon line from the transition $1 \rightarrow 3$ will disappear due to the equal population of the energy levels and the intensity of the line from transition $3 \rightarrow 4$ will increase three times, since the population difference between energy levels 3 and 4 has become greater than before. With this experiment, the population difference of a proton has been transferred to a carbon atom.

Now we apply a selective 180° pulse at the resonance frequency of transition $1 \rightarrow 3$ and invert the populations of energy levels 1 and 3 on this system (Figure 175). The new population difference is shown on the right side of the Figure 175. The proton signals preserve their intensities, but the $1 \rightarrow 3$ transition signal is now inverted. However, the signal intensities of the carbon lines are increased up to three and five times, whereas the carbon line corresponding to the transition $1 \rightarrow 2$ is inverted.

When we compare these last two experiments of the heteronuclear system with those of a homonuclear system, one can notice large differences. In the heteronuclear experiments, the intensities of the heteronuclear atom, carbon, is increased four times, whereas the intensities of the proton signals are not changed at all. In conclusion, if a sensitive nucleus, such as a proton, is directly attached to a less sensitive nucleus and if there is a spin–spin coupling between these heteronuclear nuclei, the signal intensity of the less sensitive nucleus can be increased enormously upon the population inversion of a line belonging to the sensitive nucleus. The amount of the increase in intensity of the

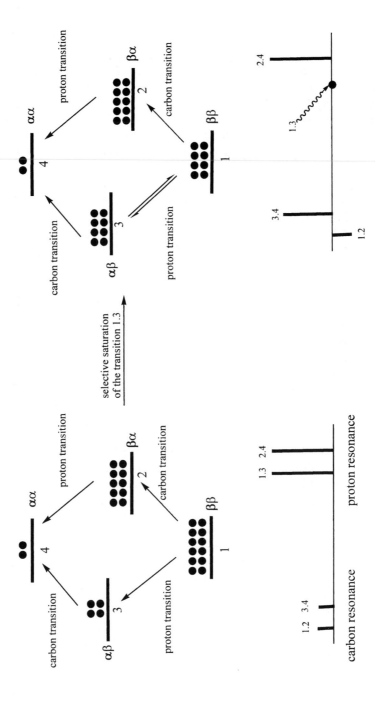

Figure 174 The energy levels and populations of a heteronuclear C–H system before and after selective saturation of the 1,2-transition and the corresponding signal intensities.

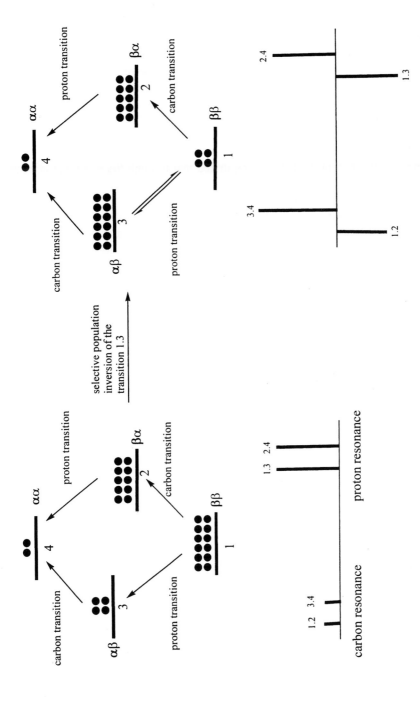

Figure 175 The energy levels and populations of a homonuclear C–H system before and after selective inversion of the 1,2-transition and the corresponding signal intensities.

coupled elements is directly given by the ratio of the gyromagnetic ratios. The larger the ratio is, the larger the increase in the signal intensity. The application of the population inversion method on the elements with smaller gyromagnetic constants makes it possible to measure them. For example, the increase in signal intensity in the case of N and Si is 10 and 5 times, respectively.

We will now turn back to Figure 175. After the population inversion of a proton signal, we obtain the doublet components with relative intensities $+5$ and -3 (as compared with direct observation). It can be very useful if we can invert the negative signal (change the phase of the negative signal) and combine these two signals with the intensities of $+5$ and $+3$ together, then we may obtain a signal where the intensity is increased up to $+8$ (Figure 176).

With an SPI experiment we have demonstrated that through the perturbation of the energy level of a single resonance in a coupled system, the other energy levels can be affected which can result in the increase of the signal intensity. However, there are some disadvantages in these experiments. First, they are selective, which means that only one line at a time in the ^{1}H spectrum can be disturbed. As mentioned above, the irradiated proton signals are not the main signals, they are ^{13}C satellites with low intensities. They are complex and can be overlapped with the other signals in most cases so that it will be very difficult to locate them. Under these circumstances it is not desirable to go through the spectrum and irradiate at the resonance frequency of all the signals one by one. Since 1980 new methods using special pulse sequences have been developed that make it possible to produce population inversion simultaneously for all of the multiplets in a molecule.

Let us go back to population inversion (Figure 175). We have applied a 180° pulse on one of the doublet lines and selectively inverted the population of levels 1 and 3. At this stage we can think about the question of can we design an experiment which can invert both of the doublet lines at the same time. A nonselective population transfer that can

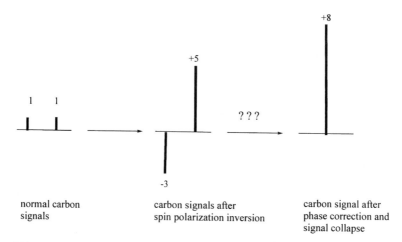

normal carbon　　　　　　carbon signals after　　　　　　carbon signal after
signals　　　　　　　　　spin polarization inversion　　　phase correction and
　　　　　　　　　　　　　　　　　　　　　　　　　　　　signal collapse

Figure 176 The new populations after inverting of the H-transitions in a heteronuclear (C–H) system and the combination of the carbon signals.

enhance the signals of all kinds of carbon atoms by a single experiment would be more desirable.

We then apply a nonselective 90° preparation pulse (proton pulse) on the sample (let us consider the doublet lines), both vectors will rotate onto the y-axis. Since the chemical shift, the Larmor frequency is exactly in the middle of these lines, one vector will rotate ahead with the frequency of $J_{CH}/2$ of the frame while the other will rotate in the opposite direction at the same rate $-J_{CH}/2$ as we have discussed before. These individual vectors will begin to spread out (Figure 177, see also Figure 164). After the time $\tau = \frac{1}{2}J$, the time required for one-fourth rotation (the time required for a single rotation is $\tau = 2/J$), the vectors will align along the $+x$- and $-x$-axis, respectively. At this moment a second 90° pulse from the y-direction will flip them onto the z-axis, one component being along the $+z$-axis, the other one along the $-z$-axis, generating the population inversion. If we would measure the carbon spectrum at this stage, it would show the characteristic polarization transfer 5: − 3 pattern which we have discussed before.

In this experiment we have applied two successive 90° pulses on the sample and generated the population inversion. The most interesting feature of this experiment is that the applied pulse is not a selective pulse. Application of this experiment on all of the protons will result in the population inversion of all protons. In this particular example we have only considered a doublet line. By the application of this experiment on all kinds of protons, the magnetic vectors of all protons will not align along the x-axis during the evolution time due to the different couplings. Therefore, any chemical shift dependence during the evolution period should be removed. This is only possible by the application of the spin-echo process in order to focus the chemical shifts. This topic will be handled in detail in the next section.

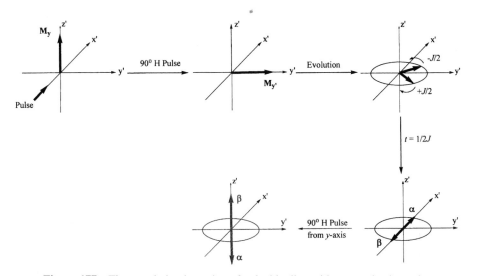

Figure 177 The population inversion of a doublet line with a nonselective pulse.

15.4   INEPT EXPERIMENT: INSENSITIVE NUCLEI ENHANCED BY POLARIZATION TRANSFER

INEPT is abbreviated from 'insensitive nuclei enhanced by polarization transfer' and it is a useful method to measure the insensitive nuclei by the simultaneous polarization transfer from the proton to the corresponding nuclei [109, 110]. It is also successfully used for multiplicity selection. The basic INEPT sequence has already been discussed in the previous section (Figure 177). We will discuss it in more detail now.

We will begin again with a heteronuclear AX system with the coupling constant J and apply a nonselective $90°$ preparation pulse (proton pulse) on the sample. The doublet lines will rotate with the frequency of $J_{CH}/2$ in the opposite direction. These individual vectors will begin to spread out (Figure 178, see also Figure 177). The delay time τ in this experiment is set to be equal at $\frac{1}{4}J_{AX}$ so that the doublet lines can move in a one-eighth rotation (the time required for a single rotation is $\tau = 2/J$). At the end of this time the vectors will have rotated $90°$ apart from one another (Figure 178).

At this instant, a $180°$ pulse is applied to the protons. After this pulse, the spins will be inverted across to the other side of the xy-plane. Since the rotational sense of their motion is unchanged, they will move towards the $-y$-axis. At the end of this second evolution period $\frac{1}{4}J_{AX}$, the magnetization vectors will be focused along the $-y$-axis. Actually, we are not interested in focusing the magnetization vectors along the $-y$-axis, however, they must be focused along the x-axis in order to generate population inversion at the next step. For that reason, we have to change the rotation directions of the vectors moving towards the $-y$-axis after the $180°$ pulse. Therefore, we apply a $180°$ carbon pulse simultaneously with the $180°$ proton pulse. After these pulses, those protons that were formerly bonded to carbons in the α-spin state are now bonded to the carbons in the β-spin state. This means that the rotation directions of the vectors are now switched and

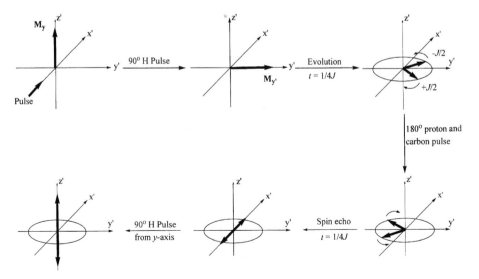

Figure 178   Population inversion during the INEPT sequence and focusing of the chemical shift.

they will move towards the *x*-axis and refocus at the end of the delay time $\frac{1}{4}J_{AX}$. With this experiment we can eliminate the chemical shift differences and focus all kinds of vectors along the *x*-axis. This situation corresponds with the desired antiphase polarization of the doublet in the SPI experiment (Figure 177). Now the system is ready to apply the 90° pulse to cause the population transfer. In the next stage of the pulse sequence we apply a 90° pulse (from the +*y*-direction) and rotate the magnetic vectors aligned along the *x*-axis into the +*z*- and −*z*-directions so that the polarization transfer to the coupled carbon is produced. The immediate 90° carbon pulse will rotate the magnetization towards the *y*-axis, and then FID is obtained in the usual way. When we record at this stage a ^{13}C spectrum, we will observe two lines with the intensities of −3:5, whereas the first line is inverted. In a normal spectrum the intensity distribution of a doublet line is actually 1:1 contributed from the natural magnetization. In order to eliminate the different signal intensities in the INEPT experiment we proceed in the following way. We alternate the phase of the last 90° pulse. If this pulse is applied from the −*y*-axis, the sense of the proton polarization is reversed. This means that the components which have previously aligned along the +*z*-axis now will be found along the −*z*-axis, and *vice versa* (Figure 179). This phase alternation cancels the natural magnetization contribution and produces an intensity pattern of −3:5 and 3: −5, respectively. Typically, the subtraction of these signals will produce an intensity pattern of 8: −8 for the observed doublets.

Pulse sequence in basic INEPT experiments is shown in Figure 180. During the acquisition time the decoupler must be off. Otherwise, we will not observe any signal because of the different phases of the doublet lines which would cancel each other out.

So far we have analyzed only the behavior of a doublet line during a basic INEPT experiment. Let us now briefly discuss triplet and quartet lines arising from CH$_2$ and CH$_3$

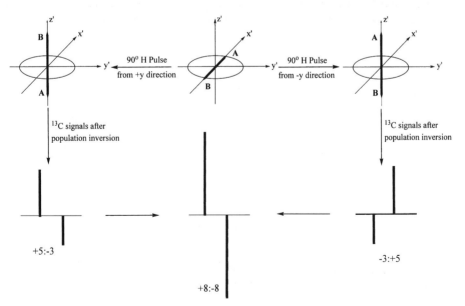

Figure 179 Increase in the signal intensities (5: − 3) of the doublet lines of a C–H system after INEPT experiment and correction of the intensity difference by an alternate 90° pulse.

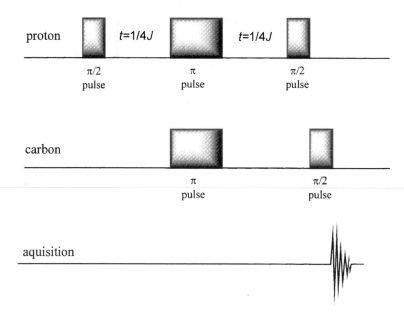

Figure 180   Schematic presentation of a basic INEPT pulse sequence.

groups. In a normal coupled ^{13}C spectrum, a CH$_2$ group resonates as a triplet with an intensity distribution of 1:2:1. Since the middle line will align along the y-axis during the INEPT experiment, the intensity of this line will be zero, however, the intensity of the outer lines will be the increased one with positive and the other one with negative amplitudes. The quartet lines will also demonstrate behavior much like doublet lines giving rise to two negative and two positive lines with the enhanced signal intensities. The quaternary carbon atoms cannot be observed during the INEPT experiment.

   In the INEPT experiments we increase the signal intensities. However, some signals appear as negative signals and some as positive signals. If we can change the phase of negative signals we may obtain two positive signals for a CH group. Furthermore, the application of a proton decoupler during the acquisition time can generate only one signal with a positive sign and enhanced intensity. How to perform this will be discussed below.

15.5 REFOCUSED INEPT EXPERIMENT

In the INEPT experiment we have demonstrated how the polarization of a sensitive element is transferred to a less sensitive element. The amplification factor of the signals in the INEPT experiment is given by the factor:

$$\frac{\gamma(^1H)}{\gamma(^{13}C)} = 4$$

As we have discussed above, the basic INEPT experiment cannot be acquired with proton decoupling, since the antiphase signal would collapse and the signal will disappear.

However, it is possible to bring the doublet lines to the same phase and switch the proton decoupler on during the data acquisition so that the spectrum will consist of a single absorption line with an enhanced signal intensity. Let us attempt to clarify how to perform this experiment.

Before starting the acquisition in normal INEPT experiments, the magnetic vectors of a doublet line are aligned along the $+y'$- and $-y'$-axis, respectively. If we record a spectrum at this stage we will obtain a normal INEPT spectrum with the intensity distribution of $-3:5$ (see Figure 181). For a completed rotation of the doublet lines the required time is $2/J$. However, we then insert a delay time of $\frac{1}{4}J$, in which they can move for a one-eighth rotation. After this time their phase difference will be reduced to 90°. Then we apply a 180° carbon and a proton pulses at the same time as we did in the first part of the INEPT experiment. The purpose of these additional 180° pulses is to eliminate the effects of chemical shift differences during this evolution time (Figure 181). After the second evolution time of $\frac{1}{4}J$ the two vectors will recombine along the x-axis. Application of the proton decoupler during the acquisition time will generate a refocused and decoupled carbon spectrum with the enhanced signal intensity. Refocused and decoupled INEPT pulse sequence is given in Figure 182.

The time required to refocus the signal in the INEPT experiments is very important. In the above experiment we have discussed the refocusing of the doublet lines and have shown that the required time is $\Delta = \frac{1}{4}J$. To refocus the other lines in a triplet, the delay time Δ has to be set equal to $\frac{1}{8}J$. The middle line has to be set to zero intensity. On the other hand, the four lines of a quartet cannot be refocused. However, maximum signal enhancement is obtained at a delay time of $\Delta = \frac{1}{8}J$. If one wishes to have a reasonable enhancement of all the signals in a sample containing CH, CH$_2$, and CH$_3$ the choice of Δ

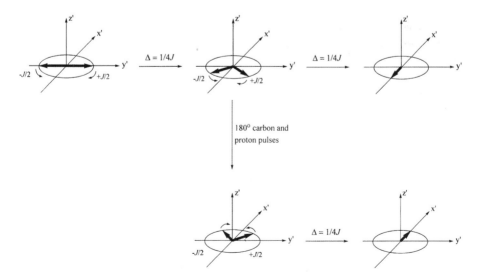

Figure 181 Vector diagram for the carbons during a refocused INEPT experiment.

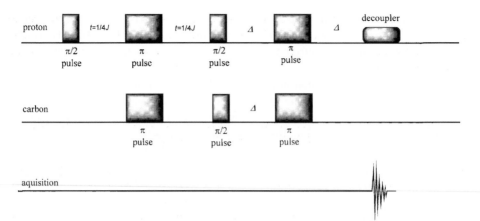

Figure 182   Schematic presentation of a refocused INEPT pulse sequence.

is essentially a compromise, as the ideal delay time is then $\Delta = \frac{1}{7}J$. Recording an INEPT spectrum at a delay time of $\Delta = \frac{1}{4}J$ will provide the observation of only tertiary carbon atoms.

The fundamental property of the INEPT experiment is to increase the sensitivity of a nonsensitive nucleus by a factor proportional to γ_A/γ_B. By the transfer of the polarization from proton to carbon, each carbon line exhibits a fourfold increase in the intensity. This is significantly important to the threefold enhancement in sensitivity obtained by the NOE by recording proton-decoupled spectra. Actually, more enhancement (sixfold, sevenfold, etc.) can be obtained by INEPT experiments since the repetition rate of the INEPT experiment is controlled by the relatively shorter proton spin-relaxation time, whereas the repetition rate in the proton-decoupled ^{13}C spectra is determined by the longer carbon relaxation times.

The major advantage of INEPT experiments arises for those nuclei with longer relaxation times and especially those with small gyromagnetic constants. Interesting INEPT experiments with metals have been carried out, to where protons were used as polarization sources. Table 15.1 shows signal enhancement ratios in NOE and INEPT experiments.

Table 15.1

A comparison of maximum signal enhancement by NOE and INEPT
experiments [110]

Nucleus	Nuclear Overhauser effect (NOE)	Polarization transfer (INEPT)
^{13}C	2.99	3.98
^{31}P	2.24	2.47
^{15}N	− 3.94	9.87
^{103}Rh	− 14.89	31.78

The INEPT experiments are not only useful for signal enhancement, but also to determine the multiplicity of the signals. A modified sequence of INEPT experiments permits the exact multiplicity selections of the signals with an enhancement of the intensities. This method is called DEPT and it is the subject of Chapter 16.

15.6   DEPT EXPERIMENTS: MULTIPLICITY SELECTION OF CH, CH$_2$, AND CH$_3$ CARBON ATOMS

The DEPT is the distortionless enhancement by polarization transfer and it is a useful method to determine the multiplicity of the carbon atoms [111]. The DEPT measurements allow the generation of subspectra for the CH, CH$_2$ and CH$_3$ groups. Before we begin to discuss the pulse sequence in a DEPT experiment, let us show a DEPT spectrum and discuss it.

Figure 183 shows the subspectra (DEPT spectra) of a bisepoxide [112]. A DEPT spectrum actually consists of several spectra, with the final data presentation depicting

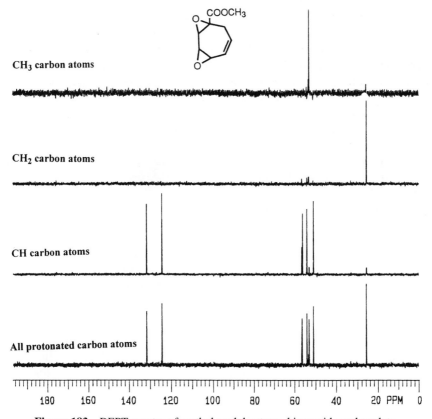

Figure 183   DEPT spectra of methyl cycloheptene–bisepoxide carboxylate.

one spectrum for each type of carbon atom. Thus, CH, CH_2 and CH_3 carbons are each printed out on separate spectra, together with a ^{13}C spectrum where all carbon types are shown. Each carbon type is thus identified unambiguously. The below spectrum in Figure 183 shows all protonated carbon atoms. The next one shows only methine carbon atoms (CH), whereas the others show the CH_2 and CH_3 carbon atoms separately. On the basis of these spectra one can easily assign carbon atoms. Nowadays, the DEPT technique has generally replaced the old-fashioned method of the off-resonance decoupling carbon spectra with reduced CH couplings from which the multiplicity could be directly determined.

The DEPT experiment is in fact an improved version of the INEPT experiment. It also results in an enhancement of the intensity of the X-nuclei by a factor of γ_X/γ_H. One of the nicest improvements of this sequence when compared with the INEPT is that the DEPT experiment does not have to deal with a variable refocusing delay, since all of the carbon signals are in phase at the beginning of the acquisition. Let us briefly describe the pulse sequence in the DEPT experiments.

The starting point of the DEPT pulse sequence, as we can see from Figure 184, is a 90° pulse on a proton. This pulse rotates the equilibrium magnetization from the z-axis to the y'-axis. The doublet lines of a CH group start to move in the xy-plane in opposite directions. At the end of the delay time $t = \frac{1}{2}J$, the vectors align along the $+x'$- and $-x'$-axis with a phase difference of 180°. A 180° pulse on the protons refocus inhomogeneity effects. Simultaneously, a 90° pulse on the carbons rotates the ^{13}C magnetization vectors from the z-axis to the y'-axis. After the subsequent delay time $t = \frac{1}{2}J$, a 90° pulse on the protons from the y-direction rotates both doublet components of proton magnetization to the $+x'$- and $-x'$-axis, respectively. This pulse sequence

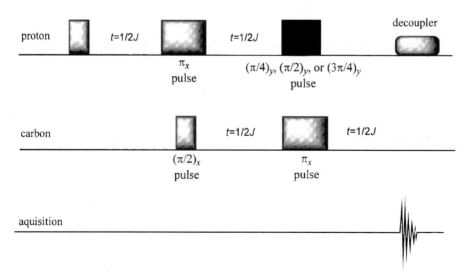

Figure 184   Schematic presentation of a basic DEPT pulse sequence.

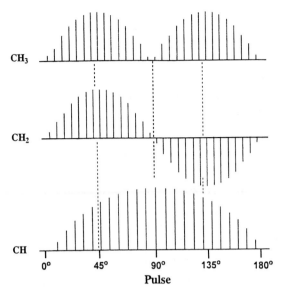

Figure 185   Intensity distribution of the CH, CH$_2$, CH$_3$ as a function of the pulse width.

causes the polarization transfer from protons to carbon atoms. A 180° carbon pulse cancels the inhomogeneity effect on carbon magnetization. At the end of the second delay time $t = \frac{1}{2}J$, a proton pulse is applied. *The pulse length of that proton pulse is very important.* This pulse forms the basis of spectrum editing with DEPT. Unlike INEPT, DEPT spectra cannot be analyzed by the simple vector model that we are using. Therefore, we will describe only briefly the consequences of the application of proton pulses with different pulse widths.

If the pulse proton length is taken at 90°, the CH signals experience a maximum polarization, whereas the CH$_2$ and CH$_3$ carbon atoms experience negligible polarization. On the other hand, CH$_2$ and CH$_3$ carbon atoms show maximum signal intensity when the length of the proton pulse is chosen to be 45° ($\pi/4$ pulse) (Figure 185). Application of a 135° proton pulse ($3\pi/4$ pulse) results in a negative maximum signal intensity for the methylene carbon atoms, whereas the CH and CH$_3$ carbon atoms appear as positive signals.

As a consequence of recording spectra with three different pulse widths (45, 90, and 135°) we can record different subspectra. 90° experiment generates a subspectrum showing only CH carbons. An additional experiment with a 135° pulse can distinguish between the CH$_2$ carbon atoms and CH/CH$_3$ carbon atoms. Differentiation between the CH and CH$_3$ carbon atoms is straightforward since the CH carbon atoms can be easily recognized from the first spectrum recorded with a 90° pulse. A 45° pulse experiment shows all kinds of protonated carbon atoms. Subtraction of this spectrum from a normal decoupled spectrum will show us all of the quaternary carbon atoms. With the help of a computer all of these spectra can be edited and the corresponding subspectra showing the C, CH, CH$_2$, and CH$_3$ carbon atoms can be drawn separately.

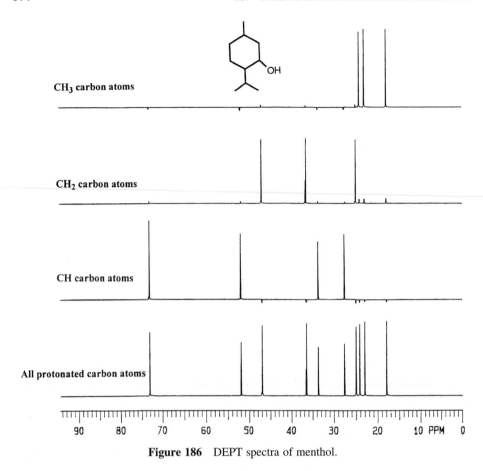

Figure 186   DEPT spectra of menthol.

Figure 186 shows the DEPT spectrum of menthol. The lower spectrum has been recorded with 256 pulses and the applied pulse was 45°. It shows all of the protonated carbon atoms. The other spectra have been obtained by the addition and subtraction of the spectra recorded by different pulse widths. The first spectrum shows the presence of three methyl resonances, whereas the next one also shows the existence of three different methylene carbons as expected from the structure of menthol. With the help of the DEPT spectra it is very easy to determine the multiplicity of the carbon signals. If one is only interested in distinguishing, for example, between CH carbons, then they are referred to HETCOR spectra.

An additional DEPT spectrum of a tetrahydroazulene derivative with the corresponding subspectra is given in Figure 187 [113].

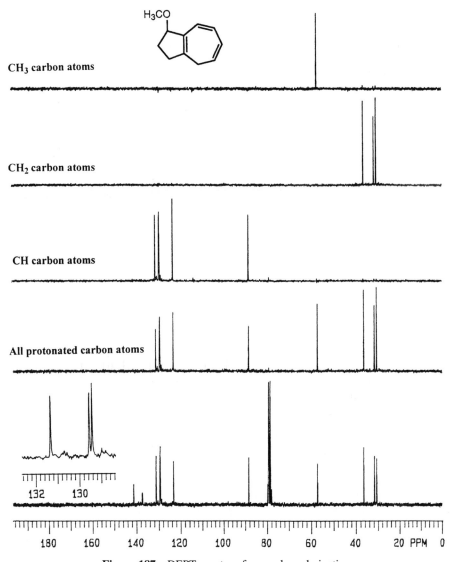

CH₃ carbon atoms

CH₂ carbon atoms

CH carbon atoms

All protonated carbon atoms

Figure 187   DEPT spectra of an azulene derivative.

Part III

Two-Dimensional (2D) NMR Spectroscopy

– 16 –

Two-Dimensional (2D) NMR Spectroscopy

INTRODUCTION

The developments in pulse technology since the 1980s form the basis of two-dimensional (2D) NMR spectroscopy [110, 114]. The two important parameters that are obtained from the NMR spectra are chemical shift and the spin–spin coupling constant. These parameters sometimes cannot be obtained from the complex spectra. Therefore, different measurement techniques are needed. With help from the new techniques discovered by chemists, it is now possible to simplify the interpretation of NMR spectra of very complex molecules. In 1991, the Nobel Prize was given to the Swiss chemist Ernst for his studies within this subject (the second Nobel Prize to be won within the field of NMR). Let us first answer the question 'What are the new facilities brought by 2D NMR spectroscopy' and move onward then to the theory of 2D spectroscopy and spectrum analysis.

(1) The spectral parameters needed for spectrum interpretation are chemical shifts and spin–spin coupling constants as mentioned above. From time to time, it is possible that these parameters cannot be extracted directly from the spectra even in the measurements with higher magnetic field instruments. J-resolved spectroscopy offers a new way to resolve even highly overlapped signals into readily interpretable signals to observe chemical shifts and coupling constants on different axes.

(2) One of the useful methods applied in the structure determination of complex molecules is the double resonance experiment (see Chapter 7). The double resonance experiment determines the coupling relation between the protons in a given molecule. This technique can be easily applied to well-resolved systems. However, it is difficult to obtain information on complex systems. Through the use of homonuclear correlation spectroscopy (COSY), one can easily obtain information concerning the spin-coupled systems.

(3) The measurement of carbon nuclei requires more time due to their low natural abundance. The assignment of carbon resonances was generally completed by selective excitation of the directly bonded protons. Because of the large number of carbon atoms in a molecule, it was applied to every carbon atom one by one and each experiment required a significant portion of time. Therefore, this method

is not frequently utilized. The relation or the connectivity between protons and carbons are determined by heteronuclear correlation spectroscopy (HETCOR, HMBC). The spectrum analysis is rather easy.

(4) The very last method that is applied by chemists when the exact configuration of a given molecule cannot be determined is the tedious and time-consuming X-ray analysis. Furthermore, X-ray analysis can be applied only to single crystals. When utilizing the NOESY and ROESY methods the configuration of the molecule in most cases can be determined by a single measurement and there is no need for the X-ray analysis.

(5) The structure of the organic compounds is mostly determined by the help of ^{1}H-NMR. ^{13}C-NMR is used to support the interpretation. Although it is possible to determine the constitution in a molecule by ^{1}H-NMR, the ^{13}C-NMR is not solely enough for the determination of constitution. The new NMR techniques, such as 2D-INADEQUATE, can enable the determination of the constitution of a molecule without measuring the ^{1}H-NMR spectrum.

Before starting to investigate 2D-NMR spectroscopy, let us examine the difference between 1D- and 2D-NMR spectra on the basis of their appearance. The 1D-NMR spectra contain two axes: The first axis is the frequency axis and the second one is the amplitude axis (Figure 188).

The frequency axis provides information about both chemical shift and spin–spin coupling constants, whereas the second axis provides information about signal intensity. All the spectra we have seen up to now are drawn on two axes. So, these spectra are actually two dimensional. However, when the parameters of the signal intensity axis are changed, there will be no change on chemical shifts and coupling constant values. Therefore, this axis is not so important and it can be ignored. Generally, the frequency

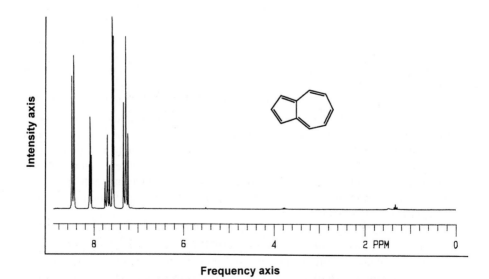

Frequency axis

Figure 188 Frequency and intensity axes in a typical 1D-NMR spectrum.

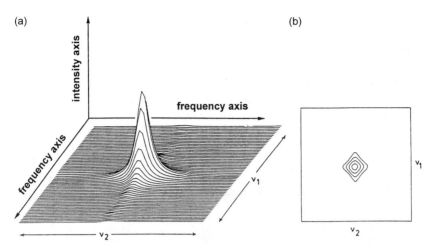

(a) (b)

Figure 189 (a) Frequency and intensity axes in a typical 2D-NMR spectrum. (b) Counter representation of the 2D-NMR spectrum. (Reprinted with permission of Elsevier Inc. from Andrew E. Derome, *Modern NMR Techniques for Chemistry Research*, 1987.)

axis is used and all spectra drawn on two axes, frequency and signal intensity, are called 1D spectra. On the 2D-NMR spectra, an additional chemical shift (homonuclear, as well as heteronuclear system) is recorded on a third axis [115]. We are then met with a three-dimensional spectrum. Since one of these axes contains signal intensity again, it is subsequently ignored and these spectra are called 2D spectra. Figure 189a shows a typical 2D spectrum obtained from chloroform. There are three axes; one of these three axes is the signal intensity axis and the others are frequency axes. The top view of this signal (counter spectrum) is shown in Figure 189b. In that spectrum, the axis containing the signal intensity is ignored.

Let us now investigate two different 2D-NMR spectra in order to obtain qualitatively some information about 2D-NMR spectra before going into more detail. In order to determine the relation between the protons in the proton 1D-NMR spectrum, double resonance experiments are applied one by one and the relation between the protons are found thereafter. Today, this and the other similar experiments can be carried out very quickly and the recorded COSY spectra contain all of the information on a single spectrum [116].

Figure 190 shows the COSY (homonuclear correlation) spectrum of a compound containing four different protons, whereas the proton H_3 couples with the protons H_1 and H_4. This spectrum provides information about proton-to-proton interactions. Both axes on the COSY spectrum are frequency axes and 1D-NMR spectrum of the corresponding compound is plotted along both the *x*- and *y*-axis. There are two different signal groups on COSY spectrum. One of them is the diagonal signal that is shown with open circles in Figure 190. These signals show the chemical shifts and do not provide any additional information. The important and new information from the COSY spectrum, however, comes from the correlation peaks or 'cross peaks' that appear off of the diagonal (shown with the filled circles), that indicate a coupling relation. If we start from any cross peak

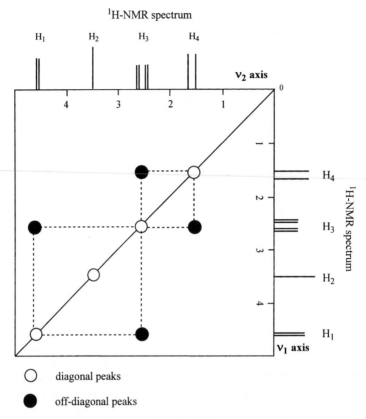

Figure 190 COSY spectrum (homonuclear 2D-NMR spectrum) of a system containing four different protons where the proton H_3 has coupling to the protons H_1 and H_4.

and draw two perpendicular lines (parallel to each spectrum axis) leading back to the diagonal, the peaks intersected on the diagonal by these lines are coupled to each other. There is a mirror reflection of these cross peaks. Only cross peaks on one side of the diagonal need to be interpreted. Interpretation of the spectrum in Figure 190 demonstrates that there are correlation peaks between the protons H_3 and H_1, as well as between H_3 and H_4. If we start from the cross peak which belongs to the proton H_3 and draw two perpendicular lines, these lines will intersect two off-diagonal peaks located at 4.6 and 1.7 ppm, respectively. This means that the proton H_3 couples to the protons H_1 and H_4. In a complicated spectrum, all kinds of coupling relations between the protons can be obtained on a single COSY spectrum.

　　In the second example, we demonstrate a heteronuclear correlation spectrum of a simple compound containing four different proton-bonded carbon atoms (Figure 191). It is not easy to assign the carbon signals from a proton-decoupled ^{13}C-NMR spectrum. Selective decoupling experiments can identify the carbon atoms. In the case of a full analysis, every proton signal should be decoupled one by one. This is a time-consuming process. This old-fashioned method is today replaced by the HETCOR. In a HETCOR spectrum, the 1D-NMR of one nucleus is plotted vs. the 1D-NMR of a different nucleus.

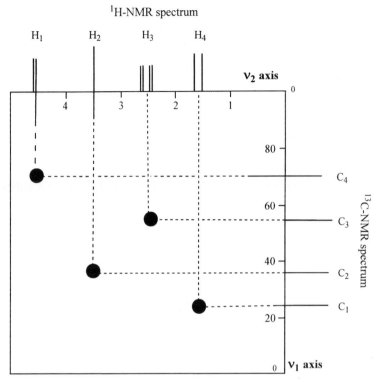

Figure 191 HETCOR spectrum (heteronuclear 2D-NMR spectrum) of a system containing four different proton-bonded carbon atoms.

In this particular case, the HETCOR experiment correlates ^{13}C nuclei with directly attached (coupled) protons. These are one-bond couplings. The ^{13}C spectrum is drawn along the F1 axis (ν_1 axis) and the ^{1}H spectrum is presented along the F2 axis (ν_2 axis). As one can easily see from the given HETCOR spectrum, there are no diagonal peaks and no symmetry. This will be always true whenever F1 and F2 represent different nuclei. The interpretation of this spectrum is straightforward. We start with a carbon resonance signal and draw a line vertically (parallel to the proton axis F2) until a cross peak is encountered. These cross peaks in a HETCOR spectrum indicate which protons are attached to which carbon atoms, or *vice versa*. If a line drawn down encounters no cross peak then this means that this carbon atom has no attached proton, i.e. a quaternary carbon atom. Of course, the full proton assignment in the proton spectrum has to be performed for the interpretation of the HETCOR spectra.

16.1 THE BASIC THEORY OF 2D SPECTROSCOPY

Before going into 2D-NMR spectroscopy in detail, let us recall the simple measurements on the 1D-NMR spectroscopy. When we place a sample in a static magnetic field, the magnetic vectors will align with the direction of the field and generate a net magnetization (Figures 112 and 113). A 90° pulse applied on the system turns the magnetization vector from the z-axis toward the y-axis to generate a component of the magnetization in the y-direction. The exponential decay of the magnetization is recorded as the FID signal. The FID signal is then Fourier transformed into a conventional frequency spectrum.

We have already seen that measurement of a spectrum consists of *preparation*, *evolution* and *detection* periods. A basic 1D FT-NMR technique operates without any evolution time. Immediately, after generation of the magnetization along the y-axis, the FID signal is detected. In some cases, like spin echo, APT etc., experiments are conducted where a constant evolution time is applied after the pulse. During this time period t, a J-modulation or a polarization transfer may evolve. By the collection of many FID signals from a sample this introduced evolution time between the pulses is kept always at a constant (see Figure 192). In the case of two-dimensional NMR we varied this evolution time t by a constant time interval (Δt) over many different experiments and collected the resulting FID signals into one overall experiment. Now we have the basis of a 2D experiment (Figure 192).

Now we will attempt to explain how the NMR spectra will be affected upon the change of that inserted evolution period. Consider a 2D-NMR experiment with a simple compound (chloroform) that contains a single proton to avoid the complication caused by

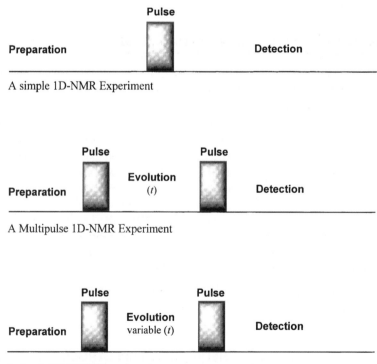

Figure 192   Schematic presentation of the pulse sequences of a simple 1D-NMR experiment, a multiple-pulse 1D-NMR experiment and a simple 2D-NMR experiment.

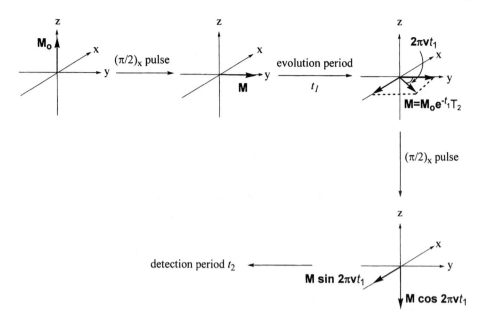

Figure 193 Schematic presentation of the pulse sequences in a simple 2D experiment with variable evolution period.

spin–spin couplings. When a chloroform sample is placed in a static magnetic field, a net magnetization along the x-axis will occur. Application of a 90° pulse from the x-direction will rotate the magnetization vector M_0 towards the y-axis. A 90° pulse is applied in order to receive a signal of maximum intensity. Magnetization vector M will start to rotate in the xy-plane about the z-axis with the Larmor frequency ν (Figure 193).

The vector will precess during the time interval t_1 through an angle $2\pi\nu t_1$. We can now calculate the components of M along the y-axis ($M \cos 2\pi\nu t_1$) and along the x-axis ($M \sin 2\pi\nu t_1$) by simple trigonometry. Application of a second 90° pulse $(\pi/2)_x$ from the x-direction will rotate the component $M_y = M \cos 2\pi\nu t_1$ downward onto the z-axis. The other component $M_x = M \sin 2\pi\nu t_1$ will remain in the xy-plane. This component will of course generate the FID signal and will determine the intensity of the NMR signal. If the NMR signal is recorded, there will actually be no difference between this signal and that recorded under the 1D-NMR conditions. Then we perform a series of experiments (n experiments) with different values of t_1, starting with $t_1 = 0$, increasing the evolution period by a constant time increment Δt_1. The receiver signal will also become dependent on t_1. Let us assume that we carry out 1024 of such experiments. Successive Fourier transformation of each of these FID signals will provide us with a series of signals. All of these signals will be chloroform resonance signals, however, the intensities of these signals will vary due to the successively changed evolution time t_1 [117] (Figure 194).

From these signals we can easily deduce that the amplitude of the chloroform peaks changes sinusoidally with the resonance frequency ν of chloroform. If we plot the amplitude of the signal as a function of t_1 we will obtain a graph showing an oscillation with the frequency ν (Figure 195). This is exactly the FID signal which we have seen earlier.

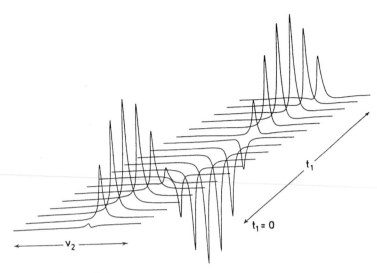

Figure 194   Schematic presentation of the chloroform signal recorded with variable evolution time t. (Reprinted with permission of Elsevier Inc. from Andrew E. Derome, *Modern NMR Techniques for Chemistry Research*, 1987.)

However, this FID signal is different to the one we know from the 1D-NMR experiments. A normal FID signal is a sinusoidal exponential decaying function in *real time* (t_2), whereas the generated FID signal (Figure 195) is a function of the variable time t_1. We can now perform a second Fourier transformation on the data orthogonal to

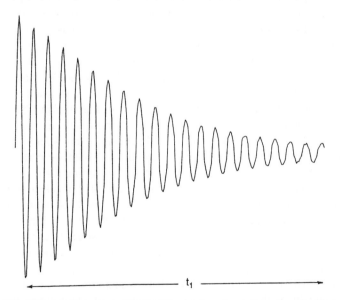

Figure 195   The FID signal generated by the Fourier transformation of each chloroform signals presented in Figure 194. (Reprinted with permission of Elsevier Inc. from Andrew E. Derome, *Modern NMR Techniques for Chemistry Research*, 1987.)

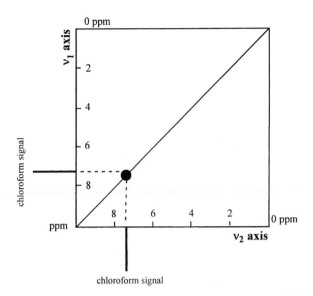

Figure 196   Graphical presentation of the 2D-NMR spectrum of chloroform.

the first. This is the end of the 2D spectra. Now, we will obtain a signal that depends on two independent time variables t_1 and t_2 (Figure 196).

The obtained signal will have two frequency axes as shown in Figure 196. Our counter diagram shows a single resonance that is drawn on two frequency axes. A closer look at these different frequency axes is given in Figure 196. The chloroform peak appears at the cross point. The frequency axis ν_1 shows the resonance frequency of the chloroform peak. The second frequency axis ν_2 shows also the resonance frequency of the chloroform peak. The diagonal signal is located at the crossing point. Actually this spectrum provides no additional information beyond the simple 1D spectrum of chloroform. All 2D experiments are performed in this way. A general pulse sequence in a 2D-NMR experiment is presented in Figure 197.

However, for useful 2D experiments, the magnetization vector M should evolve with one frequency during t_1 and with a different frequency during t_2. In the case of the chloroform experiment, the magnetization experiences identical modulation during t_1 and t_2. Therefore, the experiment produces only diagonal peaks. If the magnetization evolves with different frequencies during t_1 and t_2, the experiment will produce peaks in which ν_1 and ν_2 are different. In this case the peaks cannot be diagonal as they will be

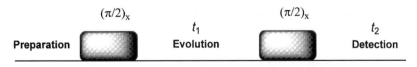

2D-NMR Experiment

Figure 197   Pulse sequence of a simple 2D-NMR experiment. The incremental delay time t_1 and the detection time (acquisition) t_2 are Fourier transferred into frequencies ν_1 and ν_2, respectively.

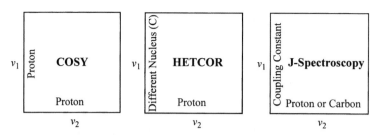

Figure 198   Classification of 2D-NMR experiments.

located out of the diagonal. These peaks are named 'off-diagonal' or cross peaks and provide very important information on the structure (see Figure 190). Now we shall discuss the axes. One of these axes always presents the nucleus detected during acquisition time (t_2). The other axis, which depends on t_1, can represent either the same nucleus (^1H–^1H or ^{13}C–^{13}C; homonuclear correlation COSY) or the other nucleus (^1H–^{13}C; heteronuclear correlation HETCOR) or the coupling constant J (J-resolved spectroscopy) (Figure 198) [118].

2D J-resolved spectroscopy was among the first 2D methods to be developed, because of the great importance of multiplicity. Nowadays, carbon multiplicities can be obtained via DEPT or APT techniques. As a result, J-resolved spectroscopy is today less popular than it was formerly. Therefore, J-resolved spectroscopy will not be discussed herein.

16.2   COSY EXPERIMENT: TWO-DIMENSIONAL HOMONUCLEAR CORRELATION SPECTROSCOPY

One of the simple 2D-NMR experiments is the COSY. In the COSY experiment, the coupled nuclei are protons. With this technique one can easily determine whether there is a coupling between two protons or not. As we have discussed earlier, the coupling relation between the protons can also be determined by double resonance experiments. However, every proton should be irradiated one by one. This is a very time-consuming method. In the case of the COSY experiment, all kinds of coupling relations between the protons in a given molecule can be determined by a single spectrum. Furthermore, in order to have a good double resonance experiment, the coupled protons must have a certain chemical shift difference. Even these kinds of problems can be solved by a COSY experiment. But, it should not be thought that COSY is a replacement of a homonuclear decoupling experiment. Of course, the protons are not decoupled during the COSY experiments.

To interpret the COSY spectra, it is not necessary to know the exact pulse sequences in the COSY experiment. To understand the mechanism of COSY experiments, interested readers are referred to Refs. [114, 116].

16.2.1   Interpretation of the COSY spectra

The COSY spectra of the following compounds will be discussed now. The 200 MHz COSY spectrum of the bicyclic endoperoxide is shown in Figure 199 [81].

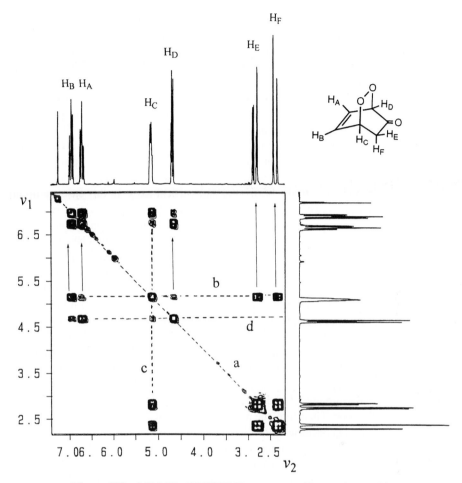

Figure 199   200 MHz COSY-NMR spectrum of ketoendoperoxide.

This spectrum shows us the correlation between coupled protons. The spectrum consists of two frequency axes (ν_1 and ν_2) where 1D spectra are displayed. These 1D spectra are not part of the COSY spectrum. They are provided solely for convenience purposes in order to determine the connectivity between protons. From the upper left to the lower right corner runs the diagonal, which has a series of signals on it (line a). These diagonal peaks do not provide any useful information beyond the 1D ^{1}H-NMR spectrum. The symmetrically disposed cross peaks on both sides of diagonal need to be interpreted. A point of the entry into a COSY spectrum is one of the important steps to obtain useful information. A peak in 1D spectrum whose assignment is relatively apparent in the 1D spectrum is a good point of reference.

Now, to start from the peak resonating at 5.2 ppm, it is likely that this peak will arise from one of the bridgehead protons. As for the next, we then draw two parallel lines to the ν_1 or ν_2 axis (we will actually obtain the same result because the cross peaks are distributed symmetrically). Both lines (b and c) will intersect five off-diagonal or cross peaks. This means that the proton resonating at 5.2 ppm is coupled with five different protons. Let us consider only the line 'b'. From here, perpendicular imaginary lines can be drawn back to its intersection with the diagonal peaks (shown as files). At their intersection we can find that diagonal peaks appearing at 4.7, 2.9 and 2.4 ppm have coupled to the selected proton. The resonance signal appearing at 2.4 and 2.9 ppm belongs to the methylenic protons H_E and H_F. Since the peak resonating at 5.2 ppm correlates with the diastereotop methylenic protons, this signal can be assigned to the proton H_C. Furthermore, the line 'b' intersects with two cross peaks appearing on the left side of the spectrum, showing couplings between the proton H_C and the olefinic protons H_A and H_B. Let us select another useful entry point: the other proton resonance at 4.7 ppm. We begin again at the diagonal peak and draw two imaginary lines 'd' and 'e' (only line 'd' is shown in Figure 199) perpendicular to the horizontal and vertical axes. In this case we do not observe any correlation between the selected proton and the methylenic protons H_E and H_F. This finding supports our assumption that the peak at 4.7 ppm belongs to the bridgehead proton H_D. On the other hand, the diagonal peak at 4.7 ppm has cross peaks with the protons resonating at 5.2, 6.7 and 7.1 ppm, respectively. Careful analysis of these cross peaks shows that some of them (cross peaks at 5.2 and 7.1 ppm) are weaker. These weak correlations are due to the small long-range couplings of $^4J_{HH}$ and $^5J_{HH}$. On the basis of this observation we can assign the resonance signal at 7.1 ppm to the proton H_B, since this proton can have a small coupling (long-range coupling with the proton H_D). We can analyze the other peaks in the same way. A parallel line drawn from the diagonal peak at 7.1 ppm to horizontal axis F_2 will intersect cross peaks at 6.7, 5.2 and 4.7 ppm, indicating that this olefinic proton H_B correlates with the olefinic proton H_A and the other bridgehead protons H_D and H_C, respectively. The cross peaks at 5.2 and 4.7 ppm show different intensities (a vicinal coupling ^3J and an allylic coupling ^4J). One at 5.2 ppm is more intense than the other one. Therefore, the resonance signal at 7.1 ppm can be ambiguously assigned to the proton H_B because of the close proximity to H_C. The weaker cross peak at 4.7 ppm shows that there is a long-range coupling between the protons H_B and H_D. Thus, from the COSY spectra we can quickly see which protons are coupled to each other. Furthermore, from the reference point of entry, we can walk around a molecule tracing the neighboring coupling relations along the given molecule.

Now let us consider the second compound, a homobenzonorbornadiene derivative whose 200 MHz COSY spectrum is given in Figure 200 [119]. The molecule has aromatic, olefinic and aliphatic protons. If we analyze the resonance signal of the aromatic protons that resonate in a very narrow range, we can see that there is no correlation (there are no cross peaks) with the olefinic and aliphatic protons as expected. The olefinic protons appear at 5.8 and 5.2 ppm as well as resolved multiplets. As a useful point of entry, let us now select the olefinic proton resonance at 5.8 ppm. If we draw two imaginary lines from the diagonal peak directly to the right (perpendicular to ν_1 axis) or directly down (perpendicular to ν_2 axis), the lines intersect four off-diagonal or cross peaks, whereas two of them are less intense. It is likely that these less intense cross peaks arise from long-range couplings. The intersected cross peaks are located at 5.2, 3.0, 2.4 and 2.0 ppm. On the basis of the number of the cross peaks and intensity of the lines, we assume that the resonance signal at 5.8 ppm belongs to the olefinic proton H_B which correlates strongly with the olefinic proton H_C (5.2 ppm) and the bridgehead proton H_A. The cross peak at 3.0 ppm arises from the coupling of H_A with H_B. The less intense peaks

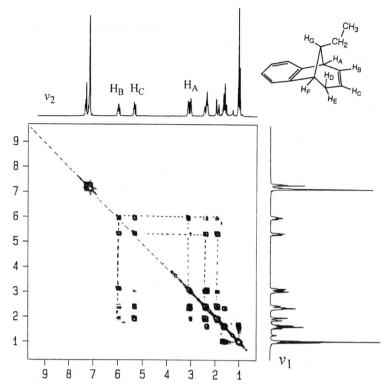

Figure 200   200 MHz COSY-NMR spectrum of ethyl homobenzonorbornadiene.

arise probably from the allylic couplings between the olefinic proton H_B and the methylene protons H_D and H_E. If we draw, similarly, the lines from the diagonal peak at 5.2 ppm (the olefinic proton H_C) we can see strong correlations with two aliphatic protons beside the olefinic proton H_B. Since the olefinic proton H_B also showed correlation with the same aliphatic protons, those resonances at 2.4 and 2.0 ppm can be assigned to the methylenic protons H_D and H_E. The diagonal peak at 3.0 ppm correlates with the olefinic proton at 5.8 ppm and the peaks appearing at 2.4 and 2.0 ppm. On the basis of this observation we can conclude that the resonance signals of the other bridgehead proton H_F appear also at 3.0 ppm together with the resonance signal of H_A as multiplet. Integration of the signals also shows the collapse of two proton resonances at 3.0 ppm. The connectivity between the other protons can be determined through similar ways.

16.3   TWO-DIMENSIONAL <u>HET</u>ERONUCLEAR <u>COR</u>RELATION SPECTROSCOPY: HETCOR EXPERIMENT

After the discussion of the COSY experiment we can now explain heteronuclear correlation spectroscopy. The word HETCOR is generated from the heading letters of the words: <u>HET</u>eronuclear <u>COR</u>relation. The HETCOR experiment is similar to the COSY experiment with the exception that it concerns two different nuclei. We will discuss only those spectra where a ^{13}C spectrum is presented along one axis and a ^{1}H spectrum is shown along the other axis. In the ^{13}C-NMR section we have discussed that the carbon nuclei can couple with directly bonded protons. Furthermore, we have shown that there are also couplings over two (^{2}J), three (^{3}J), and four (^{4}J) bond couplings. There are two different HETCOR spectra. They can be set up to demonstrate one-bond interactions between the ^{13}C and ^{1}H nuclei or to show the longer range interactions. Firstly, we will discuss how the HETCOR experiment correlates ^{13}C nuclei only with directly attached protons; these are one-bond couplings ($^{1}J_{CH}$).

16.3.1 Interpretation of the HETCOR spectra

HETCOR spectra contain no diagonal peaks, they contain only cross peaks. Furthermore, the spectra are not symmetrical. Interpretation of the HETCOR spectra is much easier than the COSY spectra. The cross peaks in a HETCOR spectrum indicate which protons are attached to which carbon atoms in a given molecule. In order to determine the correlations, we begin with any carbon atom resonance and draw a vertical line until a cross peak is encountered. In this case, we can encounter four possibilities.

(1) It is possible that the vertical drawn line will not intersect any cross peak. In that case the corresponding carbon atom has no attached proton.
(2) If the drawn line intersects one cross peak, then the corresponding carbon atom may have one, two or three protons. These protons are directly bonded to the carbon atom. If the number of the attached protons is two or three, it is likely that these protons are equal and arise from a enantiotop $-CH_2-$ or $-CH_3$ group. It should be kept in mind that two proton can be diastereotop and their resonances are overlapped.
(3) If the line intersects two cross peaks, then we have a $-CH_2-$ group bearing diastereotop protons with different chemical shifts.
(4) The fourth case is that a proton signal does not correlate with any carbon atom. In this case, this proton can be attached to a heteroatom (like $-OH$, $-NH$, etc.).

We now move on to the HETCOR spectra of those compounds whose COSY spectra have already been analyzed. After interpretation of the COSY spectra we know precisely the assignments of all protons. Figure 201 shows the HETCOR spectrum of the ketoendoperoxide. The ^{13}C-NMR spectrum is plotted along the ν_2 axis and the ^{1}H-NMR spectrum is plotted along the ν_1 axis. From the ^{1}H spectrum of this compound we have determined that the peak at 7.1 ppm belongs to the proton H_B, whereas the peak at 6.7 ppm belongs to the proton H_A. If we draw two parallel lines to the ν_2 axis starting from the proton resonances, these lines will intersect two cross peaks. Now, from these cross peaks we draw perpendicular lines to the ν_2 axis (parallel lines to the ν_1 axis), the intersected carbon signals are those carbon atoms bearing the corresponding protons. For example, in this case the proton signals at 7.1 and 6.7 ppm correlate with the carbon atoms resonating at 140 and 132 ppm, respectively. So we have assigned these peaks. It is not surprising to note that downfield carbon is attached to proton that is shifted downfield and upfield carbon signal correlates with the upfield proton. Of course, this is not normally the case. For example, the low-field proton resonance at 5.2 ppm (H_C)

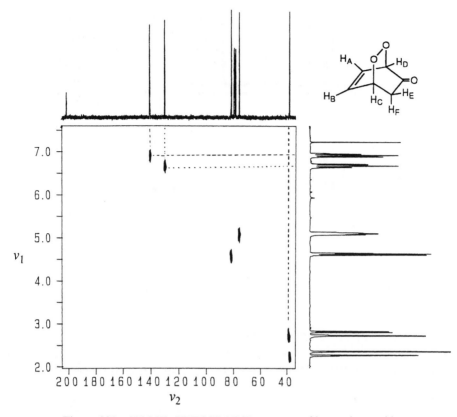

Figure 201   200 MHz HETCOR-NMR spectrum of ketoendoperoxide.

correlates with the upfield carbon resonance at 75 ppm, whereas the upfield proton resonance at 4.7 ppm (H_D) correlates with the low-field carbon resonance at 80 ppm. From the ^{1}H-NMR spectrum we have seen that the methylene protons are diastereotop and resonate separately. If we draw a perpendicular line from the carbon resonance at 39 ppm to ν_2 axis, the line intersects two cross peaks indicating that this carbon bears two unequal protons. Furthermore, the HETCOR spectrum shows that the peak is about 200 ppm which does not correlate with any proton signal. This means that it is a quaternary carbon atom which arises from a carbonyl carbon atom.

A 200 MHz HETCOR spectrum of ethyl homobenzonorbornadiene is given in Figure 202. Only seven signals can be seen at the sp^2 region, revealing that two signals are overlapped. Since the aromatic proton resonances appear in a narrow range, it is difficult to make a correlation with the corresponding carbon resonances. Carbon signals could eventually be correlated if the spectrum were recorded on an instrument with a stronger magnetic field. It is easy to assign olefinic carbon resonances since the signals are well resolved. H_B proton resonating at 5.8 ppm gives correlation with the carbon signal at 132.2; we assign it to C_B. On the other hand, the proton resonating at 5.2 ppm shows correlation with the carbon signal at 126.6 ppm which belongs to C_C. Other correlations are more straightforward. While examining the COSY spectrum of this

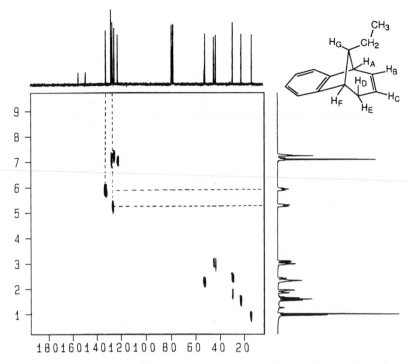

Figure 202   200 MHz HETCOR-NMR spectrum of ethyl homobenzonorbornadiene.

compound, we have determined that the signal group at 3.0 ppm belongs to two different protons (H_A and H_F). HETCOR spectrum confirms this finding. Correlation of the multiplet at around 3.0 ppm with two different carbon atoms (45.2 and 43.9 ppm) indicates that these are bridgehead carbon atoms. Similarly, the signal group observed at 2.4 ppm correlates with two different carbon atoms.

16.3.2   Two-dimensional <u>H</u>eteronuclear <u>M</u>ulti <u>B</u>ond <u>C</u>orrelation spectroscopy: HMBC experiment

In the HETCOR experiment we have shown correlations between the carbons and directly attached protons. In Chapter 14 we have discussed that there are also couplings over two, three and four bonds in the carbon NMR spectroscopy. HETCOR spectra would never show correlations over these bonds since they are eliminated. HMBC (heteronuclear <u>m</u>ulti <u>b</u>ond <u>c</u>orrelation) is a 2D experiment used to correlate, or connect, ^1H and ^{13}C peaks for atoms separated by multiple bonds (usually 2 or 3). The experiment is designed to suppress one-bond couplings, but a few are observed in most spectra. There is no way to know how many bonds separate a proton and a carbon atom when a cross peak is observed. These multi-bond correlation spectra provide us with very useful information about the structure. We obtained in an indirect way information about carbon connectivities. Furthermore, we can observe correlations between the quaternary carbons with nearby protons, so that we can locate the positions of the quaternary carbon atoms.

Major applications include the assignment of resonances of nonprotonated carbon and nitrogen nuclei and the long-range correlation of protonated carbon resonances that are separated by nonprotonated carbons and other heteronuclei.

Interpretation of HMBC spectra

In this section we will discuss HMBC spectra of two different compounds. Before starting with the HMBC spectra though, we will discuss first the HETCOR spectrum of the following bromodiketone (Figure 203) [120].

The ^{13}C-NMR spectrum of the bromodiketone is plotted along the ν_2 axis and the ^{1}H-NMR spectrum is plotted along the ν_1 axis. The ^{13}C-NMR spectrum shows

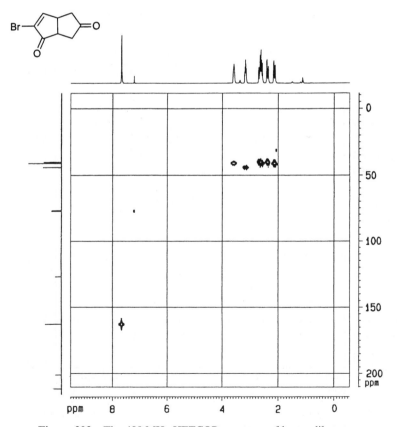

Figure 203 The 400 MHz HETCOR spectrum of bromodiketone.

the presence of seven signals. The compound must have eight resonance signals. From the higher intensity of the resonance signal at 41.1 ppm we assume that two carbon signals are overlapped. Two carbonyl carbons appear at 212.7 ppm (see Figure 204) and 201.1 ppm. The presence of two double bond carbons and four sp^3 carbons is in agreement with the proposed structure. It is expected that the carbonyl carbon peaks do not correlate with any proton signal. On the other hand, the carbon resonance at 162.7 ppm correlates with the proton resonance at 7.7 ppm arising from the olefinic proton. We can assign the carbon resonance at 162.7 ppm to the carbon atom C$_4$, and the carbon resonance at 127.1 ppm to the quaternary carbon atom C$_3$. The remaining methine and methylene protons correlate with the carbon resonances appearing between 44.1 and 40.2 ppm.

The HMBC spectrum of the bromodiketone seems similar to the HETCOR spectrum (Figure 204). The ^{13}C-NMR spectrum is plotted along the ν_2 axis and the ^{1}H-NMR spectrum is plotted along the ν_1 axis. The HMBC spectrum contains more correlations and one-bond correlations are removed. Interpretation of HMBC spectra requires a degree of flexibility, because we cannot always find what it is that we expect. For example, in some cases the three-bond correlations cannot be found.

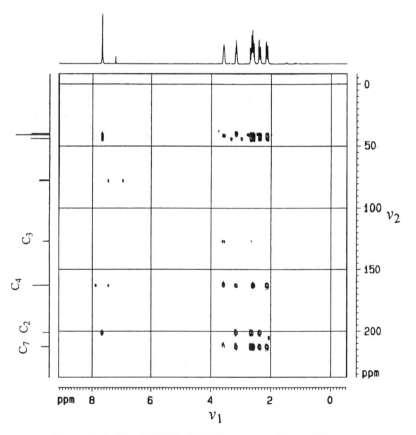

Figure 204   The 400 MHz HMBC spectrum of bromodiketone.

We can begin with the carbonyl carbon resonances. It can be easily recognized that the carbonyl carbons show many correlations (Figure 205). For example, a line drawn parallel to the ν_1 axis (^{1}H-NMR spectrum), starting from the carbonyl carbon resonance at 201.1 ppm, will intersect four cross peaks at 7.7, 3.1, 2.7 and 2.4 ppm, respectively (Figure 205). From the structure we expect that this carbonyl carbon can correlate with the methine protons (H_1 and H_5) and methylene protons H_8 and the neighboring olefinic proton H_4. On the basis of these correlations we can assign the cross peak at 3.1 ppm to the proton H_1. A correlation with the proton H_5 is not observed in this case. The three-bond correlations can be absent occasionally, whereas the two-bond correlations are always present. The other cross peaks belong to the methylene protons. The diastereotopic methylene protons (H_8) give rise to an AB system, whereas the A-part appears as doublet of doublets at 2.4 ppm. On the other hand, a parallel line drawn from the carbon resonance at 212.7 ppm intersects five cross peaks. This carbon resonance correlates with the entire proton resonances appear between 2.1 and 3.6 ppm. Careful examination of the cross peak shows that this peak actually consists of the different cross peaks, indicating the overlapping of the methylene protons (H_6 and H_8). Another useful example can be found by drawing a parallel line from

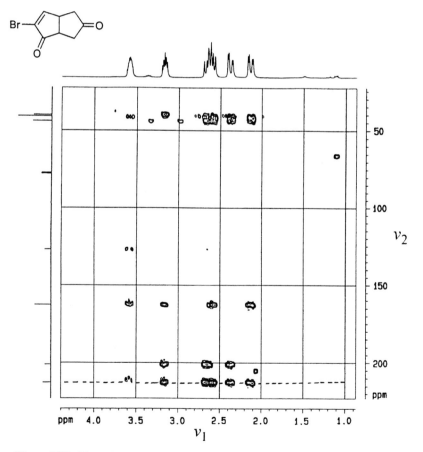

Figure 205   Expanded plot of the 400 MHz HMBC spectrum of bromodiketone.

the carbon resonance at 162.7 ppm. There are correlations to the tertiary protons H_1 and H_5 as expected. Furthermore, it correlates with the proton resonances at 2.6 and 2.1 ppm. With this observation we can assign those proton resonances to the diastereotopic methylene protons H_6. The carbon atom C_4 cannot correlate with the methylene protons attached to the carbon C_8. Other assignments can be performed in the same way. The HMBC spectrum of this bromodiketone allows one to completely confirm the suggested structure with the help of long-range carbon–proton connectivities.

As the next example, we will discuss the 400 MHz HMBC spectrum of the following oxabenzonorbornadiene derivative [121].

We will first interpret the HETCOR spectrum (Figure 206) of the oxanorbornadiene.

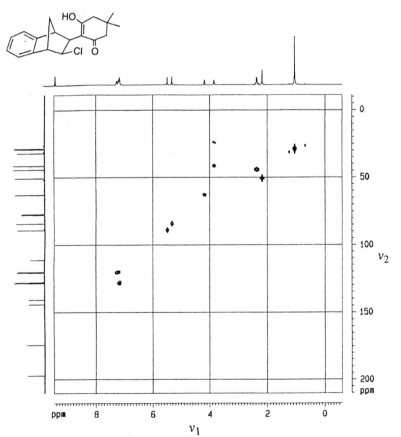

Figure 206 The 400 MHz HETCOR spectrum of oxabenzonorbornadiene derivative.

The singlet appearing at 9.8 ppm (see Figure 206) belongs to the hydroxyl group. If we draw a line from the low-resonance frequency end of the ^{1}H-NMR spectrum down, it encounters no cross peaks, indicating that this proton is not attached to a carbon atom. The aromatic proton resonances (7.1–7.3 ppm) correlate with the carbon resonances at 128.8, 128.2, 120.8, and 120.6 ppm, respectively. These carbon resonances can easily be assigned to the aromatic carbon atoms bearing protons. The singlets at 5.5 and 5.3 ppm correlate with carbon resonances at 89.1 and 84.3 ppm. These carbon resonances can be assigned to the bridgehead carbons C_8 and C_{11}. The protons H_9 and H_{10} form an AX system and resonate at 4.2 and 3.8 ppm as doublets. These peaks correlate with carbon resonances at 62.9 and 41.4 ppm. On the basis of the chemical shifts of these carbon resonances, we may assign the peak at low field (62.9 ppm) to the carbon atom C_9 bearing chlorine. The AB system appearing at 2.35 and the broad singlet at 2.2 belongs to the methylenic protons. The corresponding cross peaks appear at 50.6 and 44.2 ppm. Methyl protons at 1.05 correlate with carbon resonances at 29.1 and 28.6 ppm, respectively. Actually, there are many carbon signals that do not correlate with any proton resonance. For example, the carbon resonance at 197.1 ppm arises from the carbonyl carbon atom, whereas the peak at 174.1 ppm belongs to the β-carbon atom (C_{17}) of α,β-unsaturated carbonyl system. The carbon signals resonating at 144.2 and 140.9 belong to the quaternary aromatic carbon atoms C_2 and C_7. At this stage we cannot distinguish between those carbon atoms. A clear-cut assignment can only be performed by the analysis of the HMBC spectra. The remaining quaternary carbon atoms appear at 111.3 ppm (C_{12}) and 32.0 ppm (C_{15}).

Now we will turn back to the HMBC spectrum of this compound (Figure 207). Normal 1D ^{1}H and ^{13}C spectra are shown along the axis. In aromatic rings, the most common correlations seen in HMBC spectra are three-bond correlations because they are typically 7–8 Hz, which is the value for which the experiment is optimized. The proton-bonded aromatic carbons (resonances between 128.8 and 120.3 ppm) correlate only with the aromatic protons. The coupling constant is affected by substituents, so two-bond correlations are also sometimes observed. On the other hand, the quaternary aromatic carbon atoms C_2 and C_7 (144.2 and 140.9 ppm) correlate beside the aromatic protons with the bridgehead protons H_8 and H_{11}, as well as with the tertiary protons H_9 and H_{10}. Since the double bond carbon atom C_{12} (111.4 ppm) correlates also with the proton resonance at 4.9 ppm, this resonance can be assigned to the proton H_{11}. Furthermore, there is a strong correlation between the C_{12} carbon atom and the proton resonance at 3.8 ppm. From this correlation, this proton resonance can be assigned to the proton H_{10}. Cross peaks arising from the carbonyl carbon atom correlations also provide very important information. The carbonyl carbon resonance at 197.1 ppm has two correlations—with the proton resonances at 3.8 and 2.3 ppm. This correlation supports the assignment of the proton H_{10} and identification of the methylene protons H_{14}. The quaternary carbon atom C_{15} correlates with the both methyl and methylene protons. This finding supports the suggested structure. Correlations with the other peaks are left to the reader.

16.4 ^{13}C–^{13}C CORRELATION: INADEQUATE EXPERIMENT

The word INADEQUATE is generated from the heading letters of the words: incredible natural abundance double quantum transfer experiment which means the double quantum

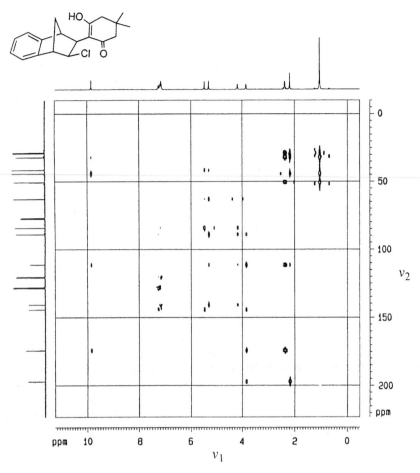

Figure 207   The 400 MHz HMBC spectrum of oxabenzonorbornadiene derivative.

transfer of an element whose natural abundance is very low. This measurement technique is applied as one-dimensional *1D-INADEQUATE* [122], as well as two-dimensional *2D-INADEQUATE*. We can explain qualitatively what type of information this measurement technique can provide us before going into further details of this subject. The 1D-INADEQUATE experiment measures the coupling constants between the carbon atoms found in the molecule, whereas the 2D-INADEQUATE experiment determines the correlation between $^{13}C-^{13}C$ nuclei. Unfortunately, the 1D version is rarely useful and the 2D version is very insensitive since INADEQUATE detects only one in 10,000 of the molecules in the NMR tube.

16.4.1   1D-INADEQUATE experiment

In the proton NMR section, we explained that the coupling constants in proton NMR provide lots of important information about the molecular structure. Similarly, coupling constants between ^{13}C–^{13}C nuclei are also expected to provide information about the molecular structure. The high abundance of protons in nature makes measurement of H–H coupling constants easier. However, the situation is completely different in the carbon nucleus. ^{12}C nuclei are inactive in NMR, only ^{13}C nuclei are measured and the natural abundance of this nucleus is about 1%. In order to measure the coupling constant between two neighboring carbon nuclei, two nuclei must be neighboring in the same molecule. The probability of this is about $0.01 \times 0.01 = 0.0001$. In order to make such a measurement, a certain carbon atom in the molecule is labeled as ^{13}C and then the satellite signals found on the right and left side of the main signal belonging to this nucleus are measured. Labeling a certain carbon atom creates synthetic problems and additionally this technique cannot measure more than one coupling constant between the labeled carbon atom and the neighboring carbon atoms. Due to these reasons, it was necessary to develop a new technique to measure the coupling constants between the carbon nuclei without labeling which has limited applicability.

Now we consider a resonating peak in a carbon NMR spectrum. As was explained above, the satellite signals which belong to the carbon nucleus will be observed on the right and left side of the main signal (Figure 208). The intensity of the satellite signal is 1% of the main signal. Since the satellite signal is resonating as a doublet, its intensity will be decreased to 0.5% of the main signal. Coupling constants through one bond ($^{1}J_{CC}$) are about 40–70 Hz. Since 1 ppm represents 100 Hz for a 100 MHz instrument (400 MHz for proton), J_{CC} couplings are observed in about 0.5 ppm region which is a very small value for carbon NMR. Here we face an important problem; since the coupling constants are too small, the satellite signals will be located very close to the main peak. In addition, when we consider that the main peaks are 200 times more intense than the satellite peaks, it would be impossible to observe the satellite signals near the main peak. However, it is very easy to observe satellite signals in proton NMR because carbon–proton couplings are large. So the only thing that can be done is to eliminate the main carbon signal. If this can be achieved, then it would be possible to observe the satellite signals and measure the ^{13}C–^{13}C coupling constants. This is possible with the 1D-INADEQUATE measurement technique.

The pulse order applied in the 1D-INADEQUATE technique is given in Figure 209. During the INADEQUATE measurement, the proton channel is always kept open. Firstly, as in other measurements, a 90° pulse is sent to the system from

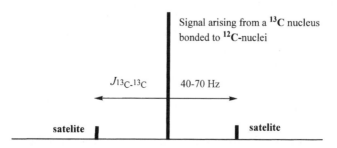

Figure 208 View of the major and satellite signals in a proton-decoupled carbon NMR spectrum.

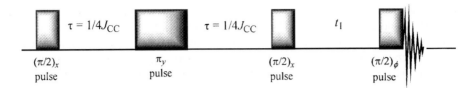

Figure 209   Pulse sequence in a 1D-INADEQUATE experiment.

the x-direction and the magnetic vector is rotated towards the y'-direction on rotating coordinates. Let us assume that the magnetic vector (S_a) belonging to the main signal (arising from $^{13}C–^{12}C$) stays constant on rotating coordinates. Both $^{13}C–^{13}C$ doublet signals will diphase with the frequency of J_{CC} until, at time $\tau = \frac{1}{4}J_{CC}/2$, they have attained a phase difference of about 90°. In order to eliminate the inhomogeneity effect during the second evolution period, a selective 180° pulse is applied from the y-direction. A difference in movement direction of magnetic vectors will not be observed. After the second evolution of the second period of $\tau = \frac{1}{4}J_{CC}$, magnetic vectors belonging to doublet lines satellites will align along the $\pm x$-axis with a phase difference of 180°. The vector belonging to the main signal will stay in the y'-axis. Up to this moment, the operations performed are not different from the ones we have previously seen. The most important stage in the INADEQUATE experiment is the second pulse which is sent to the molecule from the x-direction. How will this pulse affect the system? This pulse will not affect the magnetic vectors of the satellites, whereas the magnetic vector which belongs to the main signal will be rotated towards the $-z$-direction. In NMR, since the signal is recorded from the magnetization observed on the xy-plane, an NMR signal recorded at this stage will be formed only by the satellite signals. With this technique, one can easily determine the $^{13}C–^{13}C$ coupling constant in a natural compound.

16.4.2   2D-INADEQUATE experiment

After explaining the 1D-INADEQUATE experiment, we will continue onward with the 2D-INADEQUATE experiment. After examining the COSY spectra, here we have nothing new to add to our explanations. In COSY experiments we obtained correlation spectra between two homonuclear nuclei (proton). In the INADEQUATE experiment, since the correlating nuclei are carbon ($^{13}C–^{13}C$) we will obtain information about the directly attached (one-bond) carbon–carbon couplings. This experiment provides the direct analysis of the carbon skeleton through their J (C–C) couplings. This technique would be a method of choice to analyze complex structures, but it suffers from its very low sensitivity. Therefore, only a very concentrated solution can be analyzed. With the 2D experiment one can easily obtain the carbon skeleton of a given compound without having the proton NMR spectrum.

One way to present this experiment is as a carbon analogue COSY in which both axes (ν_1 and ν_2) are carbon axes. Diagonal peaks and also cross peaks are present on the spectrum. Similarly, by finding the cross peaks, it is directly found which carbon atom is

correlating with which carbon atom. To clarify the subject better, let us examine the
2D-INADEQUATE spectrum of menthol (Figure 210) [62].

The ^{13}C-NMR spectrum of menthol is composed of 10 carbon signals. The multiplicity
of these carbon resonances can be easily determined by DEPT spectra. For menthol
^{13}C-NMR chemical shift values and multiplicities of the peaks are shown below.

70.9 (d)   50.4 (d)   45.6 (t)   35.0 (t)   31.9 (d)

25.8 (d)   23.5 (t)   22.3 (q)   21.0 (q)   16.0 (q)

There are also diagonal peaks on the spectrum. However, these do not provide
additional information as discussed in COSY spectra. Cross peaks have to be
examined. Let us begin with the diagonal peak which belongs to the signal resonating

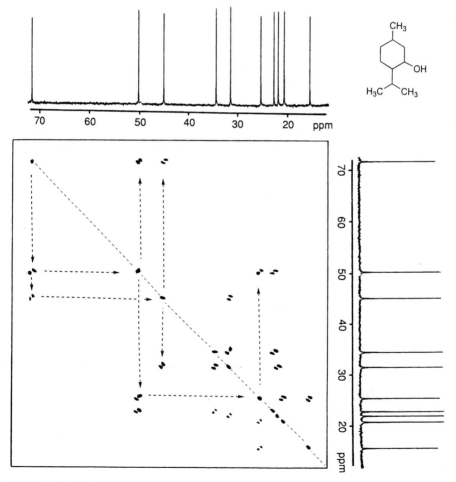

Figure 210   The 2D-INADEQUATE NMR spectrum of menthol. (Reprinted with permission
of Oxford University Press from J. Sanders *et al.*, *Modern NMR Spectroscopy: A Workbook
of Chemical Problems*, 1993, Figure 2.19.)

at 70.9 ppm. Since the spectrum is symmetrical, we can use one part of the spectrum. We draw a line parallel to the ν_2 axis; this line intersects two cross peaks at 50.4 (d) and 45.6 (t) ppm. This means that the carbon resonating at 70.9 ppm is directly connected to a ^2CH and a $^6CH_2$ carbon atom (Structure **1**).

If we draw a parallel line from the resonance signal of ^2CH carbon at 50.4 ppm, it intersects two cross peaks resonating at 25.8 (d) and 23.5 (t) indicating the connection to a ^7CH and a $^3CH_2$ group. Furthermore, this line also intersects the cross peak at 70.9 ppm showing the connectivity of ^2C with ^1C which was determined earlier. Now we end up with Structure **2**. A parallel line drawn from the resonance signal of the carbon atom 25.8 (d) intersects three cross peaks at 50.4, 21.0 (q) and 16.0 (q) which demonstrates the connection of two different methyl groups to the carbon atom ^7C (Structure **3**). This application can be continued by following the carbon atoms. When we examine the CH carbon resonating at 31.9 ppm, this carbon forms cross peaks with three different carbon atoms. These are the carbon atoms resonating at 45.6, 35.0, and 22.3 ppm. When these are added to the structure, Structure **5** is obtained which forms the skeleton of menthol.

16.5   2D-NOESY EXPERIMENT

We have already discussed 1D-NOESY experiments and shown that NOESY spectra use the nuclear Overhauser effect (NOE) to provide information about protons which are close in space, but not closely connected by chemical bonds. NOESY experiments provide very useful information about the configuration of the molecule. In order to determine the through space relation between all kinds of protons in a given molecule, all the protons have to be irradiated one by one. This is a time-consuming process and furthermore it causes some problems in the case of a proton whose chemical shifts are very close to each other. It is possible to search the relations between all protons in a single spectrum. This is possible with 2D-NOESY spectra. The NOESY spectra are distinct from COSY spectra which use the J-coupling interactions between the protons, whereas the NOESY spectra use the through space interaction. The NOESY spectra also contain diagonal peaks beside the cross peaks which give very important information in view of the structures.

Let us discuss a NOESY spectrum. Figure 211 shows the NOESY spectrum obtained from the photodimerization of nitrocinnamic acid methylester [123].

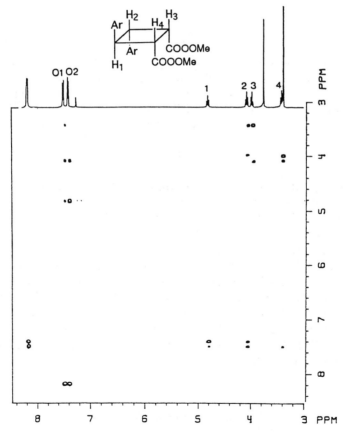

Figure 211   The NOESY spectrum of cyclobutane derivative. (Reprinted with permission of American Chemical Society from H. Ziffer, A. Bax, R.J. Highet, B. Green, *J. Org. Chem.*, 1988, **53**, 895.)

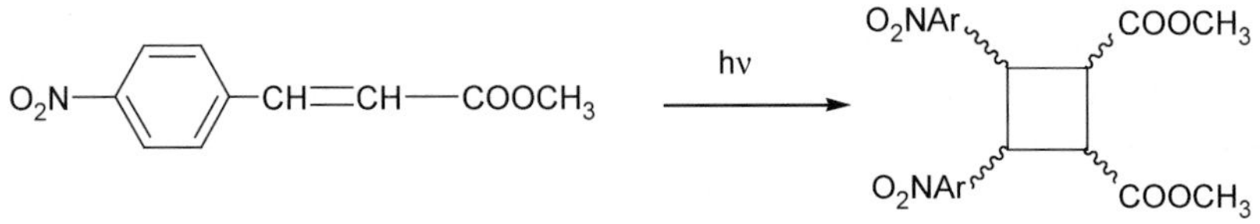

The ^{1}H-NMR spectrum shows that the formed compound is asymmetric and a head-to-head dimerization occurs. Theoretically, only two isomers (**A** and **B**) can be drawn for this dimerization as shown below. Since all cyclobutadiene protons resonate separately as triplets and vicinal cyclobutane protons have similar *cis* and *trans* couplings, it is not easy to distinguish between these suggested isomeric structures. Furthermore, the *ortho* protons in both aromatic rings also resonate separately. However, the correct stereochemistry can be assigned easily on the basis of the NOESY spectrum. The doublet arising from the *ortho*-protons of one of the aromatic rings shows cross peaks with the protons H_1, H_2, and H_4. This observation can be possible only in the case that two protons are in close proximity to the aromatic ring, i.e. they are on the same side as the aromatic ring. The interaction with the third proton can be possible if this proton is attached to the same carbon atom as the aromatic ring. This observation is in agreement only with the structure **B**. *o*-Protons of the aromatic ring attached to the carbon atom C_2 show a geminal interaction with the proton H_2 and vicinal interaction with the proton H_1. This observation also supports the structure **B**. In the case of the structure of **A**, the *ortho*-protons of the aromatic ring cannot produce three cross peaks with the ring protons.

Solutions to Exercises

The structures in this section are solutions to the exercises given at the end of Chapters 3, 5 and 8.

1   CHAPTER 3 (INCLUDE 1–30)

1) 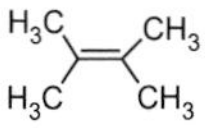   H₃C, CH₃ / H₃C, CH₃ (tetramethylethylene)

2)    (cyclohexadiene ring)

3) 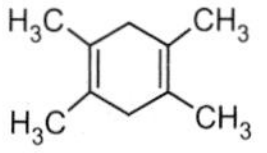   H₃C, CH₃ / H₃C, CH₃ (dimethyl-cyclohexadiene ring)

4) 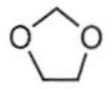   (1,3-dioxolane ring, O O)

5) 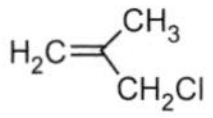   H₂C=C(CH₃)CH₂Cl

6) 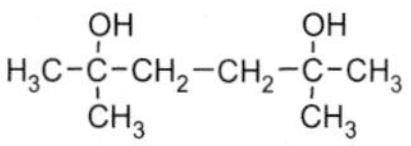   OH, OH
     H₃C-C-CH₂-CH₂-C-CH₃
          CH₃   CH₃

7) 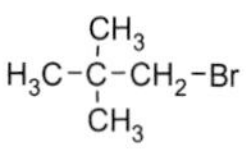   CH₃
     H₃C-C-CH₂-Br
          CH₃

8) 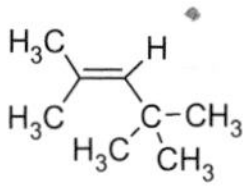   H₃C, H
     H₃C, C-CH₃
     H₃C CH₃

9) 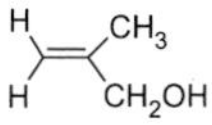   H, CH₃
     H, CH₂OH

10) 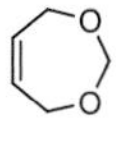 (7-membered ring with O O)

11) 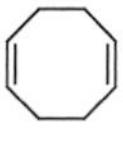 (cyclooctene ring)

12) 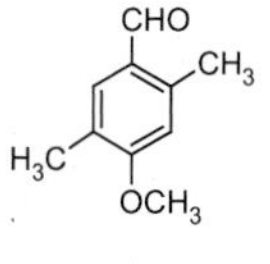 CHO
     CH₃
     H₃C
     OCH₃

13) 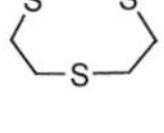 (ring with S, S, S)

14) 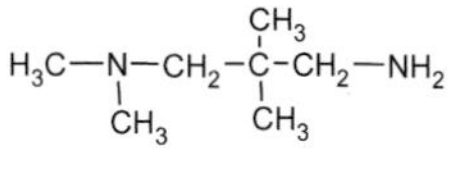 CH₃
     H₃C—N—CH₂-C-CH₂—NH₂
          CH₃   CH₃

15) 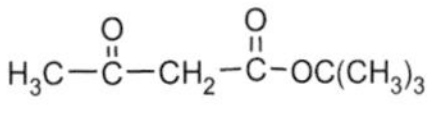 O   O
     H₃C-C-CH₂-C-OC(CH₃)₃

16) 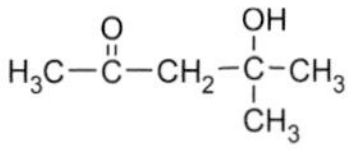 O   OH
     H₃C-C-CH₂-C-CH₃
              CH₃

17) 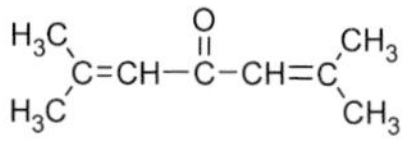 H₃C,   O   CH₃
         C=CH-C-CH=C
     H₃C   CH₃

18) 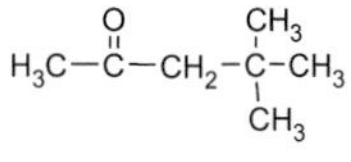 O   CH₃
     H₃C-C-CH₂-C-CH₃
              CH₃

407

19)

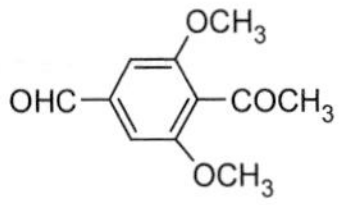

20) 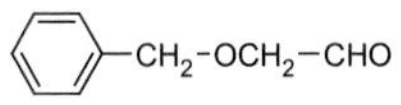 —CH$_2$–OCH$_2$–CHO

21) 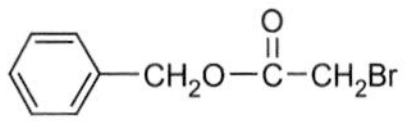 —CH$_2$O–C(=O)–CH$_2$Br

22) 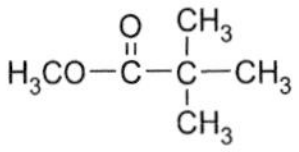 H$_3$CO–C(=O)–C(CH$_3$)(CH$_3$)–CH$_3$

23) 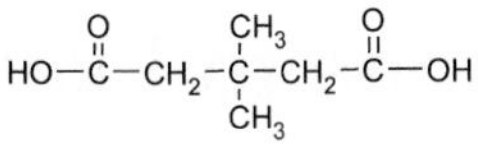 HO–C(=O)–CH$_2$–C(CH$_3$)(CH$_3$)–CH$_2$–C(=O)–OH

24)

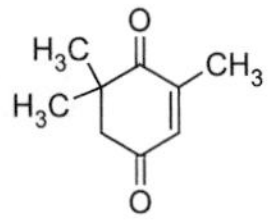

25) 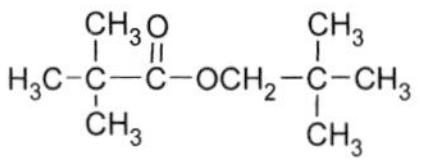 H$_3$C–C(CH$_3$)(CH$_3$)–C(=O)–OCH$_2$–C(CH$_3$)(CH$_3$)–CH$_3$

26) 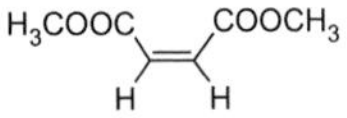 H$_3$COOC–CH=CH–COOCH$_3$ (H, H)

27) 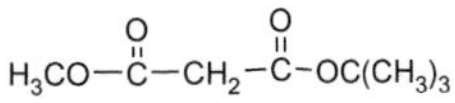 H$_3$CO–C(=O)–CH$_2$–C(=O)–OC(CH$_3$)$_3$

28)

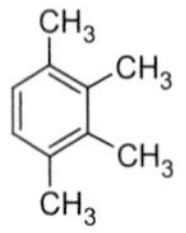

29)

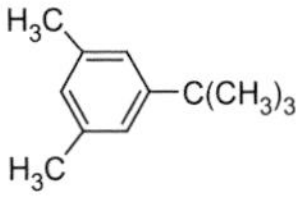

30)

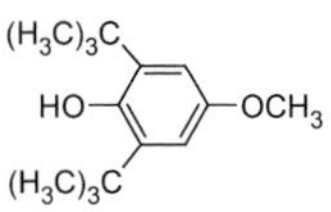

2 CHAPTER 5 (INCLUDE 31–60)

31) 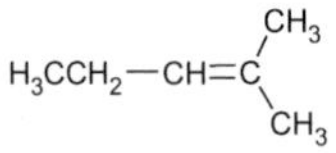 H$_3$CCH$_2$—CH=C(CH$_3$)(CH$_3$)

32) 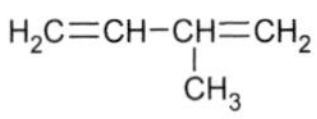 H$_2$C=CH–CH=CH$_2$ (CH$_3$)

33) 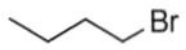 Br

34) 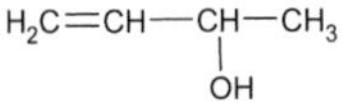 H$_2$C=CH—CH(OH)—CH$_3$

35) 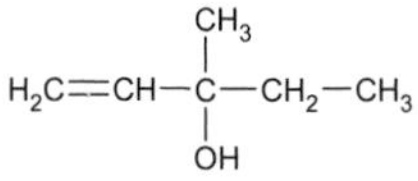 H$_2$C=CH—C(CH$_3$)(OH)—CH$_2$—CH$_3$

36)

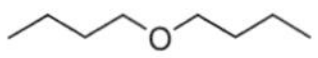

37) 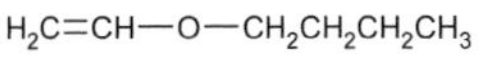 H$_2$C=CH—O—CH$_2$CH$_2$CH$_2$CH$_3$

38) 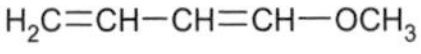 H$_2$C=CH–CH=CH–OCH$_3$

39) 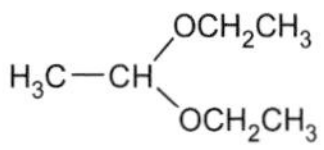 H$_3$C—CH(OCH$_2$CH$_3$)(OCH$_2$CH$_3$)

40) 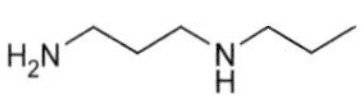 H$_2$N–CH$_2$CH$_2$CH$_2$–NH–CH$_2$CH$_2$CH$_3$

41) 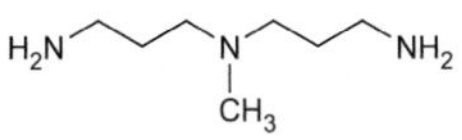 H$_2$N–CH$_2$CH$_2$CH$_2$–N(CH$_3$)–CH$_2$CH$_2$CH$_2$–NH$_2$

42) 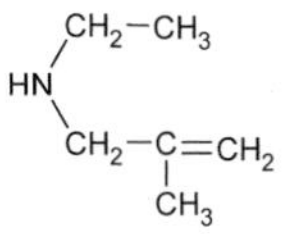 HN(CH$_2$—CH$_3$)(CH$_2$—C(CH$_3$)=CH$_2$)

43) 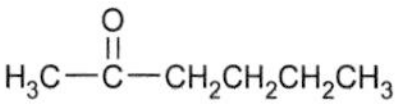 $H_3C-\overset{O}{\overset{\|}{C}}-CH_2CH_2CH_2CH_3$

44)

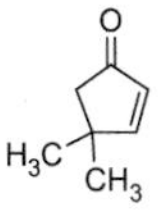

45) 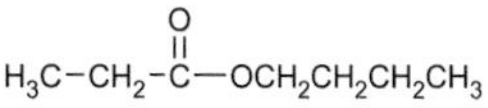 $H_3C-CH_2-\overset{O}{\overset{\|}{C}}-OCH_2CH_2CH_2CH_3$

46) 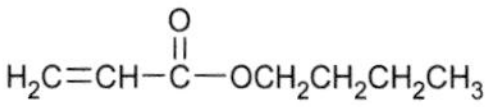 $H_2C=CH-\overset{O}{\overset{\|}{C}}-OCH_2CH_2CH_2CH_3$

47) 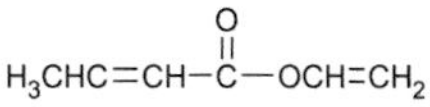 $H_3CHC=CH-\overset{O}{\overset{\|}{C}}-OCH=CH_2$

48) 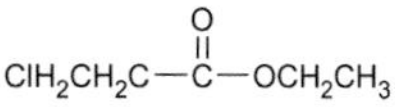 $ClH_2CH_2C-\overset{O}{\overset{\|}{C}}-OCH_2CH_3$

49) 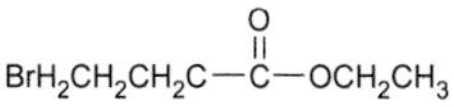 $BrH_2CH_2CH_2C-\overset{O}{\overset{\|}{C}}-OCH_2CH_3$

50)

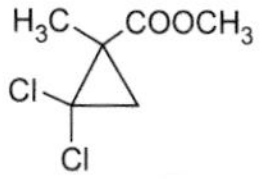

51) 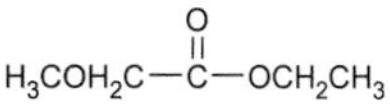 $H_3COH_2C-\overset{O}{\overset{\|}{C}}-OCH_2CH_3$

52) 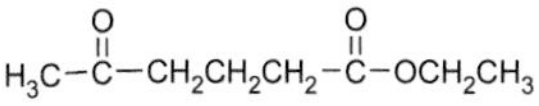 $H_3C-\overset{O}{\overset{\|}{C}}-CH_2CH_2CH_2-\overset{O}{\overset{\|}{C}}-OCH_2CH_3$

53)

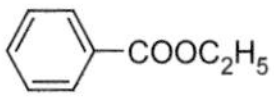

54)

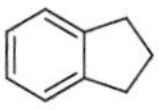

55)

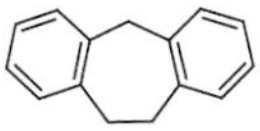

56)

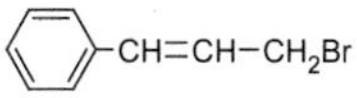

57)

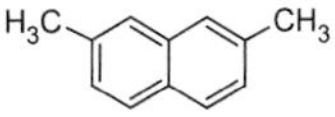

58)

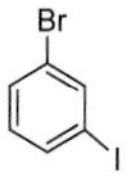

59) 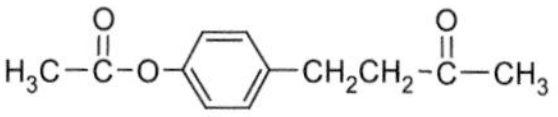 $H_3C-\overset{O}{\overset{\|}{C}}-O-\!\!\!\!\bigcirc\!\!\!\!-CH_2CH_2-\overset{O}{\overset{\|}{C}}-CH_3$

60)

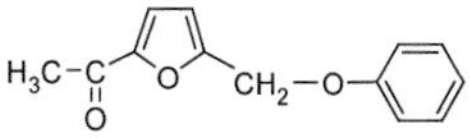

3   CHAPTER 8 (INCLUDE 61–101)

61) $H_3CH_2CH_2CH_2C-O-CH_2CH_2CH_2NH_2$

62)

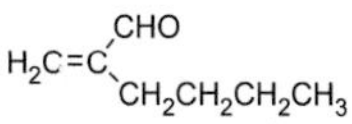

63) 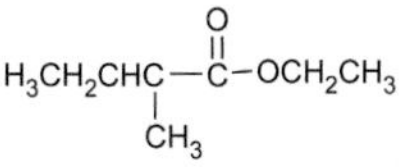 $H_3CH_2CHC-\overset{O}{\overset{\|}{C}}-OCH_2CH_3$ with CH_3 branch

64)

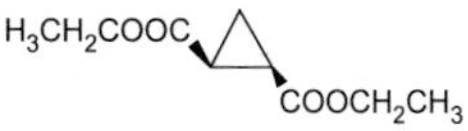

65) 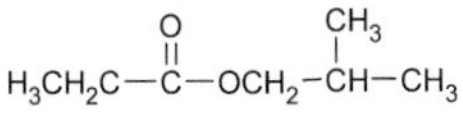 H₃CH₂C—C(=O)—O—CH₂—CH(CH₃)—CH₃

66) 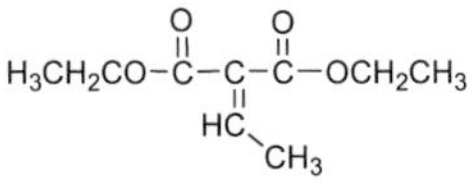 H₃CH₂CO—C(=O)—C(=CHCH₃)—C(=O)—OCH₂CH₃

67) 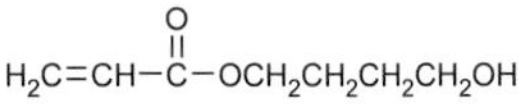 H₂C=CH—C(=O)—OCH₂CH₂CH₂CH₂OH

68) 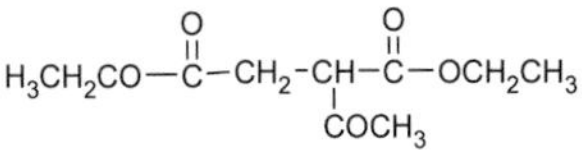 H₃CH₂CO—C(=O)—CH₂—CH(COCH₃)—C(=O)—OCH₂CH₃

69) 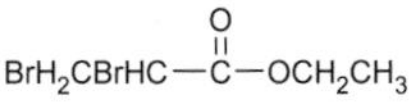 BrH₂CBrHC—C(=O)—OCH₂CH₃

70) 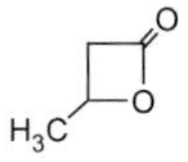 β-lactone, H₃C substituent

71) 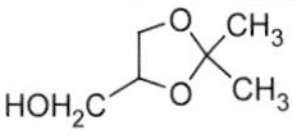 HOH₂C— (1,3-dioxolane with two CH₃)

72) 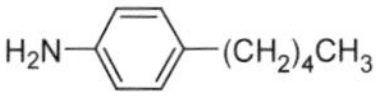 H₂N—C₆H₄—(CH₂)₄CH₃

73) 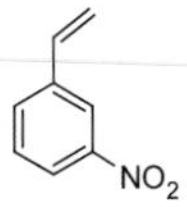 vinyl-substituted nitrobenzene (NO₂)

74) 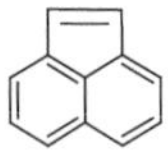 acenaphthylene

75) 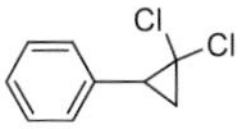 phenyl-dichlorocyclopropane (Cl, Cl)

76) 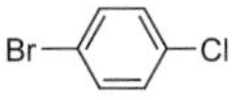 Br—C₆H₄—Cl

77) 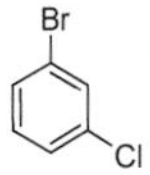 Br, Cl substituted benzene

78) 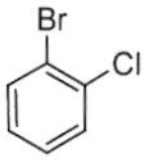 Br, Cl substituted benzene

79) 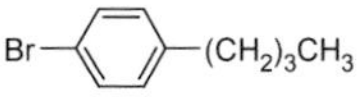 Br—C₆H₄—(CH₂)₃CH₃

80) 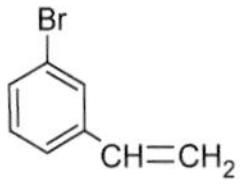 Br, CH=CH₂ substituted benzene

81) 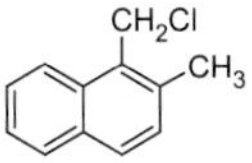 naphthalene with CH₂Cl and CH₃

82) 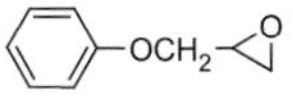 C₆H₅—OCH₂—epoxide

83) 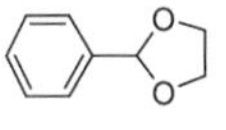 phenyl-1,3-dioxolane

84) 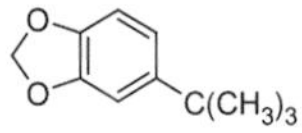 methylenedioxybenzene with C(CH₃)₃

85) 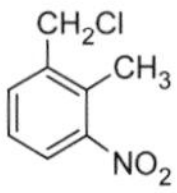 benzene with CH₂Cl, CH₃, NO₂

86) 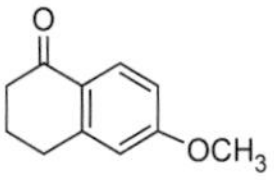 methoxy-tetralone, OCH₃

87) 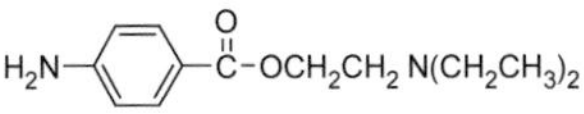 H₂N—C₆H₄—C(=O)—OCH₂CH₂N(CH₂CH₃)₂

88) 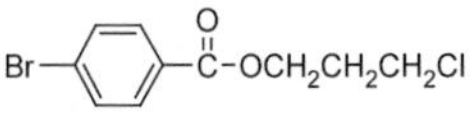 Br—C₆H₄—C(=O)—OCH₂CH₂CH₂Cl

89) 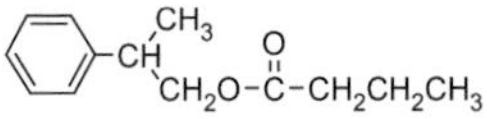 C₆H₅—CH(CH₃)—CH₂O—C(=O)—CH₂CH₂CH₃

90) 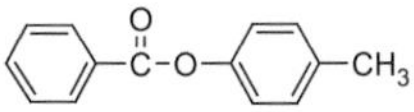 C₆H₅—C(=O)—O—C₆H₄—CH₃

91) 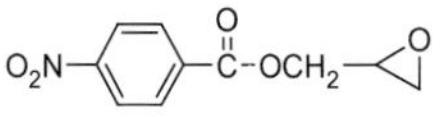 O_2N—⬡—$\overset{\displaystyle O}{\overset{\|}{C}}$-$OCH_2$—◁$O$

92) $CH_3OCH_2CH_2OCH_2CH_2Br$

93) 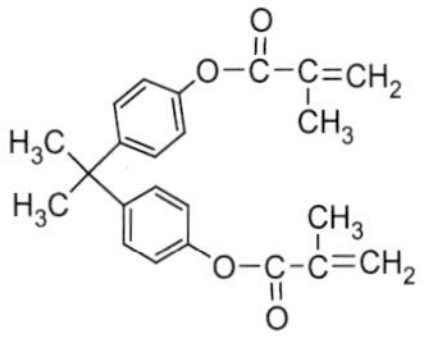 (bisphenol A dimethacrylate structure)

94) 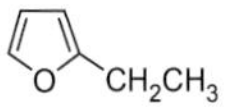 (2-ethylfuran) CH_2CH_3

95) 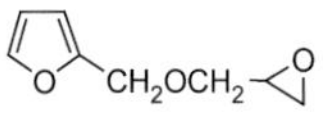 (furan) CH_2OCH_2—◁O

96) 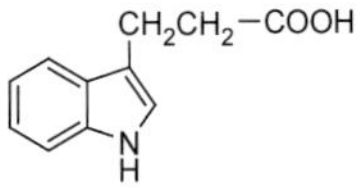 (indole-3-propanoic acid) CH_2CH_2—$COOH$

97) 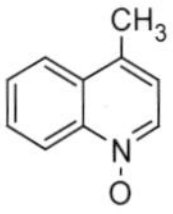 (4-methylquinoline N-oxide) CH_3

98) 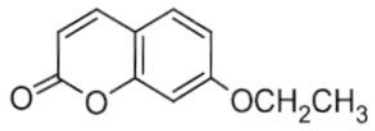 (7-ethoxycoumarin) OCH_2CH_3

99) 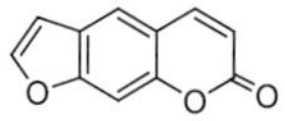 (furocoumarin)

100) 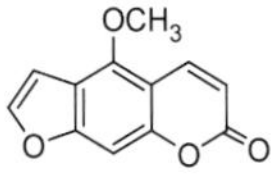 (methoxy furocoumarin) OCH_3

101) 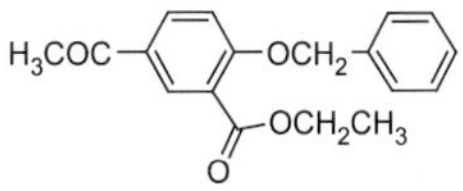 H_3COC—⬡—OCH_2—⬡, $\overset{}{C}$-OCH_2CH_3 with O

References

1. T Clerck and E Pretsch, *Kernrezonans Spektroskopie*, Akademische Verlaggesellschaft, Frankfurt, 1970, p. 52.
2. NJ Shoolerly, in *Varian Technical Information Bulletin*, **Vol. 2**, No. 3, Varian Associates, Palo Alto, CA, 1959; EC Friedrich and KG Runkle, *J. Chem. Educ.*, 1984, **61**, p. 830; HM Bell, LK Berry and EA Madigan, *Org. Magn. Reson.*, 1984, **22**, 693.
3. E Vogel and HD Roth, *Angew. Chem.*, 1964, **76**, 145; W Klug, *Substitutionsprodukte des 1.6-metano[10]annulenes*, Ph.D. Dissertation Thesis, University of Cologne, Germany, 1972.
4. DA Balc and K Conrow, *J. Org. Chem.*, 1966, **31**, 3958.
5. V Boekelheide and TH Hylton, *J. Am. Chem. Soc.*, 1970, **92**, 3669; H Blaschke, V Boekelheide, I Calder and CE Ramey, *J. Am. Chem. Soc.*, 1970, **92**, 3679.
6. F Sondheimer and Y Gaoni, *J. Am. Chem. Soc.*, 1960, **82**, 5765; F Sondheimer, R Wolowsky and Y Amiel, *J. Am. Chem. Soc.*, 1962, **84**, 274.
7. W Wagemann, *Synthese des syn-1,6/8,13-bismethano[14]annulenes*, Ph.D. Dissertation Thesis, University of Cologne, Germany, 1975.
8. E Vogel, U Haberland and H Günther, *Angew. Chem.*, 1970, **82**, 510.
9. AD Wolf, VV Kane, R Levin and M Jones, *J. Am. Chem. Soc.*, 1973, **95**, 1680; DJ Cram and JM Cram, *Acc. Chem. Res.*, 1971, **4**, 204.
10. H Günther, *NMR Spektroskopie*, Georg Thieme Verlag, Stuttgart, 1973, p. 76.
11. KG Untch and DC Wysocki, *J. Am. Chem. Soc.*, 1967, **89**, 6386.
12. D Farquar and D Leaver, *J. Chem. Soc. Chem. Commun.*, 1969, 24.
13. RH Mitchell, CE Klopfenstein and V Boekelheide, *J. Am. Chem. Soc.*, 1969, **91**, 4931.
14. LM Jackman and S Sternhell, *Application of Nuclear Magnetic Resonance Spectroscopy in Organic Chemistry*, Pergamon Press, Oxford, 1969, p. 456.
15. ME Sengül, Z Ceylan and M Balcı, *Tetrahedron*, 1997, **53**, 8522; ME Sengül, Ph.D. Dissertation Thesis, Atatürk University, Turkey, 1996.
16. H Günther, *NMR Spectroscopy*, Wiley, Stuttgart, 1980, p. 75.
17. W Adam and M Balcı, *J. Am. Chem. Soc.*, 1979, **101**, 7542.
18. M Zengin, M.Sc. Thesis, Atatürk University, Turkey, 1997.
19. H Kilic, M.Sc. Thesis, Atatürk University, Turkey, 1997; W Adam, H Kilic and M Balcı, *J. Org. Chem.*, 1998, **63**, 8544.
20. ME Sengül and M Balcı, *J. Chem. Soc. Perkin Trans.*, 1997, **1**, 2071.
21. N Simsek, M.Sc. Thesis, Atatürk University, Turkey, 1996.
22. A Dastan, Ph.D. Dissertation Thesis, Atatürk University, Turkey, 1993; A Dastan and M Balcı, *Turk. J. Chem.*, 1994, **18**, 215.
23. O Çakmak, Ph.D. Dissertation Thesis, Atatürk University, Turkey, 1990.
24. R Altundas, Ph.D. Dissertation Thesis, Atatürk University, Turkey, 1994.
25. A Altundas, M.Sc. Thesis, Atatürk University, Turkey, 1995; A Altundas, N Akbulut and M Balcı, *Helv. Chim. Acta*, 1998, **81**, 828.
26. N Horasan, M.Sc. Thesis, Atatürk University, Turkey, 1997.
27. C Pascual, J Meier and W Simon, *Helv. Chim. Acta*, 1966, **49**, 164.

28. E Vogel, W Wiedemann, H Kiefer and VF Harrison, *Tetrahedron Lett.*, 1963, 673.
29. E Vogel, W Wiedemann, HD Roth, J Eimer and H Günther, *Liebigs Ann. d. Chem.*, 1972, **759**, 1.
30. W Adam and M Balcı, *J. Org. Chem.*, 1979, **44**, 1189.
31. EC Friedrich and KG Runkle, *J. Chem. Educ.*, 1984, **61**, 830.
32. E Vogel, J Sombroek and W Wagemann, *Angew. Chem.*, 1975, **87**, 591.
33. A Menzek, M Krawiec, WH Watson and M Balcı, *J. Org. Chem.*, 1991, **56**, 6755; A Menzek, N Saracoglu, M Balcı, WH Watson and M Krawiec, *J. Org. Chem.*, 1995, **60**, 829.
34. O Çakmak and M Balcı, *J. Org. Chem.*, 1989, **54**, 181; K Tori, M Ueyama, T Tsuji, H Matsumura, H Tanida, H Iwamura, T Nishida and S Satob, *Tetrahedron Lett.*, 1974, **15**, 327.
35. M Barfield and DM Grant, Theory of nuclear spin–spin coupling in organic compounds. *Advances in Magnetic Resonance.* (ed. JS Waugh), Chapter IV, Academic Press, New York, 1965; RC Cookson, TA Crabb, JJ Frankel and J Hudec, Geminal coupling constants in methylene groups. *Tetrahedron*, 1966, **22**, Suppl. 7, 355.
36. AA Bothner-By, Geminal and vicinal proton–proton coupling constants in organic compounds. *Advances in Magnetic Resonance.* (ed. JS Waugh), Chapter V, Academic Press, New York, 1965.
37. H Hoffmann, Die Synthese von Tropylium- and Troponoid Analogen mit 10π-Elektronen, Ph.D. Dissertation Thesis, University of Cologne, Germany, 1967.
38. W Adam and M Balcı, *J. Am. Chem. Soc.*, 1979, **101**, 7537.
39. E Salamci, H Seçen, Y Sütbeyaz and M Balcı, *J. Org. Chem.*, 1997, **62**, 2453; E Salamci, Ph.D. Dissertation Thesis, Atatürk University, Erzurum, Turkey, 1997.
40. JA Pople and AA Bother-By, *J. Chem. Phys.*, 1965, **42**, 1339; M Barfield and DM Grant, *J. Am. Chem. Soc.*, 1963, **83**, 1899.
41. H Günther, *NMR Spectroscopy*, Wiley, Stuttgart, 1980, p. 103.
42. H Hopf, *Angew. Chem.*, 1972, **84**, 471; DJ Cram and RC Helgesno, *J. Am. Chem. Soc.*, 1966, **88**, 3535.
43. A Dastan and M Balcı, *Turk. J. Chem.*, 1994, **18**, 215.
44. W Grimme, H Hoffmann and E Vogel, *Angew. Chem.*, 1965, **77**, 348.
45. M Karplus, *J. Chem. Phys.*, 1959, **30**, 11; M Karplus, *J. Am. Chem. Soc.*, 1963, **85**, 2870.
46. A Dastan, B Demirci and M Balcı, Unpublished results.
47. D Cremer and H Günther, *Liebigs Ann. Chem.*, 1972, **763**, 87; H Günther, *Tetrahedron Lett.*, 1967, **8**, 2907.
48. JB Pawliczek and H Günther, *Tetrahedron*, 1970, **26**, 1755.
49. G Fraenkle, Y Asahi, MJ Mitchell and PM Cava, *Tetrahedron*, 1964, **20**, 1179; AR Katritzky and RE Reavill, *Recuil Trav. Chim. Pays-Bas*, 1964, **83**, 1230.
50. G Maier, *Angew. Chem.*, 1967, **79**, 446; G Maier, *Angew. Chem. Int. Ed. Engl.*, 1967, **6**, 402; M Balcı, *Turk. J. Chem.*, 1992, **16**, 42.
51. M Barfield and B Chakrabarti, *Chem. Rev.*, 1969, **69**, 757.
52. RG Harvey, DF Lindow and PW Rabidean, *Tetrahedron*, 1972, **28**, 2909.
53. KC Ramey and DC Lini, *J. Magn. Reson.*, 1970, **3**, 94.
54. BC Cantello and G Scholes, *J. Chem. Soc. C*, 1971, 2915.
55. M Barfield, RJ Spear and S Sternhell, *Chem. Rev.*, 1976, **76**, 593.
56. EW Garbish, *J. Chem. Soc. Chem. Commun.*, 1968, 332.
57. S Masamune, H Cuto and MG Hogben, *Tetrahedron Lett.*, 1966, 1017; WG Dauben and DL Whalen, *Tetrahedron Lett.*, 1966, 3743.
58. EA Hill and JD Roberts, *J. Am. Chem. Soc.*, 1967, **89**, 2047.
59. H Hüther and HET Brune, *Org. Magn. Reson.*, 1971, **3**, 737; S Farid, W Kothe and G Pfundt, *Tetrahedron Lett.*, 1968, **9**, 4151.
60. Y Gözel, Y Kara and M Balcı, *Turk. J. Chem.*, 1991, **15**, 274.
61. A Dastan, Y Taskesenligil, F Tümer and M Balcı, *Tetrahedron*, 1996, **2**, 14004.
62. RM Silverstein, GC Bassler and TC Morril, *Spectroscopic Identification of Organic Compounds*, Wiley, New York, 1974, p. 167.

63. ED Becker, *High Resolution NMR*. Academic Press, New York, 1969.
64. M Ceylan, Ph.D. Dissertation Thesis, Atatürk University, Erzurum, Turkey, 1995.
65. N Akbulut, A Menzek and M Balcı, *Tetrahedron Lett.*, 1987, **28**, 1689; N Akbulut, A Menzek and M Balcı, *Turk. J. Chem.*, 1991, **15**, 232.
66. D Demirci, M.Sc. Thesis, Atatürk University, Erzurum, Turkey, 1998.
67. A Dastan and M Balcı, Unpublished results.
68. RJ Abraham, *Analysis of High Resolution NMR Spectra*. Elsevier, Amsterdam, 1971; RJ Abraham and P Loftus, *Proton and Carbon-13 NMR Spectroscopy, An Integrated Approach*. Heyden, London, 1978.
69. W Adam, M Balcı and J Rivera, *Synthesis*, 1979, 807.
70. GW Buchanan and AR McCarville, *Can. J. Chem.*, 1973, **51**, 177; B Dischler, *Angew. Chem.*, 1973, **51**, 84; EW Garbisch, Jr, *J. Chem. Educ.*, 1968, **45**, 480; H Günther, *Angew. Chem.*, 1972, **84**, 907.
71. TC Morrill, *Lanthanide Shift Reagents in Stereochemical Analysis*, VCH, New York, 1988; EC McGoran, B Cutter and K Morse, *J. Chem. Educ.*, 1979, **56**, 122; Rv Ammon and RD Fischer, *Angew. Chem. Int. Ed. Engl.*, 1972, **11**, 675.
72. ME Sengül, N Simsek and M Balcı, *Eur. J. Org. Chem.*, 2000, p. 1359; N Simsek, M.Sc. Thesis, Atatürk University, Erzurum, Turkey, 1997.
73. LM Jackman and FA Cotton, *Dynamic NMR Spectroscopy*, Academic Press, New York, 1975.
74. CH Yoder and CD Schaeffer, *Introduction to Multinuclear NMR: Theory and Application*. Benjamin/Cummings, Menlo Park, CA, 1987; E Breitmeier and W Voelter, *Carbon-13 NMR Spectroscopy*, VCH, Weinheim, 1987; D Shaw, *Fourier Transform NMR Spectroscopy*. 2nd Edn., Elsevier, New York, 1984; RJ Abraham and P Loftus, *Proton and Carbon-13 Spectroscopy*, Heyden, London, 1978; TC Farrar and ED Becker, *Pulse and Fourier Transform NMR*. Academic Press, New York, 1971.
75. CP Poole and H Harach, *Relaxation in Magnetic Resonance*. Academic Press, New York, 1971; GC Levy, *Acc. Chem. Res.*, 1973, **6**, 161.
76. C Kazaz, E Sengül and M Balcı, Unpublished results.
77. C Kazaz, A Menzek and M Balcı, Unpublished results.
78. C Kazaz, A Daştan and M Balcı, Unpublished results.
79. DN Neuhaus and MP Willimson, *The Nuclear Overhauser Effect in Structural and Conformational Analysis*. VCH, New York, 1989.
80. A Menzek and M Balcı, *Tetrahedron*, 1993, **49**, 6071.
81. C Kazaz, H Kilic and M Balcı, Unpublished results.
82. M Karplus and JA Pople, *J. Chem. Phys.*, 1963, **38**, 2803.
83. DK Dalling and DM Grant, *J. Am. Chem. Soc.*, 1967, **89**, 6612; AS Perlin and HJ Koch, *Can. J. Chem.*, 1970, **48**, 2599; DE Dorman and JD Roberts, *J. Am. Chem. Soc.*, 1970, **92**, 1355.
84. HO Kalinowski, S Serger and S Sraun, *Carbon-13 NMR Spectroscopy*, Wiley, New York, 1988.
85. JB Stothers and CT Tan, *Can. J. Chem.*, 1973, **51**, 2893; AK Cheng and JB Sthoters, *Org. Magn. Reson.*, 1977, **9**, 355.
86. DM Grant and EG Paul, *J. Am. Chem. Soc.*, 1964, **86**, 2984.
87. DE Ewig, *Org. Magn. Reson.*, 1979, **12**, 499.
88. M Hesse, H Meier and B Zeeh, *Spektroskopische Methoden in der Organischen Chemie*, Georg Thieme Verlag, Stuttgart, 1984, p. 225.
89. JJ Burke and PC Lauterbur, *J. Am. Chem. Soc.*, 1964, **86**, 1870.
90. JT Clerck, E Pretsch, S Sternhell, *^{13}C-Kernresonanzspektroskopie*, Akademischer Verlaggesellschaft, Frankfurt, 1973.
91. PE Hansen, *Org. Magn. Reson.*, 1979, **12**, 239.
92. JB Stothers and PC Lauterbur, *Can. J. Chem.*, 1972, **42**, 1563.
93. GE Hawkes, K Herwig and JD Roberts, *J. Org. Chem.*, 1974, **39**, 1017.
94. LM Jackman and DP Kelly, *J. Chem. Soc. B*, 1970, 102; M Oha, J Hinton and A Fry, *J. Org. Chem.*, 1979, **44**, 3545; M Yalpani, B Modarai and E Koshdel, *Org. Magn. Reson.*, 1979, **12**, 254; FG Weigert and JD Roberts, *J. Am. Chem. Soc.*, 1970, **92**, 1347.

95. M Christl, HJ Reich and JD Roberts, *J. Am. Chem. Soc.*, 1971, **93**, 3463; PA Couperus, ADH Clague and JPCM von Dongen, *Org. Magn. Reson.*, 1980, **11**, 590; HJ Schneider, W Freitag and E Weigend, *Chem. Ber.*, 1987, **111**, 2656; DE Dorman, D Bauer and JD Roberts, *J. Org. Chem.*, 1975, **40**, 3729.

96. EL Eliel, KM Pietrusiewicz, *Topics in Carbon-13 NMR Spectroscopy*, **Vol. 3**, p. 172, Wiley, New York, 1979.

97. ER Malinowski, *J. Am. Chem. Soc.*, 1961, **83**, 4479; AW Douglass, *J. Phys. Chem.*, 1964, **40**, 2413.

98. H Kilic and M Balcı, Unpublished results.

99. JL Marshall, *Carbon–Carbon and Carbon–Proton NMR Couplings: Applications to Organic Stereochemistry and Conformational Analysis*, Verlag Chemie International, Deerfield Beach, FL, 1983.

100. FJ Weigert and JD Roberts, *J. Am. Chem. Soc.*, 1968, **90**, 3543; FJ Weigert and JD Roberts, *J. Am. Chem. Soc.*, 1969, **91**, 4940.

101. R Wasylishen and T Schafer, *Can. J. Chem.*, 1972, **50**, 3686; R Wasylishen and T Schafer, *Can. J. Chem.*, 1973, **51**, 961.

102. L Ernst, V Wray, VA Chertkow and NM Sergeyew, *J. Magn. Reson.*, 1977, **25**, 123.

103. RL Vold, JS Waugh, MP Klein and DE Phelps, *J. Chem. Phys.*, 1968, **48**, 3831; R Freeman and RC Jones, *J. Chem. Phys.*, 1970, **52**, 465.

104. C Kazaz, Ph.D. Dissertation Thesis, Atatürk University, Erzurum, Turkey, 2001.

105. C LeCocq and JY Lallemand, *J. Chem. Soc. Chem. Commun.*, 1981, 150.

106. JN Schoolerry, *APT for C NMR Spectrum Interpretation. Research and Application Notes*, Varian NMR Application Laboratory, Palo Alto, CA, 1981.

107. DL Rabenstein and TT Nakashima, *Anal. Chem.*, 1979, **51**, 1465; DW Brown, TT Nakashima and DL Rabenstein, *J. Magn. Reson.*, 1981, **45**, 302; J Wesener, P Schmitt and H Günther, *J. Am. Chem. Soc.*, 1986, **106**, 10.

108. KGR Pachler and PL Wessels, *J. Magn. Reson.*, 1973, **12**, 337; S Sorensen, RS Hansen and HJ Jacobsen, *J. Magn. Reson.*, 1974, **14**, 243; AA Chalmers, KGR Pachler and PL Wessels, *Org. Magn. Reson.*, 1974, **6**, 445.

109. AA Maudsley and RR Ernst, *Chem. Phys. Lett.*, 1977, **50**, 368; GA Morris and R Freeman, *J. Am. Chem. Soc.*, 1979, **101**, 760.

110. AE Derome, *Modern NMR Techniques for Chemistry Research*, Pergamon Press, Oxford, 1991.

111. DM Doddrell, DT Pegg and MR Bendall, *J. Magn. Reson.*, 1982, **48**, 323; DM Doddrell, DT Pegg and MR Bendall, *J. Chem. Phys.*, 1982, **77**, 2745.

112. ME Sengül and M Balcı, Unpublished results.

113. M Çelik, C Kazaz and M Balcı, Unpublished results.

114. WR Croasmun and RMK Carlson, *Two Dimensional NMR Spectroscopy: Application for Chemist and Biochemists*. VCH, New York, 1994; Atta-ur-Rahman and MI Choudhary, *Solving Problems with NMR Spectroscopy*. Academic Press, San Diego, 1996; RR Ernst, G Bodenhausen and A Wokaun, *Principles of Magnetic Resonance in One and Two Dimensions*, Clarendon Press, Oxford, 1987; JKM Sanders and BK Hunter, *Modern NMR Spectroscopy: A Guide for the Chemists*, Oxford University Press, Oxford, 1987; H Friebolin, *Basic One- and Two Dimensional NMR Spectroscopy*, VCH, Weinheim, 1991; WR Croasmum and RMK Carlson, *Two-Dimensional NMR Spectroscopy*, VCH, New York, 1987; N Chandrakumar and S Subramanian, *Modern Techniques in High Resolution FT NMR*. Springer, New York, 1987.

115. AE Derome, *Modern NMR Techniques for Chemistry Research*, Pergamon Press, Oxford, 1991, p. 186.

116. A Bax and R Freeman, *J. Magn. Reson.*, 1981, **42**, 164.

117. AE Derome, *Modern NMR Techniques for Chemistry Research*, Pergamon Press, Oxford, 1991, p. 185.

118. KR Williams and RW King, *J. Chem. Educ.*, 1990, **67A**, 125.

119. A Menzek and C Kazaz, Unpublished results.

120. T Atalar, F Algi and M Balcı, Unpublished results.

121. R Caliskan and M Balcı, Unpublished results.
122. A Bax, R Freeman and SP Kempsell, *J. Am. Chem. Soc.*, 1980, **102**, 4851.
123. H Ziffer, A Bax, RJ Highet and B Green, *J. Org. Chem.*, 1988, **53**, 895.

Index

419